Universitext

Manfred Knebusch • Claus Scheiderer

Real Algebra

A First Course

With Contributions by Thomas Unger

 Springer

Manfred Knebusch
Fakultät für Mathematik
University of Regensburg
Regensburg, Germany

Claus Scheiderer (iD)
FB Mathematik
University of Konstanz
Konstanz, Germany

Translated by
Thomas Unger
School of Mathematics and Statistics
University College Dublin
Dublin, Ireland

ISSN 0172-5939   ISSN 2191-6675   (electronic)
Universitext
ISBN 978-3-031-09799-7   ISBN 978-3-031-09800-0   (eBook)
https://doi.org/10.1007/978-3-031-09800-0

Translation from the German language edition: "Einführung in die reelle Algebra" by Manfred Knebusch et al., © Friedr. Vieweg & Sohn Verlagsgesellschaft mbH 1989. Published by Friedr. Vieweg & Sohn, Braunschweig/Wiesbaden. All Rights Reserved.

Mathematics Subject Classification: 12D15, 14P05, 14P10, 12J15, 12J10, 13J30

This Springer imprint is published by the registered company Springer Nature Switzerland AG
The registered company address is: Gewerbestrasse 11, 6330 Cham, Switzerland

Preface

More than 30 years after its publication, we are pleased and surprised with the ongoing interest in *Einführung in die reelle Algebra*. Given the vibrant development of real algebra and geometry in the previous decades, this seems by no means self-evident.

Real algebra has grown in many directions, partially from within itself, and exceedingly also through nudges and stimuli from without. While some of these developments were already discernible at the end of the 1980s, many modern ones that currently belong to the core of the area—such as the connections with tropical geometry, or with semidefinite optimization—would have been difficult to predict. Hence, it is not obvious that a text that was timely and modern in 1989 still serves current needs in a reasonable manner.

Nevertheless, we are convinced that also from today's perspective the choice and exposition of the material in *Einführung in die reelle Algebra* still constitute a solid first course on the basic principles and important techniques of real algebra. For this reason, we decided to keep the text essentially unchanged in the English translation—even though it was tempting to change the wording here and there, and to drop some sections in favour of new ones.

A small number of typos, omissions and errors that were present in the German original have been corrected, and some notation and terminology have been brought up to date. Several new examples (some in the form of exercises) have been added. All definitions and propositions have been numbered in a consistent manner. The original bibliography has been augmented with a number of newer references.

A feature of this translation is the addition of a short fourth chapter that provides a succinct overview of the most important developments and advances in those parts of real algebra that are directly related to material covered in *Einführung in die reelle Algebra*.

We are grateful to Springer Nature, and Rémi Lodh in particular, for their endorsement and friendly and professional assistance.

We received support and encouragement from many quarters, and are pleased to extend special thanks to Oliver Hien and Marcus Tressl, who made valuable remarks on the German original.

Finally, our very special and heartfelt thanks go to Thomas Unger, who translated the original text in a conscientious and competent manner, and co-authored the final chapter. It was always a great pleasure to work with him.

Regensburg, Germany   Manfred Knebusch
Konstanz, Germany   Claus Scheiderer
May 2022

Preface to *Einführung in die reelle Algebra*

Algebra textbooks that are currently in common use present real algebra only in later chapters and then usually rather tersely. This is so for the influential works of van der Waerden [113], Jacobson [55] and Lang [73], although Jacobson covers a bit more than the others. Bourbaki's substantial multivolume text *Eléments de Mathématique* shows a similar picture: the volume *Algèbre* contains just a short chapter (Chapter 6, Groupes et corps ordonnés) on real algebra. In contrast, a complete volume (currently counting nine chapters) is dedicated to commutative algebra, even though on the whole only the elementary part is covered, and not more than what is absolutely essential for the foundation of the theory, both by current standards as well as Bourbaki's own standards.

That being the case, not too many algebraists today seem to perceive real algebra as a proper branch of algebra at all. This has not always been so. In the nineteenth century real algebra flourished. The study of real zeroes of a real polynomial in one variable was at the centre of algebraic interest during the whole century, and was an indispensable part of any higher mathematical education.

Heinrich Weber's large three-volume textbook on algebra [114] provides evidence for this fact. As Weber's research interests were primarily in number theory, especially in complex multiplication and class field theory, he focussed his textbook mostly on these topics. Nevertheless he devoted well over a hundred pages in the first volume to real zeroes of real polynomials.

The twentieth century witnessed a dramatic decline in interest in real algebra. This trend only seems to have changed since the late 1970s, which is all the more astonishing since the essential seeds of a modern real algebra, as we understand it today, are already present in two works by Artin and Schreier from the 1920s [3, 4].

So, what is real algebra? Instead of giving a formal definition—which would be difficult—we rather answer the question with an analogy that puts real algebra in parallel with commutative algebra. Commutative algebra can be seen as that part of algebra that contains the algebraic foundations that are typically important for algebraic geometry (in particular in its modern, abstract form); and algebraic geometry is ultimately the study of solution sets of systems of polynomial equations $F(x_1, \ldots, x_n) = 0$ and non-equations $F(x_1, \ldots, x_n) \neq 0$. Correspondingly, real algebra provides algebraic methods that typically function as tools for real algebraic geometry, and in particular semialgebraic geometry, which is the study of the solution sets of systems of polynomial inequalities $F(x_1, \ldots, x_n) > 0$ or $F(x_1, \ldots, x_n) \geq 0$, where the coefficients classically come from the field of real numbers and more generally from any ordered field.

This analogy provides an argument for why the interest has been so much stronger in commutative than in real algebra during our century thus far. Indeed, algebraic geometry experienced a continuous and ultimately triumphant upswing in the twentieth century, while real algebraic geometry was only practiced in isolation and then mostly with transcendental methods. Thus, from the geometric point of view, there was for many decades no engine available that could have pushed real algebra forward.

It was not until 1987 that the first textbook [11] on real algebraic geometry was published. Its authors' report (in the last section of the introduction to this commendable work [11, p. 4 ff.]) on the strange sleeping beauty slumber of real algebraic geometry (insofar as it was practiced with algebraic methods) is worth considering.

In the meantime this slumber has given way to a lively development. We consider the introduction—or better: discovery—of the real spectrum Sper A of a commutative ring A by Michel Coste and Marie-Françoise Roy around the year 1979 as the most important trigger.

One can view Sper A as an analogue of the Zariski spectrum Spec A, introduced by Grothendieck. It is well-known that the Zariski spectrum is the key to Grothendieck's abstract algebraic geometry. Likewise, the real spectrum is the key to an abstract semialgebraic geometry. (There is one difference: in contrast to Spec A, the real spectrum Sper A seems to carry (at least) two important structure sheafs, the sheaf of abstract Nash functions, introduced by Coste and Roy [29], as well as the sheaf of abstract semialgebraic functions, introduced by G. Brumfiel and N. Schwartz [19, 107, 108].)

Real algebra is necessary to understand the latest developments in real algebraic geometry and to meet the future intellectual challenges in this area. This brings us to the objective of this book.

Our book is based on two insights: real algebra is a branch of algebra that is largely autonomous in its foundations, with methods specific to this branch. These foundations can be successfully taught with little more preparation than the standard background from a typical one-semester algebra course on linear algebra, group theory, field theory and ring theory.

We differentiate more precisely between an *elementary* and a *higher* real algebra. The former can be developed without any special prior knowledge and used with benefit in real algebraic geometry. The latter makes serious use of resources from other branches of mathematics, especially real algebraic geometry, commutative algebra, algebraic geometry, model theory and the theory of quadratic forms, but occasionally also algebraic topology, real analysis and complex analysis. (This list can certainly be extended.) An analogous distinction can be made in commutative algebra. A demarcation between "elementary" and "higher" in either area is not entirely objectively possible, but rather is subject to personal points of view and taste. Furthermore, the higher one goes, the more fluid and arbitrary the boundary between the two kinds of algebra and their corresponding geometries becomes.

Our book is dedicated to elementary real algebra in the above sense. Another book on higher real algebra is planned.[1] In the current book we do get by with previous knowledge of the extent sketched above. By adding another twenty to thirty pages, we could have reduced the requirements further and, for example, developed *everything* that is needed from commutative algebra and the theory of quadratic forms. We have not done so however, since students who are interested in real algebra more than likely already mastered almost all of the prerequisites for the current book, and would easily be able to find the few things they may be missing elsewhere.

One specific feature of real algebra should be pointed out in particular: the major role that general (Krull) valuation rings play. In most parts of commutative algebra, valuation rings that are not discrete are only viewed as an aid that can often be done without. In real algebra on the other hand, general valuation rings are a natural and even central concept throughout. The reason is that every convex subring of an ordered field is a valuation ring, but only in rare cases a discrete valuation ring. This fact was already observed by Krull in the introduction to his pioneering work *Allgemeine Bewertungstheorie* [67]. (It can already be found in embryonic form, without the concept of valuation ring, in the work of Artin and Schreier, cf. [4, p. 95].) Krull also identified the theory of ordered fields as an important application area of general valuation theory, but then devoted only one section— albeit substantial—of his great work to ordered fields [67, §12].

Our book is divided into three chapters. In addition to the Artin-Schreier theory of ordered fields and the elementary relationships between orderings and quadratic forms, the first chapter covers some aspects of real algebra from the nineteenth century. Various methods for determining the number of real zeroes of a real polynomial are treated (Sturm's algorithm, Hermite's method using quadratic forms, Hurwitz' Theorem). Another section is devoted to the relationship between the Cauchy index and the Hankel form, as well as the Bézoutian of a rational function.

The second chapter deals with real valuation theory. Everything we need from general valuation theory is developed from scratch. The chapter culminates in a presentation of Artin's solution of Hilbert's 17th Problem.

Finally, the third chapter is devoted to the real spectrum. After a short crash course on the Zariski spectrum, the real spectrum is examined in detail, but only as a topological space (i.e., without the introduction of a structure sheaf). In the geometric setting (affine algebras over real closed fields), the points as well as certain subsets of the real spectrum are described in terms of filter sets. In the last five sections of the chapter we come to some parts of real algebra in which the real spectrum proves to be clarifying and helpful, such as the reduced Witt ring of a field,

---

[1] This book was never written.

Positivstellensätze, preorderings of rings, convex ideals, and the holomorphy ring of a field. Many of the previously developed techniques and terms are used again.

Experts should note that the book manages without Tarski's Principle throughout. (That so much of real algebra can be developed in a meaningful way without using this principle should hopefully come as a little surprise.) Although we consider Tarski's Principle to be extremely important and by no means particularly difficult, we still count it as part of higher real algebra.

The book originates in a four-hour course by the first author in Regensburg in the 1986/87 winter semester. This course was followed by an extensive course on model theory, in which Tarski's Principle was proved fairly quickly and painlessly following Prestel [91], and then by a course on higher real algebra.

To the subject matter of the first course we added individual sections here and there, so that the book now encompasses more than can be handled in one semester. However, since some sections can be skipped without consequences for further understanding, it can still be used as a guide for a one-semester course.

We want to guard against the misunderstanding that our pair of opposites "elementary–higher" is strongly correlated with the pair "easy–difficult". We consider elementary real algebra, as presented here, not to be always easy. Beginners may need to overcome some pedagogical hurdles. Especially in the third chapter, some things may seem strange at first glance. Our text is also rather terse and requires the reader to participate intensively with paper and pencil. Our goal was to delight, motivate and enable the beginner, but not to put all the obstacles out of the way with lengthy statements, which after all could be quite tiresome.

We would like to thank the editor of the series "Aufbaukurs Mathematik", Gerd Fischer, as well as Vieweg-Verlag, and here in particular Ulrike Schmickler-Hirzebruch, for her considerable understanding and empathy during the preparation of the manuscript. Marina Franke TEXed numerous versions of the manuscript with patience and competence; Uwe Helmke, Roland Huber and Michael Prechtl proofread the manuscript and helped us a lot with comments and suggestions for improvement. We would like to take this opportunity to thank all of them.

Regensburg
January 1989

Contents

Chapter 1
Ordered Fields and Their Real Closures

Our starting point is the basic notion for the entire book, the general concept of orderings of arbitrary fields. Conceived by Artin and Schreier in their foundational 1927 paper [4], it was successfully used by Artin in his solution of Hilbert's 17th Problem in the same year [3]. We introduce real closed fields and show that they have the same algebraic properties as the field of real numbers. Moreover we prove the existence and uniqueness of a real closure for any ordered field.

The second main topic in this chapter concerns methods for counting real roots of real polynomials. Instead of trying to calculate or estimate the roots numerically, we discuss purely algebraic methods that extend from the real numbers to any real closed base field. These results were mostly found in the nineteenth century or even earlier, and are connected to famous names such as Descartes, Sturm, Sylvester and Hermite.

Parallel to these two subjects we give an introduction to the basic notions of quadratic forms over fields and their algebraic theory. They will find applications in Chaps. 2 and 3.

1.1 Orderings and Preorderings of Fields

Let K be a field.

Definition 1.1.1 An *ordering* of K is a subset P of K that satisfies the properties

(O1) $P + P \subseteq P$, $PP \subseteq P$,
(O2) $P \cap (-P) = \{0\}$,
(O3) $P \cup (-P) = K$,

where $P + P := \{a + b : a, b \in P\}$ and $PP := \{ab : a, b \in P\}$. The pair (K, P) is called an *ordered field*.

© Springer Nature Switzerland AG 2022
M. Knebusch, C. Scheiderer, *Real Algebra*, Universitext,
https://doi.org/10.1007/978-3-031-09800-0_1

Remark 1.1.2 If we assume (O1) and (O3), then (O2) is equivalent with
(O2') $-1 \notin P$.

Exercise 1.1.3 Let P be an ordering of K. Show that

$$a \leq_P b :\Leftrightarrow b - a \in P \quad (a, b \in K)$$

defines a total order relation $\leq_P$ on the set K which satisfies the properties

(i) $a \leq_P b \Rightarrow a + c \leq_P b + c$,
(ii) $a \leq_P b,\ c \geq 0 \Rightarrow ac \leq_P bc$

for all $a, b, c \in K$. Conversely, let $\leq$ be a total order relation on K which satisfies (i) and (ii). Show that $P := \{a \in K : a \geq 0\}$ is an ordering of K.

Thus we clearly have a one-one correspondence between orderings of K and total order relations on K that satisfy (i) and (ii). Usually the latter are called orderings of K as well. When P is clear we simply write $a \leq b$ instead of $a \leq_P b$.

Definition 1.1.4 A *preordering* of K is a subset T of K that satisfies the properties

(P1) $T + T \subseteq T,\ TT \subseteq T$,
(P2) $T \cap (-T) = \{0\}$,
(P3) $a^2 \in T$ for all $a \in K$.

Exercise 1.1.5 If T is a preordering of K, show that

$$a \leq_T b :\Leftrightarrow b - a \in T \quad (a, b \in K)$$

defines a *partial* order relation $\leq$ on K that is compatible with $+$ and $\cdot$.

If T is clear, we simply write $a \leq b$ instead of $a \leq_T b$.

Remarks 1.1.6

(1) If we assume (P1) and (P3), then (P2) is equivalent with
    (P2') $-1 \notin T$.
    Note that every ordering is a preordering. Indeed, (P1) is (O1), (P2) is (O2), and (P3) is a weaker version of (O3) by (P1).
(2) If $\mathcal{T} = (T_\alpha : \alpha \in I)$ is a nonempty family of preorderings of K, then $\bigcap_\alpha T_\alpha$ is also a preordering. If in addition $\mathcal{T}$ is upward directed (i.e., for all $\alpha, \beta \in I$ there exists $\gamma \in I$ such that $T_\alpha \cup T_\beta \subseteq T_\gamma$), then $\bigcup_\alpha T_\alpha$ is also a preordering. In particular, if K has a preordering, it will have a smallest one.

We introduce the notation

$$\Sigma K^2 := \{a_1^2 + \cdots + a_n^2 : n \in \mathbb{N},\ a_1, \ldots, a_n \in K\}.$$

It is clear that ΣK^2 is contained in every preordering of K, and is itself a preordering if and only if $-1 \notin \Sigma K^2$.

Definition 1.1.7 The field K is called *real* if $-1 \notin \Sigma K^2$, i.e., when -1 cannot be written as a sum of squares in K.

The previous discussion shows that a field K has a preordering if and only if it is real. In this case, ΣK^2 is the smallest preordering on K. Note that all real fields have characteristic 0.

Preorderings are precisely the intersections of orderings, as we will show next.

Lemma 1.1.8 *Let T be a preordering of K and let $a \in K$ with $a \notin T$. Then $T - aT = \{b - ac : b, c \in T\}$ is again a preordering of K.*

Proof $T - aT$ clearly satisfies (P1) and (P3). We show (P2'): assume on the contrary that $-1 \in T - aT$. Then $-1 = b - ac$ for some $b, c \in T$. But then $c \neq 0$ and so $a = c^{-2} \cdot c(1 + b) \in T$, a contradiction. $\square$

Lemma 1.1.9 *For every preordering T of K there exists an ordering P of K such that $T \subseteq P$.*

Proof Let $M := \{T' \subseteq K : T' \text{ is a preordering and } T \subseteq T'\}$. Then M is ordered by inclusion, $M \neq \varnothing$ and Zorn's Lemma applies (Remark 1.1.6(2)). Thus there exists a maximal preordering P of K with $T \subseteq P$. This P is an ordering of K. Indeed, if $a \in K, a \notin P$, then $P = P - aP$ by Lemma 1.1.8 and the maximality of P, so $-a \in P$. $\square$

Theorem 1.1.10 *Every preordering T of K is an intersection of orderings of K.*

Proof The inclusion $T \subseteq \bigcap\{P : P \text{ is an ordering and } T \subseteq P\}$ is trivial. If $a \in K$, $a \notin T$, then $T - aT$ is a preordering and thus contained in an ordering P of K (Lemma 1.1.9). It follows that $a \notin P$ since $-a \in P$ and $a \neq 0$. $\square$

Corollary 1.1.11 *K has an ordering if and only if K is real.*

(This follows already from Lemma 1.1.9.)

Corollary 1.1.12 (E. Artin [3]) *Assume char $K \neq 2$ and let $a \in K$. Then $a \geq 0$ for all orderings of K if and only if a is a sum of squares in K.*

Proof If K is real, then ΣK^2 is a preordering and the statement follows from Theorem 1.1.10. If K is not real and char $K \neq 2$, then $\Sigma K^2 = K$ since

$$a = \left(\frac{a+1}{2}\right)^2 + (-1) \cdot \left(\frac{a-1}{2}\right)^2$$

for all $a \in K$. $\square$

Naturally, we must assume that char $K \neq 2$, since otherwise ΣK^2 is a subfield of K. Next, we consider some examples of orderings. We will see many more examples later on.

Examples 1.1.13

(1) The field $\mathbb{R}$ of real numbers (and thus each of its subfields) has the usual ordering.
(2) Let $\mathbb{Q}(t)$ be the rational function field in one variable over $\mathbb{Q}$ and let $\vartheta \in \mathbb{R}$ be transcendental (over $\mathbb{Q}$). Then $f(\vartheta)$ is a well-defined real number for every $f \in \mathbb{Q}(t)$. The set

$$P = \{f \in \mathbb{Q}(t): f(\vartheta) \geq 0\}$$

is an ordering of $\mathbb{Q}(t)$. It is the ordering induced by the field embedding $\mathbb{Q}(t) \to \mathbb{R}, f \mapsto f(\vartheta)$.
(3) Let $(F, \leq)$ be an ordered field and $F(t)$ the rational function field in one variable over F. For $f \in F(t)$ the notation $f(a) \neq \infty$ signifies that $f(t)$ has no pole in $t = a$. Then $f(a) \in F$ is well-defined. The notation $f(a) = b$ for $a, b \in F$ implies in particular that $f(a) \neq \infty$. The sets

$$P_{a,+} := \{0\} \cup \{(t-a)^r f(t): r \in \mathbb{Z}, f \in F(t) \text{ with } f(a) \neq \infty \text{ and } f(a) > 0\}$$

and

$$P_{a,-} := \{0\} \cup \{(a-t)^r f(t): r \in \mathbb{Z}, f \in F(t) \text{ with } f(a) \neq \infty \text{ and } f(a) > 0\}$$

are orderings of $F(t)$ for every $a \in F$. Both orderings extend the ordering of F. The reason for this notation is that $a < t < b$ with respect to $P_{a,+}$ holds for all $b \in F$ with $b > a$. Thus $F(t)$ is ordered in such a way that the transcendental t is located "immediately to the right of a" on the "line" F. Similarly, t is located "immediately to the left of a" with respect to $P_{a,-}$.

Notation 1.1.14 If (K, P) is an ordered field, we denote the sign function with respect to P by $\text{sign}_P: K \to \{-1, 0, 1\}$. Thus, $\text{sign}_P(0) = 0$, and for $a \in K^*$ we have $\text{sign}_P(a) = 1$ if $a \in P$ and $\text{sign}_P(a) = -1$ if $a \notin P$.

As usual, $|\cdot|_P: K \to P$, $|a|_P := a \cdot \text{sign}_P(a)$ denotes the absolute value. If P is clear, the index P may be omitted.

The meaning of the generalized intervals $[a, b]_P$, $[a, b[_P$, $]a, b]_P$, $]a, b[_P$ for $a, b \in K \cup \{-\infty, +\infty\}$ is also clear, namely $[a, b[_P = \{x \in K: a \leq_P x <_P b\}$, $]a, \infty[_P = \{x \in K: x >_P a\}$, etc. Here too we will usually drop the index P, and instead often write $[a, b]_K$, etc. if several fields are considered.

1.2 Quadratic Forms, Witt Rings, Signatures

It is not possible to give an in-depth treatment of the algebraic theory of quadratic forms in this book. In the next sections we will therefore only develop what we need from the basics of the theory for later applications in real algebra. We will focus in particular on the relations between quadratic forms and orderings. For references and for a detailed treatment of quadratic form theory we recommend the books of Lam [69, 72] and Scharlau [98]. For Witt rings there is the smaller volume [62].

In this section we assume that K is a field of characteristic char $K \neq 2$. All vector spaces are assumed finite dimensional. Let V be a K-vector space. A *symmetric bilinear form* on V (over K) is a K-bilinear map $b \colon V \times V \to K$ such that $b(v, w) = b(w, v)$ ($v, w \in V$). A *quadratic form* on V (over K) is a map $q \colon V \to K$ that satisfies $q(av) = a^2 q(v)$ ($a \in K, v \in V$) and such that the map $b_q \colon V \times V \to K$, $b_q(v, w) = q(v + w) - q(v) - q(w)$ is a (symmetric) K-bilinear form. Every symmetric bilinear form b on V defines a quadratic form q_b on V via $q_b(v) := b(v, v)$. Since $2 \neq 0$, the maps $b \mapsto q_b$ and $q \mapsto \frac{1}{2} b_q$ describe mutually inverse bijections between the symmetric bilinear forms and the quadratic forms on V, and as such these concepts are usually identified.

The pair $\varphi = (V, b)$ is called a *bilinear space*, whereas the pair $\varphi = (V, q)$ is called a *quadratic space*. Two quadratic spaces (V, q) and (V', q') are called *isomorphic* (or *isometric*), denoted $(V, q) \cong (V', q')$, if there exists a vector space isomorphism $\phi \colon V \to V'$ such that $q = q' \circ \phi$. Quadratic spaces are most easily described by symmetric matrices: If $B = (b_{ij}) \in M_n(K)$ is symmetric, then B defines the quadratic space $\varphi_B = (K^n, q_B)$, where

$$q_B(x) = \sum_{i,j=1}^n b_{ij} x_i x_j \quad \left(x = (x_1, \ldots, x_n) \in K^n \right).$$

For symmetric matrices $B, B' \in M_n(K)$ it is straightforward to check that $\varphi_B \cong \varphi_{B'}$ if and only if there exists $S \in GL_n(K)$ such that $B' = SBS^t$. If $B = \operatorname{diag}(a_1, \ldots, a_n)$ is a diagonal matrix, we simply write φ_B as $\langle a_1, \ldots, a_n \rangle$. Every quadratic space is diagonalizable, i.e., isomorphic to a space $\langle a_1, \ldots, a_n \rangle$. The *hyperbolic plane* H is the quadratic space defined by $B = \left(\begin{smallmatrix} 0 & 1 \\ 1 & 0 \end{smallmatrix} \right)$. In diagonal form, $H \cong \langle 1, -1 \rangle$. We write $\langle a_1, \ldots, a_n \rangle_K$ or H_K when we want to emphasize the base field K.

Given bilinear spaces $\varphi = (V, b)$ and $\varphi' = (V', b')$, we can construct new ones, namely their *orthogonal sum* $\varphi \perp \varphi' = (V \oplus V', b \perp b')$ and their *tensor product* $\varphi \otimes \varphi' = (V \otimes V', b \otimes b')$. Here $b \perp b'$ and $b \otimes b'$ are defined by $(b \perp b')(v + v', w + w') = b(v, w) + b'(v', w')$ and $(b \otimes b')(v \otimes v', w \otimes w') = b(v, w) b'(v', w')$, respectively. If φ and φ' are represented by matrices B and B', respectively, then $\varphi \perp \varphi'$ is represented by $\left(\begin{smallmatrix} B & 0 \\ 0 & B' \end{smallmatrix} \right)$ and $\varphi \otimes \varphi'$ is represented by the Kronecker product $B \otimes B'$. In particular,

$$\langle a_1, \ldots, a_m \rangle \perp \langle b_1, \ldots, b_n \rangle \cong \langle a_1, \ldots, a_m, b_1, \ldots, b_n \rangle$$

and

$$\langle a_1, \ldots, a_m \rangle \otimes \langle b_1, \ldots, b_n \rangle \cong \perp_{i,j} \langle a_i b_j \rangle.$$

The n-fold sum $\varphi \perp \cdots \perp \varphi$ is abbreviated by $n \times \varphi$ and the n-fold tensor product $\varphi \otimes \cdots \otimes \varphi$ by $\varphi^{\otimes n}$. Finally, if $\varphi = (V, b)$, we write $-\varphi$ for $(V, -b)$.

Let $\varphi = (V, b)$ be a bilinear space and $U \subseteq V$ a subset. Then $U^\perp := \{v \in V: b(u, v) = 0$ for all $u \in U\}$ is a linear subspace of V. If $V^\perp = 0$, then φ (or b, or q) is called *nondegenerate* (or *regular*), otherwise φ (or b, or q) is called *degenerate* (or *singular*). For a symmetric matrix B, φ_B is degenerate if and only if $\det B = 0$. The space $V^\perp$ is also called the *radical* of φ, denoted $\mathrm{Rad}(\varphi)$, and $\mathrm{codim}_V(V^\perp)$ is called the *rank* of φ, denoted $\mathrm{rank}(\varphi)$. Given any quadratic space φ there exists a nondegenerate quadratic space φ' (unique up to isomorphism) such that $\varphi \cong \varphi' \perp \mathrm{Rad}(\varphi)$. For this reason one usually only considers nondegenerate spaces. Straightforward, yet important, is the following fact: If (V, q) is a quadratic space and $W \subseteq V$ a subspace such that $(W, q|_W)$ is nondegenerate, then $V = W \perp W^\perp$.

The simplest invariants of nondegenerate quadratic spaces are the *dimension* and *determinant*. We have $\det(V, q) := (\det B)K^{*2} \in K^*/K^{*2}$, where q is represented by the matrix B. Note that $\det(\varphi \perp \varphi') = \det(\varphi) \cdot \det(\varphi')$, and $\det(\varphi \otimes \varphi') = (\det \varphi)^{\dim \varphi'} \cdot (\det \varphi')^{\dim \varphi}$. Of fundamental importance is

Theorem 1.2.1 (Witt's Cancellation Theorem) *Let φ_1, φ_2 and ψ be quadratic spaces. If $\varphi_1 \perp \psi \cong \varphi_2 \perp \psi$, then $\varphi_1 \cong \varphi_2$.*

The proof can be found in any textbook on quadratic forms (e.g., [62, 69, 72, 98]), as well as in a number of algebra textbooks (e.g., [54, 55, 73]).

A quadratic form q on V *represents an element* $a \in K$ if there exists $0 \neq v \in V$ such that $q(v) = a$. If q represents all nonzero $a \in K$, then q is called *universal*. For example, the hyperbolic plane H is universal. A useful observation: If $a_1 \in K^*$ is represented by q, there exist $a_2, \ldots, a_n \in K$ such that $q \cong \langle a_1, a_2, \ldots, a_n \rangle$ ($n = \dim V$). The quadratic space (V, q) is called *isotropic* if q represents zero, and *anisotropic* otherwise. Every nondegenerate quadratic space φ has an orthogonal *Witt decomposition* $\varphi \cong \varphi_0 \perp n \times H$ where $n \geq 0$ and φ_0 is anisotropic. By Witt's Cancellation Theorem, φ_0 (up to isomorphism) and n are uniquely determined. We call φ_0 the *kernel form* and n the *Witt index* of φ. If $\varphi_0 = 0$, i.e., $\varphi \cong n \times H$, then φ is called *hyperbolic*. A quadratic space $\varphi = (V, q)$ is hyperbolic if and only if it is nondegenerate and there exists a subspace U of V such that $U = U^\perp$. From Witt decomposition it follows that every isotropic form is universal.

For the remainder of this section, all quadratic spaces are assumed to be nondegenerate, unless stated otherwise.

Definition 1.2.2 Two quadratic spaces φ, φ' are called *Witt equivalent*, denoted $\varphi \sim \varphi'$, if their kernel forms are isomorphic. We denote the class of all quadratic spaces that are Witt equivalent to φ by $[\varphi]$ (the *Witt class* of φ), and the set of all Witt classes of (nondegenerate) quadratic spaces over K by $W(K)$.

Lemma 1.2.3 *Let $\varphi, \varphi', \psi, \psi'$ be quadratic spaces.*

(a) $\varphi \sim \psi \Leftrightarrow$ *there exist $m, n \geq 0$ such that $\varphi \perp m \cdot H \cong \psi \perp n \cdot H$.*
(b) $\varphi \otimes H \cong \varphi \perp (-\varphi)$ *is a hyperbolic space.*
(c) *Witt equivalence is compatible with $\perp$ and $\otimes$, i.e., $\varphi \sim \varphi'$ and $\psi \sim \psi'$ imply $\varphi \perp \psi \sim \varphi' \perp \psi'$ and $\varphi \otimes \psi \sim \varphi' \otimes \psi'$.*

Proof (a) follows from the uniqueness of Witt decomposition.
   (b) Since $H \cong \langle 1, -1 \rangle$ we have $\varphi \otimes H \cong \varphi \perp (-\varphi)$. Since φ is diagonalizable, it suffices to settle the case $\varphi = \langle a \rangle$ $(a \in K^*)$. Assume that $\varphi \otimes H$ is represented by the matrix $\begin{pmatrix} 0 & a \\ a & 0 \end{pmatrix}$ with respect to a basis (e, f). Then it is represented by $\begin{pmatrix} 0 & 1 \\ 1 & 0 \end{pmatrix}$ with respect to the basis $(e, a^{-1}f)$. This shows $\varphi \otimes H \cong H$.
   (c) follows from (a) and (b).   □

Theorem 1.2.4 *The set $W(K)$ is made into a commutative unitary ring with unit $[\langle 1 \rangle]$ by the (well-defined) pairings*

$$[\varphi] + [\psi] := [\varphi \perp \psi] \quad \text{and} \quad [\varphi] \cdot [\psi] := [\varphi \otimes \psi].$$

$W(K)$ *is called the* Witt ring *of the field K.*

Proof That the pairings are well-defined follows from Lemma 1.2.3(c). That $[\varphi] + [-\varphi] = 0$ in $W(K)$ for all φ follows from Lemma 1.2.3(b). The other ring axioms are trivially satisfied.   □

   In order not to overload our notation, we will often write $\langle a_1, \ldots, a_n \rangle$ instead of $[\langle a_1, \ldots, a_n \rangle]$. Thus, $\langle a_1, \ldots, a_n \rangle$ may describe a quadratic form as well as its Witt class.

Example 1.2.5 With reference to Exercises 1.2.17 and 1.2.18 below we have $W(\mathbb{R}) \cong \mathbb{Z}$ and $W(\mathbb{C}) \cong \mathbb{Z}/2\mathbb{Z}$.

Exercise 1.2.6 Let K be a finite field of characteristic not 2.

(a) Show that $|K^*/K^{*2}| = 2$.
(b) Show that any 2-dimensional form is universal, and conclude that $|W(K)| = 4$.
(c) Distinguish the two cases $|K| \equiv 1$ or $-1 \mod 4$, and conclude that $W(K) \cong \dfrac{\mathbb{Z}/2\mathbb{Z}[t]}{(t^2)}$ in the first case, and $W(K) \cong \mathbb{Z}/4\mathbb{Z}$ in the second case.

Example 1.2.7 (Milnor's Exact Sequence) For every prime number p, we denote the finite field with p elements by $\mathbb{F}_p$. Then $W(\mathbb{Q})$ can be computed explicitly as an additive group by the split exact sequence

$$0 \longrightarrow \mathbb{Z} \xrightarrow{i} W(\mathbb{Q}) \xrightarrow{\oplus \partial_p} \mathbb{Z}/2\mathbb{Z} \oplus \bigoplus_{p \neq 2} W(\mathbb{F}_p) \longrightarrow 0,$$

where $i : \mathbb{Z} \to W(\mathbb{Q})$ is the unique homomorphism that sends 1 to $\langle 1 \rangle$, and ∂_2 and the ∂_p for odd primes p are certain group homomorphisms, cf. [72, VI, §4] for the details. Witt groups of number fields are considerably harder to understand, cf. [72, VI, §3]. There is a similar split exact sequence for Witt groups of rational function fields in one variable, cf. [72, IX, §3].

Next we study the relations between the Witt ring of K and orderings of K.

Definition 1.2.8 Let (K, P) be an ordered field and let $\varphi = (V, q)$ be a quadratic space over K. Then φ (or q) is called *positive definite with respect to P* if $q(v) >_P 0$ for all $0 \neq v \in V$. If $-q$ is positive definite, then q is called *negative definite*.

The following theorem is well-known:

Theorem 1.2.9 (Sylvester's Inertia Theorem) *Let φ be a possibly degenerate quadratic space over K and P an ordering of K. Then there is an orthogonal decomposition*

$$\varphi \cong \varphi_+ \perp \varphi_- \perp \mathrm{Rad}(\varphi),$$

where φ_+ is positive definite with respect to P and φ_- is negative definite with respect to P. Furthermore, $\dim \varphi_+$ and $\dim \varphi_-$ only depend on φ (but not on φ_+ and φ_-).

Proof We may assume that $\mathrm{Rad}(\varphi) = 0$. The existence of φ_+ and φ_- follows from diagonalizing φ. Assume that $\varphi \cong \varphi_+ \perp \varphi_- \cong \psi_+ \perp \psi_-$, with φ_+, ψ_+ positive definite and φ_-, ψ_- negative definite. Assume that φ (resp. $\varphi_\pm$, resp. $\psi_\pm$) is defined on V (resp. on $V_\pm$, resp. on $W_\pm$). We interpret $V_\pm$ and $W_\pm$ as subspaces of V. If $\dim V_+ < \dim W_+$, then $V_- \cap W_+ \neq 0$ since $\dim V_- + \dim W_+ > \dim V$. This is a contradiction since φ is positive and negative definite on $V_- \cap W_+$. If $\dim V_+ > \dim W_+$ we get a similar contradiction. $\square$

Definition 1.2.10 Let φ be a quadratic space over K, P an ordering of K, and $\varphi \cong \varphi_+ \perp \varphi_- \perp \mathrm{Rad}(\varphi)$ a decomposition as in Theorem 1.2.9. Then the integer

$$\mathrm{sign}_P \, \varphi := \dim(\varphi_+) - \dim(\varphi_-)$$

is called the *Sylvester signature of φ at P*.

Lemma 1.2.11 *Let (K, P) be an ordered field. Then $[\varphi] \mapsto \text{sign}_P\,\varphi$ defines a ring homomorphism sign_P from $W(K)$ to $\mathbb{Z}$.*

Proof The following equalities are easily verified:

$$\text{sign}_P(\varphi \perp \psi) = \text{sign}_P\,\varphi + \text{sign}_P\,\psi,$$

$$\text{sign}_P(\varphi \otimes \psi) = (\text{sign}_P\,\varphi) \cdot (\text{sign}_P\,\psi),$$

$$\text{sign}_P\,H = 0.$$

The statement follows.   □

Now a critical observation is that, conversely, every ring homomorphism $W(K) \to \mathbb{Z}$ arises in this manner. In other words, orderings of K and homomorphisms $W(K) \to \mathbb{Z}$ are "the same":

Theorem 1.2.12 *The Sylvester signature defines a bijection* $\text{sign}: P \mapsto \text{sign}_P$ *from the set of orderings of K to the set of ring homomorphisms $W(K) \to \mathbb{Z}$.*

Proof The map sign is injective since $P = \{0\} \cup \{a \in K^*: \text{sign}_P[\langle a \rangle] = 1\}$. Let $\phi: W(K) \to \mathbb{Z}$ be a homomorphism. Define $\chi: K^* \to \mathbb{Z}$ by $\chi(a) := \phi[\langle a \rangle]$. Then $\chi: K^* \to \{\pm 1\}$ is a group homomorphism since $\chi(1) = 1$ and $\chi(ab) = \chi(a)\chi(b)$ $(a, b \in K^*)$. We must show that $P := \{0\} \cup \text{Ker}\,\chi$ is an ordering of K. Then $\phi = \text{sign}_P$ follows immediately. Since $[\langle 1 \rangle] + [\langle -1 \rangle] = 0$ in $W(K)$, we have $\phi[\langle -1 \rangle] = \chi(-1) = -1$, i.e., $-1 \notin P$. Then $P \cup (-P) = K$ is also clear, and $PP \subseteq P$ follows since χ is a homomorphism. In order to show $P + P \subseteq P$ (which completes the proof), we use Lemma 1.2.13 below. Assume that $a, b \in P$. We may assume $a, b, a + b \neq 0$. By Lemma 1.2.13, applying ϕ to $[\langle a, b \rangle]$ gives $\chi(a) + \chi(b) = \chi(a+b) \cdot (1 + \chi(a)\chi(b))$, and $\chi(a) = \chi(b) = 1$ gives $\chi(a+b) = 1$.   □

Lemma 1.2.13 *If $a, b \in K^*$ with $a + b \neq 0$, then*

$$\langle a, b \rangle \cong \langle a + b, (a + b)ab \rangle.$$

Proof Since $a + b$ is represented by $\langle a, b \rangle$, there exists $c \in K^*$ such that $\langle a, b \rangle \cong \langle a + b, c \rangle$. Comparing determinants gives $abK^{*2} = (a + b)cK^{*2}$, i.e., $cK^{*2} = (a + b)abK^{*2}$. The statement follows.   □

Theorem 1.2.12 justifies:

Definition 1.2.14 A *signature* of a field K is a ring homomorphism $W(K) \to \mathbb{Z}$.

Remark 1.2.15 We see that a field admits a signature if and only if it is real. Note that for *every* field K (with char $K \neq 2$) there exists a ring homomorphism $e: W(K) \rightarrow \mathbb{Z}/2\mathbb{Z}$, defined by $e[\varphi] := \dim(\varphi) + 2\mathbb{Z}$. Every signature σ yields a commutative diagram

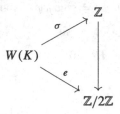

Definition 1.2.16 The map $e : W(K) \rightarrow \mathbb{Z}/2\mathbb{Z}$ is called the *dimension index*. Its kernel is denoted $I(K)$ and is called the *fundamental ideal* of $W(K)$. (Thus $I(K)$ consists of the Witt classes of even dimensional quadratic spaces.)

Exercise 1.2.17 Show that if $K^{*2} \cup \{0\} = \{a^2 : a \in K\}$ is an ordering of K, then it is the only ordering of K, and the Sylvester signature gives a ring isomorphism $W(K) \rightarrow \mathbb{Z}$.

This is the case for example when $K = \mathbb{R}$ or, more generally, when K is any real closed field, cf. Sect. 1.5.

Exercise 1.2.18 Show that if K is a quadratically closed field (i.e., $K^{*2} = K^*$) then, up to isomorphism, every nondegenerate quadratic form is determined by its dimension, and e is a ring isomorphism $W(K) \rightarrow \mathbb{Z}/2\mathbb{Z}$.

1.3 Extension of Orderings

In this section, K denotes a field of characteristic $\neq 2$.

Let L/K be a field extension. A bilinear space $\varphi = (V, b)$ over K gives rise to a bilinear space $\varphi_L = (V_L, b_L)$ over L via extension of scalars: $V_L = V \otimes_K L$ and b_L is defined by $b_L(v \otimes a, v' \otimes a') = b(v, v')aa'$ for $a, a' \in L$, $v, v' \in V$. If q is the quadratic form (over K) associated to φ, then q_L denotes the quadratic form (over L) associated to φ_L. If φ is represented by a matrix, then φ_L is represented by the same matrix, but this time considered as a matrix over L.

The form φ_L is nondegenerate if and only if the form φ is nondegenerate. If φ is hyperbolic, then φ_L is also hyperbolic, but the converse is in general false (example?). Consequently, $\varphi \mapsto \varphi_L$ induces a map on Witt classes, $i_{L/K}: W(K) \rightarrow W(L)$, which is a ring homomorphism. If M/L is a further field extension, then $i_{M/K} = i_{M/L} \circ i_{L/K}$.

Definition 1.3.1 Let L/K be a field extension.

(a) An ordering P of K *extends* to an ordering Q of L if $P \subseteq Q$ (and so, $P = K \cap Q$).
(b) A signature σ of K *extends* to a signature τ of L if $\sigma = \tau \circ i_{L/K}$. We also write $\tau \mid \sigma$.

One should convince oneself that for orderings P of K and Q of L, P extends to Q if and only if sign_P extends to sign_Q.

Proposition 1.3.2 *Let L/K be a field extension and P an ordering of K. The following statements are equivalent:*

 (i) *P extends to an ordering Q of L;*
 (ii) *-1 is not contained in the semiring T generated by P and ΣL^2 in L (i.e., the smallest subset T of L such that $P \cup \Sigma L^2 \subseteq T$, $T + T \subseteq T$ and $TT \subseteq T$);*
(iii) *every quadratic form $\langle p_1, \ldots, p_n \rangle$ with $p_1, \ldots, p_n \in P^* := P \setminus \{0\}$ is anisotropic over L.*

A further equivalent statement will be listed in Proposition 1.4.4.

Proof (i) $\Rightarrow$ (iii): If $p_1 b_1^2 + \cdots + p_n b_n^2 = 0$, with $b_1, \ldots, b_n \in L$, then $b_1 = \cdots = b_n = 0$ since Q is an ordering.

(iii) $\Rightarrow$ (ii): We can write $T = \{p_1 b_1^2 + \cdots + p_m b_m^2 : m \geq 1, \, p_i \in P^*, b_i \in L\}$. If $-1 \in T$, then $-1 = p_1 b_1^2 + \cdots + p_m b_m^2$ and the form $\langle 1, p_1, \ldots, p_m \rangle$ is isotropic over L.

(ii) $\Rightarrow$ (i): By assumption T is a preordering of L. Every ordering Q of L such that $Q \supseteq T$ (cf. Lemma 1.1.9) extends P. $\qquad\square$

Proposition 1.3.3 *Let P be an ordering of K, $d \in K$ and $L = K(\sqrt{d})$. Then P extends to L if and only if $d \in P$.*

Proof If $d \notin P$, then P has no extension to L since $d \in L^{*2}$. Assume thus that $d \in P$ and, without loss of generality, that $K \neq L$. We verify Proposition 1.3.2(ii). Assume for the sake of contradiction that there is an equality

$$-1 = \sum_{i=1}^{m} p_i (a_i + b_i \sqrt{d})^2$$

in L with $m \geq 1$, $p_i \in P$ and $a_i, b_i \in K$. It follows in particular (by comparing coefficients with respect to the K-basis $(1, \sqrt{d})$ of L) that $-1 = \sum_{i=1}^{m} p_i (a_i^2 + d b_i^2)$, and so $-1 \in P$, a contradiction. $\qquad\square$

We will see later that if $d \in P$ (and $\sqrt{d} \notin K$), then there are precisely two extensions of P to $L = K(\sqrt{d})$ (cf. Corollary 1.11.7).

Example 1.3.4 The unique ordering of $\mathbb{Q}$ has precisely two extensions, P_+ and P_-, to the field $L = \mathbb{Q}(\sqrt{2})$. They are distinguished by the fact that $\sqrt{2} \in P_+$ and $-\sqrt{2} \in P_-$. They are the only orderings of L and correspond to the two possible embeddings of L into $\mathbb{R}$.

In the following results we consider field extensions for which every ordering extends:

Proposition 1.3.5 *Let L/K be a finite field extension of odd degree. Then every ordering of K extends to L.*

By Proposition 1.3.2, this proposition follows from

Theorem 1.3.6 (T.A. Springer) *If L/K is a finite field extension of odd degree and q is an anisotropic quadratic form over K, then q_L is anisotropic over L.*

Proof We may assume without loss of generality that $L = K(\alpha)$ is a simple field extension. Let $f \in K[t]$ be the minimal polynomial of α over K. We proceed by induction on $n = [L : K] = \deg f$. Assume that $n > 1$ is odd and that the statement is true for all smaller degrees. Let $q = \langle a_1, \ldots, a_m \rangle$ be an anisotropic form over K. Assume for the sake of contradiction that q_L is isotropic. Then there exist $g_1, \ldots, g_m, h \in K[t]$ with $\deg g_i < n$ $(i = 1, \ldots, m)$ and not all $g_i = 0$, such that

$$a_1 g_1(t)^2 + \cdots + a_m g_m(t)^2 = f(t)h(t) \tag{1.1}$$

in $K[t]$. We may assume that $\gcd(g_1, \ldots, g_m) = 1$.

Let d be the maximum of the degrees of the g_i. Then $n + \deg h = 2d$ since q is anisotropic. Since $d < n$ we have $\deg h < n$ and $\deg h$ is odd. In particular, h has an irreducible factor $h_1 = h_1(t)$ of odd degree. For the extension field $E = K[t]/(h_1)$ of K we thus have $[E : K] < n$ and $[E : K]$ is odd. On the other hand, q_E is isotropic by (1.1). This contradicts the induction hypothesis.   □

Proposition 1.3.7 *If L/K is a purely transcendental field extension (not necessarily finitely generated), then every ordering of K extends to L.*

Using Zorn's Lemma, we may assume that $L = K(t)$ is simple transcendental and specify explicit extensions (cf. Example 1.1.13(3)). A different proof (again using Proposition 1.3.2) can be obtained from:

Proposition 1.3.8 *If L/K is purely transcendental and q is an anisotropic quadratic form over K, then q_L is also anisotropic.*

Proof We may assume without loss of generality that $L = K(t)$ is simple transcendental. If $q = \langle a_1, \ldots, a_m \rangle$ (with $a_i \in K^*$) and q_L is isotropic, then there exist polynomials $g_1, \ldots, g_m \in K[t]$, not all 0, with $a_1 g_1(t)^2 + \cdots + a_m g_m(t)^2 = 0$ in $K[t]$. By considering again the maximal occurring degree, it follows that q is isotropic.   □

Remark 1.3.9 From Theorem 1.3.6 and Proposition 1.3.8 it follows in particular that the homomorphism $i_{L/K} : W(K) \to W(L)$ is injective in case L/K is finite of odd degree or purely transcendental.

1.4 The Prime Ideals of the Witt Ring

In this section, K denotes a field of characteristic $\neq 2$.

We already know some of the prime ideals of the Witt ring $W(K)$, namely the fundamental ideal $I(K)$, as well as the kernels of the signatures of K. We will see that these are essentially the only prime ideals. Thus, Witt rings have a very simple and clear prime ideal structure, a fact that was observed surprisingly late (J. Leicht, F. Lorenz in [77] and D.K. Harrison in [44], both in 1970).

Theorem 1.4.1 (The Prime Ideals of $W(K)$)

(a) *If K is not real, then $I(K)$ is the only prime ideal of $W(K)$.*

(b) *If K is real, then the kernels $\mathfrak{p}_\sigma$ of the signatures σ of K are precisely the minimal prime ideals of $W(K)$. Every other prime ideal is either equal to $I(K)$, or is of the form $\mathfrak{p} = \mathfrak{p}_\sigma + p \cdot W(K)$ for some signature σ and some prime number $p \neq 2$, both uniquely determined by $\mathfrak{p}$.*

Thus, $\mathfrak{p}_\sigma + 2 \cdot W(K) = I(K)$ for every signature σ, as already observed in Sect. 1.2.

Proof Let $\mathfrak{p}$ be a prime ideal of $W(K)$, $\kappa(\mathfrak{p}) = \operatorname{Quot} W(K)/\mathfrak{p}$ and $\pi \colon W(K) \to W(K)/\mathfrak{p}$ the residue map. The composition $\phi \colon \mathbb{Z} \to W(K) \overset{\pi}{\to} W(K)/\mathfrak{p}$ is surjective since the Witt classes $\xi = [\langle a \rangle]$ ($a \in K^*$) generate $W(K)$ as a ring and $\xi^2 - 1 = 0$ in $W(K)$, hence $\xi \equiv 1$ or $\xi \equiv -1 \bmod \mathfrak{p}$.

If $\operatorname{char} \kappa(\mathfrak{p}) = 0$, then ϕ is also injective, and thus an isomorphism, and $\mathfrak{p}$ is the kernel of the signature $\phi^{-1} \circ \pi$.

If $\operatorname{char} \kappa(\mathfrak{p}) = p > 0$, then ϕ induces an isomorphism $\overline{\phi} \colon \mathbb{Z}/p\mathbb{Z} \overset{\sim}{\to} W(K)/\mathfrak{p}$. Assume that $p = 2$. Then $\mathfrak{p} = I(K)$ since there is only one ring homomorphism $W(K) \to \mathbb{Z}/2\mathbb{Z}$ (the generators $[\langle a \rangle]$, $a \in K^*$, are units in $W(K)$). Finally, assume that $p > 2$. Then $+1 \neq -1$ (in $\mathbb{Z}/p\mathbb{Z}$), and $\chi \colon a \mapsto \overline{\phi}^{-1} \circ \pi[\langle a \rangle]$ ($a \in K^*$) defines a group homomorphism $\chi \colon K^* \to \{\pm 1\}$. (Note that $\chi(K^*) \subseteq \{\pm 1\}$ holds by the argument above.) This map satisfies $\chi(-1) = -1$. Following the proof of Theorem 1.2.12 verbatim, it follows that $P := \{0\} \cup \operatorname{Ker} \chi$ is an ordering of K. Let $\sigma := \operatorname{sign}_P$. Since $\overline{\phi}$ is an isomorphism, it follows from the commutative diagram

$$
\begin{array}{ccc}
W(K) & \overset{\pi}{\longrightarrow} & W(K)/\mathfrak{p} \\
{\scriptstyle \sigma}\downarrow & \overset{\phi}{\nearrow} & \cong \uparrow {\scriptstyle \overline{\phi}} \\
\mathbb{Z} & \longrightarrow & \mathbb{Z}/p\mathbb{Z}
\end{array}
$$

that $\mathfrak{p} = \operatorname{Ker}\left(W(K) \overset{\sigma}{\to} \mathbb{Z} \to \mathbb{Z}/p\mathbb{Z} \right)$, i.e., that $\mathfrak{p} = \mathfrak{p}_\sigma + p \cdot W(K)$. It is clear that σ and p are uniquely determined by $\mathfrak{p}$. This establishes the proof. $\qquad\square$

Since in every (commutative) ring the nilradical (i.e., the set of all nilpotent elements) is equal to the intersection of all prime ideals ([55, 68, 73], see also Proposition 3.1.11), we obtain immediately

Corollary 1.4.2 *Let φ be a nondegenerate quadratic space over K. The following statements are equivalent:*

(i) *$\varphi^{\otimes n}$ is hyperbolic for some $n \in \mathbb{N}$;*
(ii) *$\dim \varphi$ is even, and $\mathrm{sign}_P \varphi = 0$ for every ordering P of K.*   □

If K is ordered, the parity condition on $\dim \varphi$ can of of course be omitted from (ii).

Applying Corollary 1.4.2 to $\varphi = \langle 1, 1 \rangle = 2 \times \langle 1 \rangle$ gives

Corollary 1.4.3 *If K is not real, then there exists $n \in \mathbb{N}$ such that $2^n \cdot W(K) = 0$.*   □

Having determined the prime ideals of $W(K)$, we can state a further criterion for when orderings extend:

Proposition 1.4.4 *Let L/K be a field extension. A signature of K extends to L if and only if it vanishes on the kernel of $i_{L/K} \colon W(K) \to W(L)$.*

For the proof we require a simple fact from commutative algebra:

Lemma 1.4.5 *If $\phi \colon A \to B$ is an injective ring homomorphism and $\mathfrak{p}$ a minimal prime ideal of A, then there exists a prime ideal $\mathfrak{q}$ of B such that $\mathfrak{p} = \phi^{-1}(\mathfrak{q})$.*

Proof Since $S := \phi(A \setminus \mathfrak{p})$ does not contain zero, we have that $S^{-1}B \neq 0$ and every prime ideal $\mathfrak{q}'$ of $S^{-1}B$ satisfies $\mathfrak{p} = \phi^{-1}\left(j^{-1}(\mathfrak{q}')\right)$, where $j \colon B \to S^{-1}B$ denotes the canonical homomorphism.   □

Proof of Proposition 1.4.4 Let $I := \mathrm{Ker}\, i_{L/K}$ and let σ be a signature of K. If σ has an extension τ, then $\sigma = \tau \circ i_{L/K}$, i.e., $\sigma(I) = 0$.

Conversely, assume that $\sigma(I) = 0$, i.e., $I \subseteq \mathfrak{p}_\sigma := \mathrm{Ker}\, \sigma$. Then $\mathfrak{p}_\sigma/I$ is a minimal prime ideal of $W(K)/I$ by Theorem 1.4.1. By Lemma 1.4.5 (applied to $W(K)/I \hookrightarrow W(L)$) there exists a prime ideal $\mathfrak{q}$ of $W(L)$ such that $\mathfrak{p}_\sigma = i_{L/K}^{-1}(\mathfrak{q})$. Since $\mathbb{Z} \cong W(K)/\mathfrak{p}_\sigma \hookrightarrow W(L)/\mathfrak{q}$ and again by Theorem 1.4.1, it follows that $W(K)/\mathfrak{p}_\sigma \to W(L)/\mathfrak{q}$ is an isomorphism, i.e., $\mathfrak{q}$ determines an extension τ of σ.   □

In Sect. 1.3 we proved the injectivity of the map $i_{L/K}$ for extensions L/K that are of odd degree or that are purely transcendental. Next, we will determine the kernel of $i_{L/K}$ for quadratic extensions (and so with Proposition 1.4.4 give a new proof of Proposition 1.3.3):

Proposition 1.4.6 *Let $a \in K^* \setminus K^{*2}$ and $L = K(\sqrt{a})$. Then the kernel of $i_{L/K}$ is the ideal of $W(K)$ generated by $\langle 1, -a \rangle$.*

A more precise statement is:

Proposition 1.4.7 *Let q be an anisotropic quadratic form over K. There exist quadratic forms q', q'' over K with $q \cong q' \perp \langle 1, -a \rangle \otimes q''$ such that q'_L is anisotropic.*

Proof By induction on $\dim q$. The case $\dim q = 1$ is clear. Assume thus that $\dim q \geq 2$ and without loss of generality that q_L is isotropic. If (V, b) denotes the bilinear space associated to q, then there exist $v, w \in V$, not both zero, such that

$$0 = q_L(v + \sqrt{a}\, w) = q(v) + a \cdot q(w) + \sqrt{a} \cdot b(v, w).$$

It follows that $q(v) + a\, q(w) = 0 = b(v, w)$. Since q is anisotropic, $q(v), q(w) \neq 0$. Let $W := Kv + Kw \subseteq V$. Then $\dim W = 2$ (if v, w were linearly dependent, then we would have $b(v, w) \neq 0$), and $q|_W \cong \langle -ac, c \rangle = \langle 1, -a \rangle \otimes \langle c \rangle$ for $c := q(w)$. In particular, $q|_W$ is nondegenerate. Since $V = W \perp W^\perp$, we can apply the induction hypothesis to $q|_{W^\perp}$ and so conclude the proof.   □

1.5 Real Closed Fields and Their Field Theoretic Characterization

After our excursion into the theory of quadratic forms we return to real algebra in the narrower sense. This section is fundamental for everything that follows.

Definition 1.5.1 A field is called *real closed* if it is real and has no proper algebraic extension that is real.

The field $\mathbb{R}$ of real numbers is a well-known example of a real closed field.

Proposition 1.5.2 *Let K be a field. The following statements are equivalent:*

 (i) *K is real closed;*
 (ii) *there exists an ordering P of K that cannot be extended to any proper algebraic extension of K;*
(iii) *$K^{*2} \cup \{0\} = \{a^2 : a \in K\}$ is an ordering of K, and every polynomial (in one variable) of odd degree over K has a zero in K.*

Moreover, if these conditions are satisfied, then $K^{*2} \cup \{0\}$ is the only ordering of K.

Proof The final comment is clear by (iii) (every ordering P satisfies $K^{*2} \cup \{0\} \subseteq P$).

(i) $\Rightarrow$ (ii) is clear since a field has an ordering if (and only if) it is real.

(ii) $\Rightarrow$ (iii): If it were true that $K^{*2} \cup \{0\} \neq P$, then there would exist $d \in P$ with $\sqrt{d} \notin K$, and P would extend to $K(\sqrt{d})$ (Proposition 1.3.3), a contradiction. Furthermore, K has no proper extensions of odd degree by Proposition 1.3.5.

(iii) $\Rightarrow$ (i): Let $L \supset K$ be a finite proper field extension of K. We will show that L is not real. By (iii), $[L : K]$ is a power of 2. Since K has an ordering by (iii), it follows that char $K = 0$ and in particular that L/K is separable. Let L_1 be the Galois closure of L over K, $G = \mathrm{Gal}(L_1/K)$ and $H = \mathrm{Gal}(L_1/L)$. Then H is a subgroup of G, and $\bigcap_{g \in G} H^g = \{1\}$ by Galois theory, where $H^g := g^{-1}Hg$. By elementary group theory it follows that $|G|$ divides $\prod_{g \in G}[G : H^g]$. By assumption (iii) this number is a power of 2, and so G is a 2-group. From Sylow's Theorems (elementary group theory again) it follows that there exists a subgroup $\widetilde{H}$ of G of index 2 with $H \subset \widetilde{H}$. Let F be the fixed field of $\widetilde{H}$. Then $K \subset F \subset L$ and $[F : K] = 2$. There exists $a \in K$ with $F = K(\sqrt{a})$. Since $K^{*2} \cup \{0\}$ is an ordering of K and $a \notin K^{*2} \cup \{0\}$, there exists $b \in K^*$ with $a = -b^2$. Hence $-1 = (\sqrt{a}/b)^2$ is a square in F, and so F is not real. We conclude that L is not real either.   □

Remark 1.5.3 Condition (iii) in Proposition 1.5.2 is clearly equivalent to

(iii') K is real, a or $-a$ is a square in K for every $a \in K$, and every polynomial of
        odd degree over K has a zero in K.

In the literature it is common to denote real closed fields with capital letters $R, S, \ldots$. We will usually follow this convention.

Theorem 1.5.4 (Fundamental Theorem of Algebra) *If R is a real closed field, then $R(\sqrt{-1})$ is algebraically closed.*

Proof (Gauss) We use Proposition 1.5.2(iii). Every finite extension of R (and so also every finite extension of $C := R(i)$, $i := \sqrt{-1}$) has 2-power degree. Assume for the sake of contradiction that $C = R(i)$ is not algebraically closed. As in the proof of Proposition 1.5.2 it follows that C has an extension F with $[F : C] = 2$. Thus, there exist $a, b \in R$ such that $\alpha := a + bi$ is not a square in C. Hence, $b \neq 0$. Since R^2 is an ordering of R, there exists $c \in R$ with $a^2 + b^2 = c^2$ and $c > 0$, as well as $x, y \in R$ with $x > 0$, $\text{sign}\, y = \text{sign}\, b$ and $x^2 = (c + a)/2$, $y^2 = (c - a)/2$. (Note that $c \pm a$ are positive and that all signs are of course understood to be with respect to the unique ordering of R.) Now we have $(x + iy)^2 = (x^2 - y^2) + 2xyi$, and also $x^2 - y^2 = a$ and $(2xy)^2 = b^2$. Since $\text{sign}\,(xy) = \text{sign}\, b$ we have $2xy = b$, and it follows that $(x + iy)^2 = \alpha$, contradicting $\sqrt{\alpha} \notin C$.   □

A very strong converse to this theorem is given by Theorem 1.6.1.

Corollary 1.5.5 *Let R be a real closed field and let $K \subseteq R$ be a subfield. Then K is real closed if and only if K is (relatively) algebraically closed in R.*

Proof Assume that K is algebraically closed in R and let $i = \sqrt{-1}$. Then $K(i)$ is also algebraically closed in $R(i)$. Indeed, if $\alpha = a + bi \in R(i)$ is algebraic over $K(i)$ $(a, b \in R)$, then also over K, and the same holds for $\overline{\alpha} = a - bi$, thus also for $a = (\alpha + \overline{\alpha})/2$ and $b = i(\overline{\alpha} - \alpha)/2$. It follows that $a, b \in K$, hence $\alpha \in K(i)$. Thus $K(i)$ is algebraically closed by Theorem 1.5.4 and it follows immediately that K is real closed.

The converse direction is trivial by definition.   □

Another characteristic of real closed fields is their relative rigidity:

Proposition 1.5.6 *Let R be a real closed field.*

(a) *Every (ring) endomorphism φ of R is order preserving, i.e., $a \leq b \Rightarrow \varphi(a) \leq \varphi(b)$ for all $a, b \in R$.*

(b) *If $K \subseteq R$ is a subfield and R/K is algebraic, then the identity is the only K-automorphism of R.*

Proof (a) This follows immediately from $R^2 = \{a \in R : a \geq 0\}$.

(b) Let $\varphi \in \text{Aut}_K(R)$ and $a \in R$. Since R/K is algebraic, $\{\varphi^n(a) : n \in \mathbb{N}\}$ is a finite set. If it were true that $\varphi(a) \neq a$, thus for instance $\varphi(a) > a$, it would follow from (a) by induction that $a < \varphi(a) < \varphi^2(a) < \cdots$, a contradiction. □

If R is real closed, we will henceforth denote the uniquely determined ordering of R by $\leq$ without further comment (as we already did several times above).

1.6   Galois Theoretic Characterization of Real Closed Fields

Let K be a field of arbitrary characteristic with algebraic closure $\overline{K}$. This section is dedicated to a proof of the following beautiful result of Artin and Schreier [4]:

Theorem 1.6.1 (E. Artin, O. Schreier, 1927) *If $K \neq \overline{K}$ and $[\overline{K} : K] < \infty$, then K is real closed and $\overline{K} = K(\sqrt{-1})$.*

Proof The elementary proof that we present here goes back to J. Leicht [75]. We break it down into several steps. Let $n = [\overline{K} : K]$.

(1) *The field K is perfect.*

Assume K is not perfect. Then char $K = p > 0$, and there exists $a \in K$ with $\alpha := a^{1/p} \notin K$. But then also $\alpha^{1/p} \notin K(\alpha)$. Indeed, if $\alpha = \beta^p$ with $\beta \in K(\alpha)$, then $\left(N_{K(\alpha)/K}(\beta)\right)^p = N_{K(\alpha)/K}(\alpha) = (-1)^{p-1}a = a$ since $(-1)^{p-1} = 1$. Iterating this step, we obtain field extensions of K of degree $p, p^2, p^3, \ldots$, contradicting $[\overline{K} : K] < \infty$.

It follows that $\overline{K}/K$ is a finite Galois extension. Let L be a maximal intermediate field of $\overline{K}/K$, different from $\overline{K}$, and let $q := [\overline{K} : L]$. (Note that q is prime!)

(2) char $K \neq q$.

Assume for the sake of contradiction that char $K = p = q$. By Artin–Schreier theory for Galois extensions of degree p in characteristic p we have that $\overline{K}$ is L-isomorphic to $L[t]/(t^p - t - a)$ for some $a \in L$ (see, for example, [96, Theorem 12.2.1], [73, VI, Theorem 6.4] or [55, Vol. II, §8.11]). We only need the consequence that $\psi(L) \neq L$ for the map $\psi : \overline{K} \to \overline{K}$ defined by $\psi(b) := b^p - b$.

Note that ψ is additive, i.e., $\psi(b + b') = \psi(b) + \psi(b')$. Consider the trace $\mathrm{tr} = \mathrm{tr}_{\overline{K}/L} \colon \overline{K} \to L$. Then $\mathrm{tr} \circ \psi = (\psi|_L) \circ \mathrm{tr}$. Indeed, if $b \in \overline{K}$ and if $b_1, \ldots, b_p$ are the L-conjugates of b in $\overline{K}$, then $\psi(b_1), \ldots, \psi(b_p)$ are the L-conjugates of $\psi(b)$, and it follows that $\mathrm{tr}\,(\psi(b)) = \sum_i \psi(b_i) = \psi(\sum_i b_i) = \psi(\mathrm{tr}\,b)$. Hence the diagram

$$\begin{array}{ccc} \overline{K} & \xrightarrow{\mathrm{tr}} & L \\ \downarrow{\psi} & & \downarrow{\psi|_L} \\ \overline{K} & \xrightarrow{\mathrm{tr}} & L \end{array}$$

commutes. Since ψ and tr are surjective (the second map since $\overline{K}/L$ is separable), the map $\psi|_L \colon L \to L$ is also surjective, a contradiction.

(3) $q = 2$ *(and* char $K \neq 2$*)*.

L contains the q-th roots of unity since $[\overline{K} : L] = q$ and the q-th cyclotomic polynomial is of degree $q - 1$. Hence, since char $K \neq q$, there exists $a \in L$ with $\overline{K} = L(a^{1/q})$ (see for instance [55, Vol. I, §4.7] or [73, VI, §6]). For $\beta := a^{1/q^2}$ we obtain as in (1) that $(N_{\overline{K}/L}(\beta))^q = (-1)^{q-1}a$. Since a is not a q-th power in L, q must be even, thus $q = 2$.

(4) $\overline{K} = L(\sqrt{-1})$.

Let $i = \sqrt{-1} \in \overline{K}$, and let $a \in L$ and $\beta \in \overline{K}$ as in (3), thus with $\beta^4 = a$ and $\overline{K} = L(\beta^2)$. Since $N_{\overline{K}/L}(\beta)^2 = -\beta^4$, we have $\beta^2/N_{\overline{K}/L}(\beta) = \sqrt{-1}$, and so $\overline{K} = L(\sqrt{-1})$.

(5) $L = K$, *thus* $\overline{K} = K(i)$.

If $K \neq L$, then also $K(i) \neq \overline{K}$. Carrying out steps (1) to (4) for $K(i)$ instead of K leads to a contradiction by (4).

(6) K *is real (and thus real closed)*.

Since $i \notin K$ it suffices to show that the sum of two squares in K is a square (Proposition 1.5.2). Thus, let $a, b \in K^*$. Since $\overline{K}$ is algebraically closed, there are $c, d \in K$ with $a + bi = (c + di)^2$. Hence, $a - bi = (c - di)^2$ and $a^2 + b^2 = (c^2 + d^2)^2$. This proves the theorem. □

Let K_s denote the separable closure of K.

Corollary 1.6.2 *If* char $K \neq 0$ *and* $[K_s : K] < \infty$, *then* $K = K_s$.

Proof Steps (2)–(6) of the proof can be carried out for K_s just as for $\overline{K}$. □

We finish this section with a consequence of the theorem of Artin and Schreier for absolute Galois groups of fields. These are profinite groups (i.e., projective limits of finite groups). For the basic aspects of infinite Galois theory, [54, Vol. III, §IV.2], [55, Vol. II, §8.6] or [12, Ch. V, §10], for example, can be consulted, but for understanding what follows this is not required.

Corollary 1.6.3 *Let K be a field with separable algebraic closure K_s and absolute Galois group $\Gamma = \mathrm{Gal}\,(K_s/K)$. All elements of finite order > 1 in Γ are involutions (i.e., have order 2), and the map $\tau \mapsto \mathrm{Fix}\,(\tau)$ (the fixed field of τ in $K_s = \overline{K}$) is a bijection from the set of involutions in Γ to the set of real closed overfields of K in $\overline{K}$. In particular, Γ contains elements of finite order > 1 if and only if K is real.* □

1.7 Counting Real Zeroes of Polynomials (without Multiplicities)

One of the oldest problems in real analysis consists of determining the number and location of the real zeroes of a polynomial $f(t)$ with real coefficients. The first general answer was formulated by J. C. F. Sturm (1803–1855) in the form of an algorithm, the so-called Sturm sequence, that determines the number of zeroes of f in any given interval, counted without multiplicities. Another solution of the problem, based on the examination of certain quadratic forms, was found by Ch. Hermite (1822–1901). Both methods will be presented in this section.

It should be remarked that these results used to be part of the mathematics curriculum at the turn of the nineteenth century (see for example H. Weber's *Lehrbuch der Algebra* [114]).

Sturm's solution, found in 1829, initiated an historically very interesting and to some extent stormy period in real algebra. (Sturm published the full proof only in 1835.) Hermite's theorem (Theorem 1.7.17) was anticipated by C.G.J. Jacobi (1804–1851) and his student and friend C.W. Borchardt (1817–1880) (Theorem 1.7.15), and was proved in 1853 by Hermite and—essentially independently— also by J.J. Sylvester (1814–1897). In addition, Sylvester, A. Cayley (1821–1895) and Sturm established interesting links between the theorems of Sturm and Hermite. In addition to the original literature, the reader can find full particulars of this history in the substantial text [65] of Krein and Naimark, as well as in [39].

We start with a number of classical results, well-known from real analysis, that stay valid for rational functions over arbitrary real closed fields.

In the remainder of this section R denotes a real closed field.

Proposition 1.7.1 *Let $f(t) = t^n + a_1 t^{n-1} + \cdots + a_n \in R[t]$ be a monic polynomial. Then all real zeroes of f (i.e., all zeroes of f in R) are in the interval $[-M, M]$, where $M := \max\{1, |a_1| + \cdots + |a_n|\}$.*

Proof If $0 \neq a \in R$ is such that $f(a) = 0$, then $a = -(a_1 + a_2 a^{-1} + \cdots + a_n a^{1-n})$, and so $|a| \leq 1$ or $|a| \leq |a_1| + \cdots + |a_n|$ by the triangle inequality. □

Proposition 1.7.2 (Intermediate Value Theorem) *Let* $f \in R(t)$ *and* $a, b \in R$ *with* $a < b$, *such that* f *has no pole in* $[a, b]$. *If* $f(a)f(b) < 0$, *then* f *has an odd number of zeroes in* $]a, b[$, *counted with multiplicities*.

Proof If $f = g/h$ with polynomials g, h, where h is nonzero on $[a, b]$, then we may replace f by $fh^2 = gh$, and thus assume that $0 \neq f \in R[t]$. Let $a_1 \leq \cdots \leq a_r$ be the real roots of f, counted with multiplicities $(r \geq 0)$. By the Fundamental Theorem of Algebra (Theorem 1.5.4) there exist $c \in R^*$ and irreducible monic quadratic polynomials $p_1, \ldots, p_s$ $(s \geq 0)$ with $f(t) = c \cdot (t - a_1) \cdots (t - a_r) p_1(t) \cdots p_s(t)$. Since the p_k only take positive values, we have

$$-1 = \text{sign}\frac{f(a)}{f(b)} = \prod_{j=1}^{r} \text{sign}\frac{a - a_j}{b - a_j}.$$

The number of j with $a < a_j < b$ is therefore odd.   □

Lemma 1.7.3 *Let* $0 \neq f \in R(t)$ *be a rational function and* $a \in R$ *a zero of* f. *The logarithmic derivative* f'/f *of* f *changes sign from minus to plus at* $t = a$ *(it has a pole at* $t = a$), *i.e., there exists* $\varepsilon > 0$ *such that* f'/f *is negative on* $]a - \varepsilon, a[$ *and positive on* $]a, a + \varepsilon[$.

Proof There exist $g, h \in R[t]$ with $g(a)h(a) \neq 0$ and $f(t) = (t - a)^d g(t)/h(t)$ for some $d \in \mathbb{N}$. Then

$$\frac{f'(t)}{f(t)} = \frac{d}{t - a} + \frac{g'(t)}{g(t)} - \frac{h'(t)}{h(t)},$$

from which the statement follows.   □

Proposition 1.7.4 *Let* $0 \neq f \in R(t)$ *and* $a, b \in R$ *with* $a < b$ *and* $f(a) = f(b) = 0$. *If* f *has neither poles nor zeroes in* (a, b), *then the number of zeroes of* f' *in* $]a, b[$, *counted with multiplicities, is odd*.

Corollary 1.7.5 (Rolle's Theorem) *If* $f \in R(t)$, $f(a) = f(b) \in R$ *for* $a, b \in R$, $a < b$, *and if* f *has no poles in* $[a, b]$, *then* f' *has a zero in* $]a, b[$.

Proof of Proposition 1.7.4 Let $\varepsilon > 0$ be such that f' neither has zeroes in $]a, a + \varepsilon]$, nor in $[b - \varepsilon, b[$, and such that $a + \varepsilon < b - \varepsilon$. By Lemma 1.7.3 we have

$$\frac{f'(a + \varepsilon)}{f(a + \varepsilon)} \cdot \frac{f'(b - \varepsilon)}{f(b - \varepsilon)} < 0,$$

and by Proposition 1.7.2 the number of zeroes of f'/f in $]a + \varepsilon, b - \varepsilon[$ is odd. This number equals the number of zeroes of f' in $]a, b[$.   □

In what follows we assume that $f \in R[t]$ is a fixed non-constant polynomial and that $a, b \in R$ with $a < b$ and $f(a)f(b) \neq 0$. Unless stated otherwise, zeroes will be counted *without multiplicities*.

We consider

Sturm's Problem *Determine the number of distinct real zeroes of the polynomial f in the interval $[a, b]$.*

Definition 1.7.6 The *Sturm sequence* of f is the sequence $(f_0, f_1, \ldots, f_r)$ of polynomials, defined recursively as follows: let $f_0 = f$, $f_1 = f'$ and

$$f_0 = q_1 f_1 - f_2,$$
$$f_1 = q_2 f_2 - f_3,$$
$$\vdots$$
$$f_{r-2} = q_{r-1} f_{r-1} - f_r,$$
$$f_{r-1} = q_r f_r,$$

where $r \geq 1$, $f_0, \ldots, f_r, q_1, \ldots, q_r \in R[t]$, $f_i \neq 0$ and $\deg f_i < \deg f_{i-1}$ for $i = 2, \ldots, r$. These conditions determine r, as well as the f_i (and q_i) uniquely.

If we change the minus signs on the right hand side into plus signs, we obtain the usual version of Euclid's algorithm for the pair f, f'. This sign difference is essential for the determination of the Sturm sequence, but plays no role in the computation of the greatest common divisor. In particular, $f_r = \gcd(f, f')$.

Notation 1.7.7 For $(c_0, \ldots, c_r) \in R^{r+1}$ we denote by $\mathrm{Var}(c_0, \ldots, c_r)$ the number of sign changes in the given sequence, after cancelling all zeroes. We let $\mathrm{Var}(0, \ldots, 0) := -1$. For example, for the sequence $(1, 0, -2, 1, 0, 4, -1, 1, 0)$ we obtain $\mathrm{Var} = 4$. For $x \in R$ we let $V(x) := \mathrm{Var}(f_0(x), f_1(x), \ldots, f_r(x))$, where $(f_0, \ldots, f_r)$ is the Sturm sequence of f.

Theorem 1.7.8 (Sturm, 1829) *Let $f \in R[t]$ be a non-constant polynomial, and $a, b \in R$ with $a < b$ and $f(a)f(b) \neq 0$. Then the number of distinct zeroes of f in the interval $]a, b[$ equals $V(a) - V(b)$.*

Proof We proceed in two steps.

(1) Let $(g_0, \ldots, g_r)$ be a sequence in $R[t] \setminus \{0\}$ that satisfies properties (a)–(d) below:

(a) $g_0(a)g_0(b) \neq 0$;
(b) $g_r(a)g_r(b) \neq 0$, and g_r is semidefinite on $[a, b]$ (i.e., either $g_r(x) \geq 0$ or $g_r(x) \leq 0$ for all $x \in [a, b]$);
(c) if $c \in [a, b]$ and $g_0(c) = 0$, then $g_0(x)g_1(x)$ changes sign from minus to plus at $x = c$;
(d) if $c \in [a, b]$ and $g_i(c) = 0$, $0 < i < r$, then $g_{i-1}(c)g_{i+1}(c) < 0$.   □

For $x \in R$ let $W(x) := \mathrm{Var}\,(g_0(x), \ldots, g_r(x))$. We claim that g_0 has exactly $W(a) - W(b)$ distinct zeroes in $[a, b]$.

To prove the claim, we first note that for every $c \in R$ with $g_0(c) \cdots g_r(c) \neq 0$ the function $W(x)$ is constant on a neighbourhood of $x = c$. Let $g := g_0 \cdots g_r$, and let $c \in [a, b]$ be a zero of g. We distinguish three cases:

If $g_0(c) = 0$, then $a < c < b$, and the function $x \mapsto \mathrm{Var}\,(g_0(x), g_1(x))$ has value 1 on $]c - \varepsilon, c[$ and value 0 on $[c, c + \varepsilon[$, for some $\varepsilon > 0$. This follows from (c).

If $g_i(c) = 0, 0 < i < r$, then by (d) there exists $\varepsilon > 0$, such that $g_{i-1}(x)g_{i+1}(x)$ is negative on $]c - \varepsilon, c + \varepsilon[$. Hence, $x \mapsto \mathrm{Var}\,(g_{i-1}(x),\, g_i(x),\, g_{i+1}(x))$ is constant equal to 1 on $]c - \varepsilon, c + \varepsilon[$ (note that for every $\alpha \in R$, $\mathrm{Var}(-1, \alpha, 1) = \mathrm{Var}(1, \alpha, -1) = 1$!).

If $g_r(c) = 0$, and $g_0(c) \neq 0$ or $r > 1$, then $g_{r-1}(c) \neq 0$ by (d), and $x \mapsto \mathrm{Var}\,(g_{r-1}(x), g_r(x))$ is constant on $\{x : 0 < |x - c| < \varepsilon\}$ for some $\varepsilon > 0$, since g_r is semidefinite on $[a, b]$.

From these three cases we obtain for all $c \in [a, b]$: If $g_0(c)\, g_r(c) \neq 0$, then $W(x)$ is constant on a neighbourhood of $x = c$; if $g_0(c) \neq 0$, then $W(x)$ is constant on a punctured neighbourhood of $x = c$; if $g_0(c) = 0$, there exists $\varepsilon > 0$ and $N \in \mathbb{Z}$ with $W(x) = N$ for $c - \varepsilon < x < c$ and $W(x) = N - 1$ for $c < x < c + \varepsilon$.

The claim then follows since g_0 and g_r do not vanish at the boundary points a, b.

(2) Now let f be as in the statement of the theorem and let $(f_0, \ldots, f_r)$ be the Sturm sequence of f. We let $g_i := f_i / f_r \in R[t]$ $(i = 0, \ldots, r)$ and $W(x) := \mathrm{Var}\,(g_0(x), \ldots, g_r(x))$ $(x \in R)$. Then the sequence $(g_0, \ldots, g_r)$ satisfies (a)–(d) from (1). Indeed, since $g_{i-1} = h_i g_i - g_{i+1}$ $(i = 1, \ldots, r - 1)$ and $g_{r-1} = h_r g_r$ for suitable $h_i \in R[t]$, and since $g_r = 1$, it follows that g_{i-1} and g_i have no zeroes in common $(i = 1, \ldots, r)$, and so (d) follows recursively. (a) and (b) are clear, and (c) follows from Lemma 1.7.3 since $g_0 g_1 = f f' / f_r^2$.

Furthermore, g_0 and $f_0 = f$ have the same zeroes (if $f_r(c) = 0$, then also $f(c) = 0$, and the zero c occurs with larger multiplicity in f than in f_r!). It thus follows from (1) that f has precisely $W(a) - W(b)$ distinct zeroes in $[a, b]$. The proof then follows from the observation that $V(a) = W(a)$ and $V(b) = W(b)$ since $f_r(a)\, f_r(b) \neq 0$. $\square$

We state a more general version of Sturm's Theorem, also due to Sturm. We make use of

Definition 1.7.9 Let $f \in R[t]$ and $a, b \in R$ with $a < b$ and $f(a)f(b) \neq 0$. A *generalized Sturm sequence* of f on $[a, b]$ is a sequence $(f_0, \ldots, f_r)$ in $R[t] \setminus \{0\}$ with $r \geq 1$ that satisfies the properties

(1) $f_0 = f$;
(2) $f_r(a)f_r(b) \neq 0$, and f_r is semidefinite on $[a, b]$;
(3) for $0 < i < r$ and $c \in [a, b]$ with $f_i(c) = 0$, we have $f_{i-1}(c)f_{i+1}(c) < 0$.

(Warning: *The* Sturm sequence of f is usually not a generalized Sturm sequence of f in the sense of Definition 1.7.9!)

Let $f \in R[t]$ and a, b as in Definition 1.7.9. Let $(f_0, \ldots, f_r)$ be a generalized Sturm sequence of f on $[a, b]$ and, once again, $V(x) := \mathrm{Var}(f_0(x), \ldots, f_r(x))$ $(x \in R)$.

Theorem 1.7.10 (Sturm)

$$V(a) - V(b) = \#\Big\{c \in \,]a, b[\, : \, f(c) = 0 \text{ and } ff_1 \text{ changes sign from } - \text{ to } + \text{ at } c\Big\}$$

$$- \#\Big\{c \in \,]a, b[\, : \, f(c) = 0 \text{ and } ff_1 \text{ changes sign from } + \text{ to } - \text{ at } c\Big\}.$$

The statement is clear by the proof of Theorem 1.7.8! (We only have to count the different types of sign changes of ff_1.)   □

Remark 1.7.11 From the proof we see that simplifications of the original Sturm algorithm are possible. For example:

(a) The original Sturm sequence $f_0 = f$, $f_1 = f', \ldots$ may be terminated at f_s provided that $f_s(a) \, f_s(b) \neq 0$ and f_s is semidefinite on $[a, b]$.
(b) For a remainder f_{i+1} in the modified Euclidean algorithm (Definition 1.7.6) it is allowed to leave out those factors that are semidefinite on $[a, b]$ and that do not vanish at a, b.

Exercise 1.7.12 Let $(K, \leq)$ be an ordered field, and let

$$f = t^n + \sum_{i=1}^{n} a_i t^{n-i} = \prod_{j=1}^{n} (t - \xi_j)$$

be a monic polynomial in $K[t]$ that splits over K (with $a_i, \xi_j \in K$). Prove that

$$\xi_1, \ldots, \xi_n \geq 0 \quad \Leftrightarrow \quad (-1)^i a_i \geq 0 \text{ for } i = 1, \ldots, n,$$

and similarly with strict instead of non-strict inequalities.

Next, we discuss Hermite's method. Let K be an arbitrary field and $f(t) = t^n + a_1 t^{n-1} + \cdots + a_n \in K[t]$ a monic polynomial $(n \geq 1)$. Let $\alpha_1, \ldots, \alpha_n$ be the zeroes of f in the algebraic closure of K. For $r = 0, 1, 2, \ldots$, let $s_r = s_r(f) := \alpha_1^r + \cdots + \alpha_n^r$. Then $s_r(f) \in K$ since $s_r(f)$ is symmetric in the α_i. For instance, $s_0(f) = n$, $s_1(f) = -a_1$, $s_2(f) = a_1^2 - 2a_2$, $s_3(f) = -a_1^3 + 3a_1 a_2 - 3a_3$, and so on.

Exercise 1.7.13 Prove the *Newton identity*

$$s_k + s_{k-1} a_1 + s_{k-2} a_2 + \cdots + s_1 a_{k-1} + k a_k = 0$$

for all $k \geq 0$, where we put $a_k = 0$ for $k > n$. Conclude for $r \geq 0$ that $s_r = s_r(a_1, \ldots, a_n)$ is a polynomial in $a_1, \ldots, a_n$ with integer coefficients.

Definition 1.7.14 The *Sylvester form* $S(f)$ of f is the n-dimensional quadratic form over K defined by the $n \times n$-matrix

$$\left(s_{j+k-2}(f)\right)_{1 \le j,k \le n}.$$

In other words, $S(f)(x_1, \ldots, x_n) = \sum\limits_{j,k=1}^{n} s_{j+k-2}(f) x_j x_k.$

Theorem 1.7.15 (Jacobi, Borchardt, Hermite) *Let R be a real closed field and $f \in R[t]$ a monic non-constant polynomial. Then the rank of $S(f)$ equals the number of distinct roots of f in the algebraic closure $C = R(\sqrt{-1})$ of R, and the signature of $S(f)$ equals the number of distinct real roots of f.*

Note that therefore $S(f)$ always has nonnegative signature!

Proof Let $\beta_1, \ldots, \beta_r$ be the pairwise distinct real roots of f, and $\gamma_1, \overline{\gamma_1}, \ldots, \gamma_s, \overline{\gamma_s}$ the pairwise distinct non-real roots of f, where $r, s \ge 0$ and $\alpha \mapsto \overline{\alpha}$ denotes the nontrivial R-automorphism of C. Denote the multiplicity of β_j by m_j and the multiplicity of γ_j by n_j. Let $i = \sqrt{-1} \in C$, and let $\mathrm{Re}\,\alpha$, $\mathrm{Im}\,\alpha$ for $\alpha \in C$ have their usual meaning. Then

$$S(f)(x_1, \ldots, x_n) = \sum_{1 \le j,k,l \le n} \alpha_j^{k+l-2} x_k x_l$$

$$= \sum_{j=1}^{n} \left(\sum_{k=1}^{n} \alpha_j^{k-1} x_k \right)^2$$

$$= \sum_{j=1}^{r} m_j \left(\sum_k \beta_j^{k-1} x_k \right)^2 + \sum_{j=1}^{s} n_j \left[\left(\sum_k \gamma_j^{k-1} x_k \right)^2 + \left(\sum_k \overline{\gamma_j}^{k-1} x_k \right)^2 \right]$$

$$= \sum_{j=1}^{r} m_j \left(\sum_k \beta_j^{k-1} x_k \right)^2$$

$$\quad + 2 \sum_{j=1}^{s} n_j \left[\left(\sum_k \mathrm{Re}(\gamma_j^{k-1}) x_k \right)^2 - \left(\sum_k \mathrm{Im}(\gamma_j^{k-1}) x_k \right)^2 \right],$$

and we obtain $S(F)$ in diagonal form. Indeed, letting

$$u_j(x) := \sum_k \beta_j^{k-1} x_k \quad (j = 1, \ldots, r),$$

$$v_j(x) := \sum_k \mathrm{Re}(\gamma_j^{k-1}) x_k, \quad w_j(x) := \sum_k \mathrm{Im}(\gamma_j^{k-1}) x_k \quad (j = 1, \ldots, s),$$

we have

$$S(f) = \sum_{j=1}^{r} m_j\, u_j^2 + 2\sum_{j=1}^{s} n_j(v_j^2 - w_j^2),$$

and the linear forms $u_1, \ldots, u_r,\ v_1, w_1, \ldots, v_s, w_s$ are linearly independent. (The transition matrix arises from the Vandermonde matrix associated to $\beta_1, \ldots, \beta_r, \gamma_1, \overline{\gamma_1}, \ldots, \gamma_s, \overline{\gamma_s}$ via two elementary transformations.) It follows that

$$S(f) \cong r \times \langle 1 \rangle \perp s \times H \perp (n - r - 2s) \times \langle 0 \rangle,$$

from which we immediately see that rank $S(f) = r + 2s$, sign $S(f) = r$. $\square$

The following modification makes it possible to also count real zeroes in intervals:

Definition 1.7.16 Let K be a field, $f \in K[t]$ a monic polynomial of degree $n \geq 1$, and $\lambda \in K$. The *Sylvester form of f with parameter λ* is the quadratic form

$$(x_1, \ldots, x_n) \mapsto \sum_{1 \leq j,k \leq n} \left(\lambda\, s_{j+k-2}(f) - s_{j+k-1}(f) \right) x_j x_k$$

over K, denoted by $S_\lambda(f)$.

Theorem 1.7.17 (Hermite, Sylvester, 1853) *Let R be real closed and $f \in R[t]$ a monic polynomial of degree $n \geq 1$. For $\lambda \in R$ we have*

$$\text{rank } S_\lambda(f) = \#\{a \in R(\sqrt{-1})\colon f(a) = 0 \text{ and } a \neq \lambda\},$$

$$\text{sign } S_\lambda(f) = \#\{a \in R\colon f(a) = 0,\ a < \lambda\} - \#\{a \in R\colon f(a) = 0,\ a > \lambda\}.$$

Proof We use the notation from the previous proof.

$$S_\lambda(f)(x_1, \ldots, x_n) = \sum_{1 \leq j,k,l \leq n} (\lambda - \alpha_j)\alpha_j^{k+l-2} x_k x_l$$

$$= \sum_{j=1}^{n} (\lambda - \alpha_j) \left(\sum_{k} \alpha_j^{k-1} x_k \right)^2$$

$$= \sum_{j=1}^{r} m_j(\lambda - \beta_j) u_j^2$$

$$+ \sum_{j=1}^{s} n_j \left[(\lambda - \gamma_j)(v_j + i w_j)^2 + (\lambda - \overline{\gamma_j})(v_j - i w_j)^2 \right].$$

We rewrite $[\cdots]$ as follows:

$$
\begin{aligned}
[\cdots] &= (\lambda - \gamma_j + \lambda - \overline{\gamma_j})(v_j^2 - w_j^2) + 2i(\lambda - \gamma_j - \lambda + \overline{\gamma_j})v_j w_j \\
&= 2\,(\lambda - \mathrm{Re}\,\gamma_j)(v_j^2 - w_j^2) + 4\,\mathrm{Im}\,(\gamma_j)v_j w_j \\
&=: 2\varphi_j(v_j, w_j).
\end{aligned}
$$

Hence,

$$
S_\lambda(f) = \sum_{j=1}^{r} m_j(\lambda - \beta_j)u_j^2 + \sum_{j=1}^{s} 2n_j\,\varphi_j(v_j, w_j).
$$

The forms φ_j are hyperbolic ($j = 1, \ldots, s$). Indeed, if $\gamma_j = a_j + ib_j$ with $a_j, b_j \in R$, then φ_j is represented by the matrix

$$
\begin{pmatrix} \lambda - a_j & b_j \\ b_j & -(\lambda - a_j) \end{pmatrix}
$$

which has negative determinant since $b_j \neq 0$. We conclude that

$$
S_\lambda(f) \cong \langle \lambda - \beta_1, \ldots, \lambda - \beta_r \rangle \perp s \times H \perp (n - r - 2s) \times \langle 0 \rangle. \qquad \square
$$

1.8   Conceptual Interpretation of the Sylvester Form

Let K be a field of characteristic not 2.

In this section we will present a more conceptual interpretation of the Sylvester form of a polynomial, namely as the trace form of a certain algebra. We start with generalities about trace forms of algebras.

By a *K-algebra* we mean a commutative (and associative) unital finite-dimensional algebra over K. Let A be a K-algebra. For $a \in A$, let $\lambda(a)$ be the K-linear endomorphism $b \mapsto ab$ of A, and $\mathrm{tr}_{A/K}(a)$ the trace of $\lambda(a)$ (over K). The K-linear form $\mathrm{tr}_{A/K} : A \to K$ is called the *trace* of A over K.

The following rules are easily verified:

Lemma 1.8.1

(a) *If K'/K is a field extension and $A' = A \otimes_K K'$, then*

$$
\mathrm{tr}_{A/K}(a) = \mathrm{tr}_{A'/K'}(a \otimes 1) \quad \textit{for all } a \in A.
$$

(b) *If $A_1, \ldots, A_r$ are K-algebras and $A = A_1 \times \cdots \times A_r$, then*

$$\mathrm{tr}_{A/K}(a_1, \ldots, a_r) = \sum_{i=1}^{r} \mathrm{tr}_{A_i/K}(a_i) \quad \textit{for all } (a_1, \ldots, a_r) \in A.$$

(c) $\mathrm{tr}_{A/K}(\mathrm{Nil}\,A) = 0$ *(where* $\mathrm{Nil}\,A = \{a \in A : a^n = 0 \text{ for some } n \in \mathbb{N}\}$ *is the nilradical of A).* □

Assume now that $A = K[t]/(f)$, where $f \in K[t]$ is a monic polynomial of degree $n \geq 1$. We denote the image of t in A by τ. Let $\alpha_1, \ldots, \alpha_n$ be the roots of f in the algebraic closure $\overline{K}$ of K.

Proposition 1.8.2 *For $g \in K[t]$ we have* $\mathrm{tr}_{A/K}\,(g(\tau)) = \sum\limits_{j=1}^{n} g(\alpha_j)$.

Proof By Lemma 1.8.1(a) we may assume $K = \overline{K}$. Let $\beta_1, \ldots, \beta_r$ be the *distinct* roots of f and $e_1, \ldots, e_r$ their multiplicities, i.e., $f(t) = (t - \beta_1)^{e_1} \cdots (t - \beta_r)^{e_r}$. Let $A_i = K[t]/(t-\beta_i)^{e_i}$ $(i = 1, \ldots, r)$. Then $A \to A_1 \times \cdots \times A_r$ is an isomorphism by the Chinese Remainder Theorem, and by Lemma 1.8.1(b) we may assume $r = 1$, i.e., $f(t) = (t - \beta)^e$. There exists $h \in K[t]$ such that $g(t) = g(\beta) + (t - \beta)h(t)$. Since $(\tau - \beta)h(\tau) \in \mathrm{Nil}\,A$, and using Lemma 1.8.1(c), we conclude $\mathrm{tr}_{A/K}\,(g(\tau)) = \mathrm{tr}_{A/K}\,(g(\beta)) = e \cdot g(\beta)$. □

Notation 1.8.3

(a) Let A be a K-algebra and $a \in A$. Then $[A, a]_K$ denotes the quadratic space over K defined by the quadratic form $x \mapsto \mathrm{tr}_{A/K}(ax^2)$ on A.
(b) For polynomials $f, g \in K[t]$, $f \neq 0$, let

$$\mathrm{Syl}_K(f; g) := \left[K[t]/(f), g(\tau) \right]_K$$

and

$$\mathrm{Syl}_K(f) := \mathrm{Syl}_K(f; 1).$$

We usually drop the index K in the notation.

Lemma 1.8.1 immediately yields the following computational rules for these forms:

Lemma 1.8.4 *Let $A, A', A_1, \ldots, A_r$ be K-algebras.*

(a) *If $\varphi \colon A \to A'$ is a K-algebra isomorphism, then*

$$[A, a] \cong \left[A', \varphi(a) \right] \quad \textit{for all } a \in A.$$

(b) *If K'/K is a field extension then, for all $a \in A$,*

$$[A, a]_K \otimes_K K' \cong [A \otimes_K K', a \otimes 1]_{K'} \quad (over\ K').$$

(c) $\left[A_1 \times \cdots \times A_r, (a_1, \dots, a_r) \right] \cong [A_1, a_1] \perp \cdots \perp [A_r, a_r] \quad (a_i \in A_i).$
(d) $\operatorname{Nil} A \subseteq \operatorname{Rad}[A, a]$ *for all* $a \in A$. $\qquad\qquad\qquad\qquad\qquad\qquad\quad\square$

Lemma 1.8.5 *Let* $f, g \in K[t]$ *with* $f \neq 0$.

(a) *If* K'/K *is a field extension, then*

$$\operatorname{Syl}_K(f; g) \otimes_K K' \cong \operatorname{Syl}_{K'}(f; g) \quad (over\ K').$$

(b) *Let* $f_1, \dots, f_r \in K[t]$ *be nonzero and pairwise relatively prime, then*

$$\operatorname{Syl}(f_1 \cdots f_r; g) \cong \operatorname{Syl}(f_1; g) \perp \cdots \perp \operatorname{Syl}(f_r; g).$$

(c) $\operatorname{Syl}(f^e; g) \cong \langle e \rangle \otimes \operatorname{Syl}(f; g) \perp \langle 0, \dots, 0 \rangle \quad (e \in \mathbb{N})$.
(d) $\operatorname{Syl}(af; g) = \operatorname{Syl}(f; g)$ *and* $\operatorname{Syl}(f; ag) \cong \langle a \rangle \otimes \operatorname{Syl}(f; g)$ *for any* $a \in K^*$.
(e) $\operatorname{Syl}(t - a; g(t)) = \langle g(a) \rangle$ *for any* $a \in K$.

Proof (a) and (b) follow from Lemma 1.8.4, and (d) and (e) are clear. We show (c): let $A = K[t]/(f^e)$, $A' = K[t]/(f)$, and $\pi \colon A \to A'$ the residue class map. By Proposition 1.8.2 we have $\operatorname{tr}_{A/K}(a) = e \cdot \operatorname{tr}_{A'/K}(\pi a)$ for all $a \in A$, from which the proof follows. $\qquad\qquad\qquad\qquad\qquad\qquad\qquad\qquad\qquad\qquad\qquad\qquad\square$

Corollary 1.8.6 *Let* $f, g \in K[t]$ *with* $f \neq 0$. *The following statements are equivalent:*

 (i) $\operatorname{Syl}_K(f; g)$ *is nondegenerate;*
(ii) *f is separable, and f and g are relatively prime.*

Proof Since (i) and (ii) are not affected by field extensions, we may assume that $K = \overline{K}$ is algebraically closed. Let $\alpha_1, \dots, \alpha_r$ be the distinct zeroes of f, with respective multiplicities $e_1, \dots, e_r \in \mathbb{N}$. By Lemma 1.8.5 we have

$$\operatorname{Syl}(f; g) \cong \perp_{j=1}^{r} \operatorname{Syl}\left((t - \alpha_j)^{e_j}; g(t) \right)$$

$$\cong \perp_{j=1}^{r} \left[\langle e_j \rangle \otimes \operatorname{Syl}\left(t - \alpha_j; g(t) \right) \perp (e_j - 1) \times \langle 0 \rangle \right]$$

$$\cong \perp_{j=1}^{r} \left[\langle e_j g(\alpha_j) \rangle \perp (e_j - 1) \times \langle 0 \rangle \right],$$

from which the statement follows. $\qquad\qquad\qquad\qquad\qquad\qquad\qquad\qquad\qquad\qquad\square$

Proposition 1.8.7 *Let $f \in K[t]$ be a monic polynomial of degree $n \geq 1$. The Sylvester form $S(f)$ of f satisifies $S(f) \cong \mathrm{Syl}(f)$, whereas the Sylvester form $S_\lambda(f)$ of f with parameter λ satisfies $S_\lambda(f) \cong \mathrm{Syl}(f; \lambda - t)$.*

Proof Let $\alpha_1, \ldots, \alpha_n$ be the zeroes of f in $\overline{K}$ and let $g \in K[t]$ be arbitrary. With respect to the basis $1, \tau, \ldots, \tau^{n-1}$ of $A := K[t]/(f)$, the quadratic space $\mathrm{Syl}(f; g)$ is represented by the matrix $(b_{jk})_{1 \leq j,k \leq n}$, with $b_{jk} = \mathrm{tr}_{A/K}\left(\tau^{j+k-2}g(\tau)\right)$. We have $b_{jk} = \sum_{i=1}^{n} \alpha_i^{j+k-2} g(\alpha_i)$ by Proposition 1.8.2, from which the statements follow with $g = 1$ and $g = \lambda - t$, respectively.   □

We finish this section by showing that the theorems of Sylvester and Hermite (Theorems 1.7.15 and 1.7.17) are quite easily obtained by means of trace form calculus.

To this end, assume that $K = R$ is real closed, and that $f \in R[t]$ is monic of degree $n \geq 1$ with distinct real roots $\beta_1, \ldots, \beta_r$ and distinct non-real roots $\gamma_1, \overline{\gamma_1}, \cdots, \gamma_s, \overline{\gamma_s}$. Assume again that the β_j and γ_j have multiplicities m_j and n_j, respectively. We let $q_j(t) := (t - \gamma_j)(t - \overline{\gamma_j})$ $(j = 1, \ldots, s)$. Then

$$f(t) = \prod_{j=1}^{r}(t - \beta_j)^{m_j} \prod_{j=1}^{s} q_j(t)^{n_j},$$

and by Lemma 1.8.5 it follows that for all $g \in R[t]$,

$$\mathrm{Syl}(f; g) \underset{(b)}{\cong} \perp_{j=1}^{r} \mathrm{Syl}\left((t - \beta_j)^{m_j}; g\right) \perp \perp_{j=1}^{s} \mathrm{Syl}(q_j^{n_j}; g)$$

$$\underset{(c)}{\cong} \perp_{j=1}^{r} \mathrm{Syl}(t - \beta_j; g) \perp \perp_{j=1}^{s} \mathrm{Syl}(q_j; g) \perp \langle 0, \ldots, 0 \rangle$$

$$\underset{(e)}{\cong} \langle g(\beta_1), \ldots, g(\beta_r) \rangle \perp \perp_{j=1}^{s} \mathrm{Syl}(q_j; g) \perp \langle 0, \ldots, 0 \rangle.$$

The forms $\mathrm{Syl}(q_j; g)$ are all hyperbolic (or zero in case $q_j \mid g$) since $R[t]/(q_j)$ and $C = R(\sqrt{-1})$ are isomorphic as R-algebras, and $[C; \alpha]_R \cong \langle 1, -1 \rangle = H$ for all $\alpha \in C^*$. (Choosing $\beta \in C$ with $2\alpha\beta^2 = i$, the form $[C; \alpha]_R$ is represented by the matrix $\left(\begin{smallmatrix} 0 & 1 \\ 1 & 0 \end{smallmatrix}\right)$ with respect to the basis $\beta, -i\beta$.)

It follows that

$$\mathrm{Syl}(f; g) \cong \langle g(\beta_1), \ldots, g(\beta_r) \rangle \perp t \times H \perp (n - r - 2t) \times \langle 0 \rangle$$

for some $0 \leq t \leq s$, and in particular that

$$S(f) \cong \mathrm{Syl}(f) = \mathrm{Syl}(f; 1) \cong r \times \langle 1 \rangle \perp s \times H \perp (n - r - 2s) \times \langle 0 \rangle$$

and

$$\text{Syl}(f; \lambda - t) \cong \langle \lambda - \beta_1, \ldots, \lambda - \beta_r \rangle \perp s \times H \perp (n - r - 2s) \times \langle 0 \rangle,$$

from which the statements of the theorems of Sylvester and Hermite can be read off directly.

In addition to the works [65] and [39], mentioned before, the reader may want to compare the current topic and related themes with the article [10] that contains interesting historical remarks about Sylvester.

1.9   Cauchy Index of a Rational Function, Bézoutian and Hankel Forms

This section, and the next one, will not be used in the remainder of this book and can therefore be skipped during a first reading. Having said that, its content is a classical part of real algebra, and is essential for understanding the historical development of mathematics in the nineteenth century. Furthermore, it has become important in many applications such as the stability of motion or control theory.

We will follow the paper [65] of Krein and Naimark, but can only cover a small part for lack of space. The interested reader may therefore want to consult the original source.

R will always denote a real closed field.

Definition 1.9.1 Let $\varphi(t) = g(t)/f(t) \in R(t)$ be a rational function with g and f relatively prime polynomials over R.

(a) Let $\alpha \in R$ be a pole of φ (i.e., a zero of f). The *Cauchy index* $\text{ind}_\alpha(\varphi)$ of φ at α is defined as follows:

$$\text{ind}_\alpha(\varphi) := \begin{cases} +1 & \text{if the value of } \varphi(t) \text{ jumps from } -\infty \text{ to } +\infty \text{ at } t = \alpha \\ -1 & \text{if the value of } \varphi(t) \text{ jumps from } +\infty \text{ to } -\infty \text{ at } t = \alpha \\ 0 & \text{otherwise} \end{cases}$$

(b) The Cauchy index of φ in an open interval $]a, b[\subseteq R$ (a and b are allowed to be $\pm\infty$), denoted $I_a^b(\varphi)$, is defined as the sum of the Cauchy indices of φ at those poles of φ that are in $]a, b[$. If $a = -\infty$ and $b = +\infty$, we speak of the *global Cauchy index* of φ. (Note that if φ has a pole at ∞, this pole is ignored for the Cauchy index, no matter what a and b are.)

The Cauchy index of $\varphi(t) = f(t)/g(t)$ at the pole $t = \alpha$ is thus $+1$, -1 or 0, depending on whether the polynomial $f(t)g(t)$ changes sign from minus to plus, from plus to minus, or not at all, at $t = \alpha$.

Examples 1.9.2

(1) Let $f \in R[t]$ be a non-constant polynomial with logarithmic derivative $\varphi = f'/f$. Let $\alpha_1, \ldots, \alpha_r$ be the distinct real zeroes of f with respective multiplicities $m_1, \ldots, m_r$. Then

$$\varphi(t) = \psi(t) + \sum_{i=1}^{r} \frac{m_i}{t - \alpha_i},$$

for some rational function ψ without real poles. Thus $\varphi = f'/f$ has precisely r real poles $\alpha_1, \ldots, \alpha_r$ and its Cauchy index at each one of them is $+1$.

(2) More generally: let $f, h \in R[t]$ with $f \neq 0$. For $\varphi = hf'/f$ we have

$$I_{-\infty}^{+\infty}(\varphi) = \#\{\alpha \in R : f(\alpha)=0 \text{ and } h(\alpha) > 0\} - \#\{\alpha \in R : f(\alpha)=0 \text{ and } h(\alpha) < 0\},$$

as can easily be verified.

How can we determine the integer $I_a^b(\varphi)$ from the coefficients of φ? A solution was already presented in Sect. 1.7: Carry out Sturm's algorithm for the polynomials f, g (instead of f, f') and apply Theorem 1.7.10 to the generalized Sturm sequence thus obtained. This procedure was already developed in the nineteenth century as a practical method for finding $I_a^b(\varphi)$. For an important special case, the so-called Routh–Hurwitz Problem, we refer to [40, §16].

Next we will establish a connection between the global Cauchy index and certain quadratic forms. This will require some preparation. For $\varphi = g/f$ we may always assume that $\deg g \leq \deg f$ since the Cauchy indices do not change after adding a polynomial.

Consider thus two polynomials

$$f(t) = a_0 t^n + a_1 t^{n-1} + \cdots + a_n \quad \text{and} \quad g(t) = b_0 t^n + b_1 t^{n-1} + \cdots + b_n$$

with coefficients in a field K (char $K \neq 2$) and let $a_0 \neq 0$. To the pair (f, g) we associate two quadratic forms over K as follows.

Let x and y be new variables. The polynomial $f(x)g(y) - f(y)g(x)$ is divisible by $x - y$. More precisely,

$$\frac{f(x)g(y) - f(y)g(x)}{x - y} = \sum_{i,j=0}^{n-1} c_{ij} x^i y^j,$$

where

$$c_{ij} = \sum_{k=0}^{i} d_{k,i+j+1-k} \qquad \text{with} \quad d_{ij} := a_{n-j}b_{n-i} - a_{n-i}b_{n-j}.$$

Independently from these explicit formulas (for which we will have no further use), it is clear that $c_{ij} = c_{ji}$ for all $i, j = 0, \ldots, n - 1$.

Definition 1.9.3 The symmetric $n \times n$-matrix $(c_{ji})_{0 \le i, j \le n-1}$ is called the *Bézout matrix* of f and g, and is denoted $B(f, g)$. The associated quadratic form over K is called the *Bézoutian* of f, also denoted $B(f, g)$. Thus,

$$B(f, g)(x_0, \ldots, x_{n-1}) = \sum_{i,j=0}^{n-1} c_{ij} x_i x_j.$$

Since $\deg g \le \deg f$ we can write the rational function $\varphi(t) = g(t)/f(t)$ as a formal power series in t^{-1}:

$$\varphi(t) = s_{-1} + s_0 t^{-1} + s_1 t^{-2} + \cdots.$$

Definition 1.9.4 The symmetric $n \times n$-matrix $(s_{i+j})_{0 \le i, j \le n-1}$ is called the *Hankel matrix* of f and g and is denoted $H(f, g)$. (In general, any quadratic matrix (a_{ij}) whose entry a_{ij} only depends on $i + j$ is called a Hankel matrix.) Again we interpret $H(f, g)$ as a quadratic form,

$$H(f, g)(x_0, \ldots, x_{n-1}) = \sum_{i,j=0}^{n-1} s_{i+j} x_i x_j,$$

and then refer to it as the *Hankel form* of f and g. For $1 \le p \le n$ the truncated matrix $(s_{i+j})_{0 \le i, j \le p-1}$ is denoted $H_p(f, g)$.

Remark 1.9.5 $H(f, g)$ clearly only depends on n and the rational function $\varphi = g/f$. More generally, a Hankel matrix $H_n(\varphi)$ as in Definition 1.9.4 can be associated to any formal power series $\varphi(t) \in K[\![t^{-1}]\!]$ in t^{-1} and any $n \ge 0$. A well-known (elementary) theorem of Frobenius says that $\varphi(t)$ is rational (i.e., $\varphi(t) \in K(t)$) if and only if there exists $n \ge 0$ such that $\det H_m(\varphi) = 0$ for all $m \ge n$. Furthermore, in this case the smallest such n is the degree of the denominator of φ, written as an irreducible fraction, and every matrix $H_m(\varphi)$ with $m > n$ has rank n. See, for example, [40, §16.10].

Example 1.9.6 We will compute the Hankel form $H(f, f')$, where $f \in K[t]$ is a non-constant polynomial. Let $f(t) = a(t - \alpha_1) \cdots (t - \alpha_n)$ be the factorization of f over the algebraic closure of K, then (geometric series!)

$$\frac{f'(t)}{f(t)} = \sum_{i=1}^{n} \frac{1}{t - \alpha_i} = \sum_{j=0}^{\infty} s_j t^{-j-1} \quad \text{with} \quad s_j = \sum_{k=1}^{n} \alpha_k^j.$$

Hence, $H(f, f')$ is the Sylvester form $S(f)$ of f, considered in Sect. 1.7.

Remark Concerning the Terminology We will see shortly that $H(f, g)$ is isometric to $B(f, g)$. Some modern papers and books (e.g., [11]) therefore refer to the form $S(f) = H(f, f')$ as the "Bézoutian of f". We feel that this usage is unfortunate, as it is not in harmony with the terminology developed in the nineteenth century. Back then, computations were more concrete than is common nowadays and there was a strict distinction between a quadratic form and its isometry class. In order to avoid such clashes we called $S(f)$ the Sylvester form of f, which is certainly justified given Sylvester's pioneering work in this area.

Proposition 1.9.7 *The Bézoutian $B(f, g)$ is isometric to the Hankel form $H(f, g)$.*

Proof As usual, let $\varphi(t) = g(t)/f(t)$. Then

$$\sum_{i,j=0}^{n-1} c_{ij} x^i y^i = f(x) f(y) \frac{\varphi(y) - \varphi(x)}{x - y}$$

$$= f(x) f(y) \sum_{i=0}^{\infty} s_{i-1} \frac{y^{-i} - x^{-i}}{x - y}$$

$$= f(x) f(y) \sum_{i=1}^{\infty} \sum_{j=0}^{i-1} s_{i-1} x^{j-i} y^{-j-1}$$

$$= f(x) f(y) \sum_{k,l=0}^{\infty} s_{k+l} x^{-(k+1)} y^{-(l+1)}$$

$$= \sum_{k,l=0}^{\infty} s_{k+l} (a_0 x^{n-k-1} + \cdots + a_n x^{-k-1})(a_0 y^{n-l-1} + \cdots + a_n y^{-l-1}).$$

Since there are no negative powers of x or y on the left hand side, we may omit all such terms on the right hand side and obtain

$$\sum_{i,j=0}^{n-1} c_{ij} x^i y^j = \sum_{k,l=0}^{n-1} s_{k+l} (a_0 x^{n-k-1} + \cdots + a_{n-k-1})(a_0 y^{n-l-1} + \cdots + a_{n-l-1}).$$

Via the transformation of variables

$$u_0 := \quad a_0 x_{n-1} + a_1 x_{n-2} + \cdots + a_{n-1} x_0$$

$$u_1 := \qquad\qquad a_0 x_{n-2} + \cdots + a_{n-2} x_0$$

$$\vdots \qquad\qquad\qquad \vdots$$

$$u_{n-1} := \qquad\qquad\qquad\qquad a_0 x_0$$

we then get

$$B(f, g)(x_0, \ldots, x_{n-1}) = H(f, g)(u_0, \ldots, u_{n-1}). \tag{1.2}$$

Since $a_0 \neq 0$ the theorem is proved. $\square$

Corollary 1.9.8 *Consider the minors*

$$B_p := \det(c_{ij})_{n-p \leq i, j \leq n-1} \qquad (1 \leq p \leq n)$$

of the Bézout matrix $B(f, g)$ (the "principal minors from the bottom right up").
Then

$$B_p = a_0^{2p} \det H_p(f, g). \tag{1.3}$$

Proof In the identity (1.2) we let $x_0 = \cdots = x_{n-p-1} = 0$, i.e., $u_p = \cdots = u_{n-1} = 0$, and then read the new identity as an equation of symmetric matrices. Taking determinants finishes the proof. $\square$

Theorem 1.9.9 (Hurwitz)

$$B_p = \begin{vmatrix} a_0 & b_0 & 0 & 0 & \cdots & 0 & 0 \\ a_1 & b_1 & a_0 & b_0 & \cdots & 0 & 0 \\ \vdots & \vdots & \vdots & \vdots & & \vdots & \vdots \\ a_{2p-1} & b_{2p-1} & a_{2p-2} & b_{2p-2} & \cdots & a_p & b_p \end{vmatrix}, \tag{1.4}$$

where $a_i = b_i = 0$ for $i > n$.

Proof From the identity $g(t) = f(t)\varphi(t)$ we obtain the recursion formulas

$$b_i = a_0 s_{i-1} + a_1 s_{i-2} + \cdots + a_i s_{-1} \tag{1.5}$$

for all $i \geq 0$. These can be used to transform the identity (1.3) as follows:

$$
a_0 B_p \underset{(1.3)}{=} (-1)^{\frac{p(p-1)}{2}} a_0^{2p+1}
\begin{vmatrix}
1 & 0 & \cdots & 0 & 0 & \cdots & 0 \\
0 & 1 & \cdots & 0 & 0 & \cdots & 0 \\
\vdots & \vdots & \ddots & \vdots & \vdots & & \vdots \\
0 & 0 & \cdots & 1 & 0 & \cdots & 0 \\
0 & s_{-1} & \cdots & s_{p-2} & s_{p-1} & \cdots & s_{2p-2} \\
0 & 0 & \cdots & s_{p-3} & s_{p-2} & \cdots & s_{2p-3} \\
\vdots & \vdots & & \vdots & \vdots & & \vdots \\
0 & 0 & \cdots & s_{-1} & s_0 & \cdots & s_{p-1}
\end{vmatrix}
$$

$$
= (-1)^p \cdot
\begin{vmatrix}
1 & 0 & 0 & 0 & & \cdots & & 0 \\
0 & s_{-1} & s_0 & s_1 & & \cdots & & s_{2p-2} \\
0 & 1 & 0 & 0 & & \cdots & & 0 \\
0 & 0 & s_{-1} & s_0 & & \cdots & & s_{2p-3} \\
\vdots & \vdots & \vdots & \vdots & & & & \vdots \\
0 & 0 & 0 & 0 & \cdots & 1 & 0 & \cdots & 0 \\
0 & 0 & 0 & 0 & \cdots & 0 & s_{-1} & \cdots & s_{p-1} \\
0 & 0 & 0 & 0 & \cdots & 0 & 1 & \cdots & 0
\end{vmatrix}
\cdot
\begin{vmatrix}
a_0 & a_1 & a_2 & \cdots & a_{2p} \\
0 & a_0 & a_1 & \cdots & a_{2p-1} \\
0 & 0 & a_0 & \cdots & a_{2p-2} \\
\vdots & \vdots & \vdots & \ddots & \vdots \\
0 & 0 & 0 & \cdots & a_0
\end{vmatrix}.
$$

Performing the matrix multiplication gives

$$
a_0 B_p \underset{(1.5)}{=} (-1)^p
\begin{vmatrix}
a_0 & a_1 & a_2 & \cdots & a_{2p} \\
0 & b_0 & b_1 & \cdots & b_{2p-1} \\
0 & a_0 & a_1 & \cdots & a_{2p-1} \\
0 & 0 & b_0 & \cdots & b_{2p-2} \\
\vdots & \vdots & \vdots & & \vdots \\
0 & 0 & 0 & \cdots & b_p \\
0 & 0 & 0 & \cdots & a_p
\end{vmatrix}
= (-1)^p a_0
\begin{vmatrix}
b_0 & b_1 & \cdots & b_{2p-1} \\
a_0 & a_1 & \cdots & a_{2p-1} \\
0 & b_0 & \cdots & b_{2p-2} \\
\vdots & \vdots & & \vdots \\
0 & 0 & \cdots & b_p \\
0 & 0 & \cdots & a_p
\end{vmatrix}.
$$

Permuting the rows and transposition then yield (1.4). $\qquad\square$

Definition 1.9.10 The *resultant* $R(f, g)$ of f and g is defined to be the determinant (with $2n$ rows)

$$\begin{vmatrix} a_0 & a_1 & \cdots & a_{n-1} & a_n & 0 & 0 & \cdots & 0 \\ 0 & a_0 & \cdots & a_{n-2} & a_{n-1} & a_n & 0 & \cdots & 0 \\ \vdots & \vdots & \ddots & \vdots & \vdots & \vdots & \vdots & & \vdots \\ 0 & 0 & \cdots & a_0 & a_1 & a_2 & a_3 & \cdots & a_n \\ b_0 & b_1 & \cdots & b_{n-1} & b_n & 0 & 0 & \cdots & 0 \\ 0 & b_0 & \cdots & b_{n-2} & b_{n-1} & b_n & 0 & \cdots & 0 \\ \vdots & \vdots & \ddots & \vdots & \vdots & \vdots & \vdots & & \vdots \\ 0 & 0 & \cdots & b_0 & b_1 & b_2 & b_3 & \cdots & b_n \end{vmatrix}.$$

If $b_0 \neq 0$, this is the resultant of f and g in the usual algebra sense (cf. [55, 73, 113]). In contrast, if $\deg(g) = m < n = \deg(f)$, then $R(f, g)$ is obtained from the usual resultant via multiplication with the (unimportant) factor a_0^{n-m}.

In case $n = p$, Theorem 1.9.9 says

$$\det B(f, g) = (-1)^{\frac{n(n-1)}{2}} R(f, g). \tag{1.6}$$

Since the resultant of two polynomials vanishes if and only if they have a common divisor (cf. [113], ...), we obtain

Corollary 1.9.11 *The Bézoutian of f and g is nondegenerate if and only if f and g have no common divisor.* □

Denoting the quadratic space $(K^n, B(f, g))$ by $V(f, g)$, we have more generally

Proposition 1.9.12 *Let $D = t^q + d_1 t^{q-1} + \cdots + d_q$ be the greatest common divisor of f and g, and let $f = f^0 \cdot D$, $g = g^0 \cdot D$. Then the radical $\mathrm{Rad}\, V(f, g)$ of $V(f, g)$ (see Sect. 1.2) has dimension q, and the quadratic space $V(f, g)/\mathrm{Rad}\, V(f, g)$ is isometric to $V(f^0, g^0)$.*

Proof Let $(c_{ij}^0) := B(f^0, g^0)$. Then

$$\sum_{i,j=0}^{n-1} c_{ij} x^i y^j = \frac{f(x)g(y) - f(y)g(x)}{x - y}$$

$$= D(x)D(y) \frac{f^0(x)g^0(y) - f^0(y)g^0(x)}{x - y}$$

$$= D(x)D(y) \sum_{i,j=0}^{n-q-1} c_{ij}^0 x^i y^j$$

$$= \sum_{i,j=0}^{n-q-1} c_{ij}^0 \, (d_0 x^{q+i} + \cdots + d_q x^i)\, (d_0 y^{q+j} + \cdots + d_q y^j),$$

where $d_0 := 1$. Assume that $x_0, \ldots, x_{n-1}$ are independent variables. We form $n - q$ new independent variables

$$u_0 = d_q x_0 + d_{q-1} x_1 + \cdots + d_0 x_q$$

$$u_1 = \qquad d_q x_1 + \cdots + d_1 x_q + d_0 x_{q+1}$$

$$\vdots$$

$$u_{n-q-1} = \qquad d_q x_{n-q-1} + \qquad \cdots \qquad + d_0 x_{n-1}$$

($d_0 := 1$). From the above identity we then read off that

$$B(f, g)(x_0, \ldots, x_{n-1}) = B(f^0, g^0)(u_0, \ldots, u_{n-q-1}).$$

The statement now follows since $B(f^0, g^0)$ is nondegenerate by Corollary 1.9.11.

$\square$

From now on we assume again that $K = R$ is real closed. As before, let f and g be polynomials of degree n and degree at most n, respectively, and let $\varphi := g/f$.

Theorem 1.9.13 *The global Cauchy index $I_{-\infty}^{+\infty}(\varphi)$ is equal to the signature of the Hankel form $H(f, g)$, and thus also equal to the signature of the Bézoutian $B(f, g)$ of f and g.*

Proof By Proposition 1.9.12 we may assume without loss of generality that f and g are relatively prime and thus that $H(f, g)$ is nondegenerate.

We use complex analysis to prove the case $R = \mathbb{R}$. Specifically, we will compute the residues of the differential form $\omega = \varphi(z) \theta(z)^2 dz$, where z is a complex variable,

$$\theta(z) = x_0 + x_1 z + \cdots + x_{n-1} z^{n-1}$$

and $x_0, \ldots, x_{n-1}$ are independent real parameters. In order to compute the residue of ω at $z = \infty$, we introduce the local coordinate $\zeta = z^{-1}$ at ∞. Then we have

$$\omega = \varphi(z) \theta(z)^2 \left(-\zeta^{-2}\right) d\zeta,$$

and so $-\operatorname{Res}_\infty(\omega)$ is the coefficient of ζ in the Laurent expansion of $\varphi(z) \theta(z)^2$ in terms of ζ. We see from

$$\varphi(z) \theta(z)^2 = (s_{-1} + s_0 \zeta + s_1 \zeta^2 + \cdots) \cdot \sum_{j,k=0}^{n-1} x_j x_k \zeta^{-j-k}$$

that

$$- \mathrm{Res}_\infty(\omega) = \sum_{j,k=0}^{n-1} s_{j+k} x_j x_k = H(f, g)(x_0, \ldots, x_{n-1}). \qquad (1.7)$$

By the Residue Theorem this expression equals the sum of the residues of ω at the simple poles of ω, in other words at the zeroes of f.

Assume thus that $z = \alpha$ is a (complex) zero of $f(z)$ of multiplicity m. There is a Laurent expansion

$$\varphi(z + \alpha) = c_0 z^{-m} + c_1 z^{-m+1} + \cdots$$

with complex coefficients c_j and $c_0 \neq 0$. If $\alpha \in \mathbb{R}$, then the c_j are also real. Furthermore,

$$\theta(z + \alpha) = x_0 + x_1(z + \alpha) + \cdots + x_{n-1}(z + \alpha)^{n-1}$$
$$= u_0 + u_1 z + \cdots + u_{n-1} z^{n-1},$$

where $u_0, \ldots, u_{n-1}$ are linear combinations of $x_0, \ldots, x_{n-1}$ whose coefficients are integer (universal) polynomials in α. It follows that

$$\mathrm{Res}_\alpha(\omega) = c_{m-1} u_0^2 + 2c_{m-2} u_0 u_1 + \cdots + c_0(u_0 u_{m-1} + \cdots + u_{m-1} u_0).$$

If $m = 2r$ is even, we can write

$$\mathrm{Res}_\alpha(\omega) = u_0 v_0 + u_1 v_1 + \cdots + u_{r-1} v_{r-1} \qquad (A)$$

where the v_j are appropriate linear combinations of $x_0, \ldots, x_{n-1}$ whose coefficients are integral polynomials in $\alpha, c_0, \ldots, c_{m-1}$.

If $m = 2r + 1$ is odd, on the other hand, we can write

$$\mathrm{Res}_\alpha(\omega) = u_0 v_0 + \cdots + u_{r-1} v_{r-1} + c_0 u_r^2 \qquad (B)$$

where the v_j are linear combinations as before.

If $\alpha \in \mathbb{R}$, then all linear combinations u_j, v_j have real coefficients. Assume now that $\alpha \notin \mathbb{R}$, more precisely $\mathrm{Im}(\alpha) > 0$. Then $\overline{\alpha}$ is also a zero of f, of the same multiplicity as α, and precisely the complex conjugate linear combinations $\overline{u}_j, \overline{v}_j$ appear in the residue $\mathrm{Res}_{\overline{\alpha}}(\omega)$. In order to split the u_j, v_j in real and imaginary parts we write

$$u_j = p_j + \sqrt{-1}\, p'_j, \qquad v_j = q_j - \sqrt{-1}\, q'_j,$$

for $0 \le j \le r - 1$, where the p_j, p'_j, q_j, q'_j are real linear combinations of $x_0, \ldots, x_{n-1}$; when $m = 2r + 1$ we write

$$u_r = p_r + \sqrt{-1}\, p'_r$$

where p_r, p'_r are real linear combinations.

If $m = 2r$ we then obtain

$$\text{Res}_\alpha(\omega) + \text{Res}_{\overline{\alpha}}(\omega) = \sum_{j=0}^{r-1} (u_j v_j + \overline{u}_j \overline{v}_j) = 2 \sum_{j=0}^{r-1} (p_j q_j + p'_j q'_j), \qquad \text{(C)}$$

and if $m = 2r + 1$ we obtain

$$\text{Res}_\alpha(\omega) + \text{Res}_{\overline{\alpha}}(\omega) = 2 \sum_{j=0}^{r-1} (p_j q_j + p'_j q'_j) + 2c_0(p_r^2 - p'^2_r). \qquad \text{(D)}$$

In total precisely n real linear forms occur in the identities (A), (B) (for real α) and (C), (D) (for nonreal α) since f has precisely n zeroes, counted with multiplicities. Using the Residue Theorem, (1.7) yields a representation of $H(f, g)$ as a quadratic form in those n linear forms. These must be linearly independent since otherwise $H(f, g)$ is degenerate. Furthermore, $H(f, g)$ is the orthogonal sum of the quadratic forms determined by (A), (B) for $\alpha \in \mathbb{R}$ and (C), (D) for $\text{Im}(\alpha) > 0$. The forms coming from (A), (C) and (D) are clearly hyperbolic, while the form coming from (B) is the orthogonal sum of a hyperbolic form and the form $c_0 u_r^2$, and so its signature is the sign of c_0. Observe now that for $\alpha \in \mathbb{R}$ the Cauchy index $\text{ind}_\alpha(\varphi) = 0$ if m is even, but $\text{ind}_\alpha(\varphi) = \text{sgn}\, c_0$ if m is odd. We conclude that

$$\text{sign}\, H(f, g) = \sum_{\alpha \in \mathbb{R},\, f(\alpha)=0} \text{ind}_\alpha(\varphi) = I_{-\infty}^{+\infty}(\varphi).$$

There are now two different options for proving Theorem 1.9.13 for an arbitrary real closed field R. The first option is to use a purely algebraic residue theorem for rational differential forms over $R(\sqrt{-1})$. Indeed, for the form ω introduced above one can determine formal Laurent series at all poles over any algebraically closed field, use them to define the residues of ω, and then show that their sum is zero. The proof above then remains valid word for word. See also [109, p. 31].

The other option consists of applying *Tarski's Principle* (cf. [90, §5] or [91, §4.2]). This theorem from model theory states among other things that every "elementary sentence" in the language of ordered fields that is valid over $\mathbb{R}$ is also valid over *every* real closed field. Even if one has had only very little exposure to model theory, one readily recognizes that Theorem 1.9.13 is such an elementary sentence. Tarski's Principle plays an important role in higher real algebra (as mentioned in the Preface), but will not be considered any further in this book.

In this way we declare the theorem proven. □

From this result we immediately obtain a method for expressing Cauchy indices $I_a^b(\varphi)$ on finite open intervals (a, b) as signatures of "modified" Bézoutians. Without loss of generality we may assume $\deg g < \deg f$ and define, for each $\lambda \in R$, the quadratic form

$$B_\lambda(f, g) := B\left(f, (\lambda - t)g\right).$$

Its signature is denoted $\rho(\lambda)$. If λ is at most a pole of order one of φ, then

$$\rho(\lambda) = I_{-\infty}^{+\infty}\left((\lambda - t)\varphi\right) = I_{-\infty}^{\lambda}(\varphi) - I_{\lambda}^{\infty}(\varphi),$$

by Theorem 1.9.13. Thus, if a, b are at most poles of order one of φ and if $a < b$, then

$$\rho(b) = I_{-\infty}^{a}(\varphi) + \operatorname{ind}_a(\varphi) + I_a^b(\varphi) - I_b^{\infty}(\varphi),$$

$$\rho(a) = I_{-\infty}^{a}(\varphi) - I_a^b(\varphi) - \operatorname{ind}_b(\varphi) - I_b^{\infty}(\varphi).$$

This proves

Corollary 1.9.14 *If φ has at most simple poles at a and b and if $a < b$, then*

$$2I_a^b(\varphi) = \rho(b) - \rho(a) - \operatorname{ind}_a(\varphi) - \operatorname{ind}_b(\varphi). \qquad \square$$

In our proof of Theorem 1.9.13 we only used the Hankel form $H(f, g)$. The Bézoutian $B(f, g)$ was considered only indirectly in the sense that we showed via the theory of Bézoutians that f and g may be assumed relatively prime and so $H(f, g)$ may be assumed nondegenerate. One could thus get the impression that the Hankel form is the central object, and that the Bézoutian only plays an auxiliary role.

In fact, there are two reasons for casting the Bézoutian in the central role. The first one is of a formal nature: $B(f, g)$ can be defined for arbitrary polynomials f and g without restriction on the degrees, as in Definition 1.9.3, whereas for $H(f, g)$ the natural range of definition is $\deg g \le \deg f$ (or even $\deg g < \deg f$). Furthermore, the Bézoutian $B(f, g)$, beloved in the nineteenth century, but then mostly forgotten, has in the meantime been shown to have fascinating properties, not shared with $H(f, g)$. See [46, 47] and the cited literature.

The second reason is that $B(f, g)$ has practical advantages over $H(f, g)$ since the coefficients of $B(f, g)$ can be obtained quickly from those of f and g (see the formulas at the beginning of the section), whereas the Taylor coefficients of g/f that are needed for $H(f, g)$ depend on f and g in a complicated manner. (On the other hand, the matrix $H(f, g)$ has a simpler form than $B(f, g)$.) Hence, when f and g vary it is easier to study the change in $B(f, g)$ than the change in $H(f, g)$, which is important for applications in control theory.

Moreover, Hurwitz' Theorem 1.9.9 gives us tight control over the signature of $B(f, g)$. As an illustration we mention

Theorem 1.9.15 (Hurwitz) *Let f and g be polynomials over R as above (deg $g \leq$ deg $f = n$). The following statements are equivalent:*

(i) *All the roots of f are real and simple, and between any two such roots there is a root of g;*
(ii) *The entries in the sequence of determinants*

$$1, \quad \begin{vmatrix} a_0 & b_0 \\ a_1 & b_1 \end{vmatrix}, \ldots, \quad \begin{vmatrix} a_0 & b_0 & 0 & 0 & \cdots & 0 & 0 \\ a_1 & b_1 & a_0 & b_0 & \cdots & 0 & 0 \\ \vdots & \vdots & \vdots & \vdots & & \vdots & \vdots \\ a_{2n-1} & b_{2n-1} & a_{2n-2} & b_{2n-2} & \cdots & a_n & b_n \end{vmatrix}$$

are either all positive, or alternating positive and negative.

Proof Condition (i) signifies $I_{-\infty}^{+\infty} (g/f) = \pm n$, whereas by Theorem 1.9.9 the condition on the determinants says that sign $B(f, g) = \pm n$. The claim then follows from Theorem 1.9.13.   □

1.10 An Upper Bound for the Number of Real Zeroes (with Multiplicities)

In this section R denotes a real closed field and, in contrast to previous practice, all zeroes will be counted *with* multiplicities.

Let $f \in R(t)$ be a non-constant rational function without pole in the interval $[a, b]$ $(a, b \in R, a < b)$. We fix $n \geq 1$ such that $f^{(n)}(a) f^{(n)}(b) \neq 0$ and denote by N_i the number of real zeroes of the i-th derivative $f^{(i)}$ in $]a, b]$ $(i = 0, 1, \ldots, n)$; in particular we let $N := N_0$ denote the number of zeroes of f in $]a, b]$. For $x \in [a, b]$ we let $V(x) := \mathrm{Var}\left(f(x), f'(x), \ldots, f^{(n)}(x)\right)$ (cf. Sect. 1.7).

Theorem 1.10.1 (Hurwitz) *There exists $v \in \mathbb{N}_0$ such that*

$$N = N_n + V(a) - V(b) - 2v.$$

Corollary 1.10.2 (Budan–Fourier) *If $f \in R[t]$ is a polynomial of degree $n \geq 1$, then there exists $v \in \mathbb{N}_0$ such that*

$$N = V(a) - V(b) - 2v.$$

Before proving the theorem, we will derive a number of interesting consequences from it (more precisely: from Corollary 1.10.2). These show that without any computations one can already draw remarkable conclusions about the real roots of a polynomial from the sign sequence of its coefficients alone!

Thus, let $f(t) = c_0 t^n + c_1 t^{n-1} + \cdots + c_n \in R[t]$, where $n \geq 1$ and $c_0 \neq 0$. We introduce the following notation:

$$p := \text{number of strictly postive roots of } f,$$

$$p' := \text{number of strictly negative roots of } f,$$

$$V := \text{Var}(c_0, \ldots, c_n),$$

$$V' := \text{Var}\left(c_0, -c_1, c_2, \ldots, (-1)^n c_n\right).$$

Corollary 1.10.3

(a) *(Rule of Descartes) There exist* $v, v' \in \mathbb{N}_0$ *such that* $p = V - 2v$ *and* $p' = V' - 2v'$.
(b) *If* $c_n \neq 0$ *and if all roots of* f *are real, then* $p = V$ *and* $p' = V'$.
(c) *If* $c_n \neq 0$ *and if* $c_{i-1} = c_i = 0$ *for some* i *with* $1 < i < n$, *then* f *has non-real roots.*

Proof (a) $V(0) = \text{Var}(c_n, 1! c_{n-1}, \ldots, n! c_0) = V$ and $V(b) = 0$ for $b \gg 0$, together with Corollary 1.10.2, give the first formula. For the second one, simply replace $f(t)$ by $f(-t)$.

(b) For $x \in R$, let

$$W(x) := \text{Var}(c_0 + x, \ldots, c_n + x), \quad \text{and}$$

$$W'(x) := \text{Var}\left(c_0 + x, -(c_1 + x), \ldots, (-1)^n (c_n + x)\right).$$

Then, for all $x \in R$ such that

$$0 < x < \min\{|c_i| : c_i \neq 0\},$$

we have $W(x) + W'(x) = n$, as well as $W(x) \geq V$ and $W'(x) \geq V'$. Thus, $V + V' \leq n$ and it follows from (a) that $p + p' \leq V + V'$. On the other hand $p + p' = n$ by the assumption in (b), and so $p = V$, $p' = V'$.

(c) Omit positions $i - 1$ and i in the sequences and repeat the argument in the proof of (b). $\qquad\qquad\square$

For the proof of Hurwitz' Theorem, we provide two auxiliary results.

Proposition 1.10.4 (Taylor) *Let* $0 \neq f \in R(t)$ *and let* $a \in R$ *be a zero of* f *of multiplicity* $m \geq 0$. *There exists* $g \in R(t)$ *such that* $g(0) \neq \infty$ *and*

$$f(a + t) = \tfrac{1}{m!} f^{(m)}(a) t^m + t^{m+1} g(t).$$

Proof There exists $h \in R(t)$ with the properties

$$h(0) \neq \infty, \ h(0) \neq 0 \quad \text{and} \quad f(a+t) = t^m h(t).$$

The Leibniz rule gives $f^{(m)}(a) = m! \, h(0)$. There exists $g \in R(t)$ such that $g(0) \neq \infty$ and $h(t) = h(0) + t \, g(t)$, from which the statement follows.   □

Proposition 1.10.5 *Let $a, b \in R$, $a < b$, and $f \in R(t)$ without a pole in $[a, b]$.*

(a) *If f has no zeroes in $[a, b[$, $f(b) = 0$ and $f'(a) \neq 0$, then the number of zeroes of f' in $]a, b[$ is*

$$\begin{cases} even & \text{if } f(a) f'(a) < 0 \\ odd & \text{if } f(a) f'(a) > 0 \end{cases}.$$

(b) *If f has no zeroes in $]a, b]$, $f(a) = 0$ and $f'(b) \neq 0$, then the number of zeroes of f' in $]a, b[$ is*

$$\begin{cases} odd & \text{if } f(b) f'(b) < 0 \\ even & \text{if } f(b) f'(b) > 0 \end{cases}.$$

Proof On grounds of symmetry it suffices to prove (a). By Lemma 1.7.3 we have $f(b - \varepsilon) \, f'(b - \varepsilon) < 0$ for small $\varepsilon > 0$. It follows from the assumptions that $f(a) f(b - \varepsilon) > 0$. If $f(a) f'(a) > 0$ (resp. < 0), then $f'(a)$ and $f'(b - \varepsilon)$ have distinct (resp. equal) signs, and so the number of zeroes of f' in $[a, b - \varepsilon]$ is odd (resp. even).   □

We have arrived at the proof of Theorem 1.10.1 and adopt the assumptions and notation in its formulation.

Proof of Theorem 1.10.1 For all $x \in R$ with $f(x) \neq \infty$ we define

$$w(x) := \tfrac{1}{2}\Big(1 - \text{sign}\big(f(x) \, f'(x)\big)\Big),$$

and also $(Wf)(x) := 0$ if $f(x) = 0$ and $(Wf)(x) := \tfrac{1}{2}\Big(1 - \text{sign}\big(f(x) f^{(k)}(x)\big)\Big)$ if $f(x) \neq 0$ and $k \geq 1$ is minimal such that $f^{(k)}(x) \neq 0$.

Lemma 1.10.6 *Assume that $f(a) \, f'(a) \, f(b) \, f'(b) \neq 0$. Then there exists $v \in \mathbb{N}_0$ such that $N - N_1 = w(a) - w(b) - 2v$.*

Proof Let $c_1, \ldots, c_r$ be the distinct zeroes of f in $[a, b]$, with respective multiplicities $m_1, \ldots, m_r$, and assume $a < c_1 < \cdots < c_r < b$. By Proposition 1.7.4 f' has an odd number of zeroes in $]c_{i-1}, c_i[$ $(i = 2, \ldots, r)$, and in addition an $(m_i - 1)$-fold zero at c_i $(i = 1, \ldots, r)$. Thus f' has precisely

$$(r - 1) + 2u + (m_1 - 1) + \cdots + (m_r - 1) = N + 2u - 1$$

zeroes in $[c_1, c_r]$ for some $u \in \mathbb{N}_0$. Furthermore, the number of zeroes of f' in $]a, c_1[$, resp. in $]c_r, b[$, equals $1 - w(a) + 2u_1$, resp. $w(b) + 2u_2$, for some $u_1, u_2 \in \mathbb{N}_0$ (Proposition 1.10.5). It follows that

$$N_1 = (N + 2u - 1) + (1 - w(a) + 2u_1) + (w(b) + 2u_2),$$

i.e.,

$$N_1 = N - w(a) + w(b) + 2v \quad \text{with} \quad v := u + u_1 + u_2. \qquad \square$$

(The assumption in Lemma 1.10.6 is now revoked.)

Lemma 1.10.7 *For all $x \in R$ with $f(x) \neq \infty$ we have $(Wf)(x) = w(x + h)$ for all sufficiently small $h > 0$.*

Proof We distinguish three cases:
1) If $f(x) = 0$, then $(Wf)(x) = 0$, and $w(x + h) = 0$ for a small $h > 0$ by Lemma 1.7.3.
2) If $f(x) \neq 0$ and $f'(x) = 0$, let $k \geq 1$ be minimal such that $f^{(k)}(x) \neq 0$. By Proposition 1.10.4 there exists $g \in R(t)$ such that $g(0) \neq \infty$ and

$$f'(x + h) = \frac{f^{(k)}(x)}{(k - 1)!} \cdot h^{k-1} + h^k g(h).$$

It follows that sign $f'(x + h) = $ sign $f^{(k)}(x)$ for a small $h > 0$, i.e., $w(x + h) = (Wf)(x)$ for a possibly even smaller $h > 0$.
3) If $f(x) f'(x) \neq 0$, then $(Wf)(x) = w(x) = w(x + h)$ for a small $h > 0$. $\square$

Lemma 1.10.8 *There exists $v_0 \in \mathbb{N}_0$ such that $N - N_1 = (Wf)(a) - (Wf)(b) - 2v_0$.*

Proof Let $h > 0$ be sufficiently small such that $f \cdot f'$ has no zeroes nor poles in $]a, a + h] \cup]b, b + h]$ and thus such that N, resp. N_1, is the number of zeroes of f, resp. f', in $[a + h, b + h]$. From Lemma 1.10.6 it follows that $N - N_1 = w(a + h) - w(b + h) - 2v_0$ for some $v_0 \geq 0$, and the statement follows from Lemma 1.10.7. $\square$

This result can now be iterated. Since the assumptions made about f are also valid for $f', \ldots, f^{(n-1)}$, we obtain

$$N - N_1 = (Wf)(a) - (Wf)(b) - 2v_0,$$
$$N_1 - N_2 = (Wf')(a) - (Wf')(b) - 2v_1,$$

$$\vdots$$

$$N_{n-1} - N_n = (Wf^{(n-1)})(a) - (Wf^{(n-1)})(b) - 2v_{n-1}$$

for natural numbers $v_0, \ldots, v_{n-1} \geq 0$, and so

$$N - N_n = W(a) - W(b) - 2v,$$

where $W := Wf + Wf' + \cdots + Wf^{(n-1)}$ and $v := v_0 + v_1 + \cdots + v_{n-1}$.

However, if $f^{(n)}(x) \neq 0$, then $W(x) = \mathrm{Var}\left(f(x), f'(x), \ldots, f^{(n)}(x)\right) = V(x)$ (and we are done!). Indeed, if $k \geq 1$ is minimal such that $f^{(k)}(x) \neq 0$ then, by definition,

$$(Wf)(x) = \begin{cases} 0 & \text{if } f(x)\, f^{(k)}(x) \geq 0 \\ 1 & \text{otherwise} \end{cases}.$$

Let $0 \leq k_1 < \cdots < k_r = n$ be precisely the indices $k \in \{0, \ldots, n\}$ such that $f^{(k)}(x) \neq 0$. Among the $\left(Wf^{(i)}\right)(x)$ $(i = 0, \ldots, n)$ at most the $\left(Wf^{(k_j)}\right)(x)$ can be nonzero, and $\left(Wf^{(k_j)}\right)(x) = 1$ if and only if $f^{(k_j)}(x)\, f^{(k_{j+1})}(x) < 0$ $(j = 1, \ldots, r-1)$. Therefore,

$$(Wf)(x) = \mathrm{Var}\left(f^{(k_1)}(x), \ldots, f^{(k_r)}(x)\right) = \mathrm{Var}\left(f(x), f'(x), \ldots, f^{(n)}(x)\right) = V(x),$$

which concludes the proof of Theorem 1.10.1. $\qquad\square$

In our proof of the Budan–Fourier Theorem via Hurwitz' Theorem we followed the book [85] of N. Obreschkoff [loc.cit. §15], which also contains a substantially different approach to the theorem of Budan–Fourier [loc.cit. §14] as well as further investigations on this topic. We recommend in particular a number of interesting theorems of Laguerre [loc.cit. §16], which generalize the Rule of Descartes in a way different from the Budan–Fourier Theorem.

1.11   The Real Closure of an Ordered Field

Definition 1.11.1   Let K be a field.

(a) Let P be an ordering of K. A *real closure of* (K, P) is a real closed overfield R of K that is algebraic over K and whose ordering extends P.
(b) A *real closure of* K is an algebraic extension of K that is real closed.

It follows from Proposition 1.5.2 (the equivalence (i)⇔(ii)) and Zorn's Lemma that every ordered field has a real closure. It is uniquely determined (similar to the algebraic closure of a field), up to *unique* isomorphism (which is not the case for the algebraic closure):

Theorem 1.11.2   *Let* (K, P) *be an ordered field, R a real closure of (K, P) and* $\varphi \colon K \to S$ *an order preserving homomorphism (with respect to P) into another real closed field S. Then there exists a unique homomorphism* $\psi \colon R \to S$ *that extends* φ.

Corollary 1.11.3   *Let R/K and R'/K be real closures of (K, P). Then there exists a unique K-isomorphism $R \to R'$.*

Because of Corollary 1.11.3 we will say *the* real closure of (K, P) from now on.

Proposition 1.11.4   *Let L/K be a finite field extension such that $K \subseteq L \subseteq R$. Then φ has an order preserving extension $\psi_L \colon L \to S$. (Here "order preserving" is with respect to the ordering of L induced by R.)*

Proof   Let $\alpha \in L$ be such that $L = K(\alpha)$, and let $f \in K[t]$ be the minimal polynomial of α. By Hermite's Theorem (Theorem 1.7.15) the Sylvester form $S(f)$ of f has positive signature over R, and thus also over S. Again by this theorem, f thus also has a zero in S and so there is at least one extension $\chi \colon L \to S$ of φ. Let $\chi_1, \ldots, \chi_r$ be all such extensions ($r \geq 1$). Suppose that none of the χ_i is order preserving. Then there exist $a_1, \ldots, a_r \in L$ with $a_i > 0$ (in R), but $\chi_i(a_i) < 0$ (in S). Consider $L' := L(\sqrt{a_1}, \ldots, \sqrt{a_r}) \subseteq R$. Then L'/K is finite, but there are actually no extensions $\chi' \colon L' \to S$ of φ to L' (since $\chi'|_L$ must be one of the χ_i, i.e., $\chi'\left((\sqrt{a_i})^2\right) < 0$ for some i), contradicting Hermite's Theorem.   □

Proof of Theorem 1.11.2   By Zorn's Lemma there exists an intermediate field K'/K of R/K and a maximal order preserving extension $\psi' \colon K' \to S$ of φ. However, it follows from Proposition 1.11.4 that $K' = R$, which establishes the existence of ψ. Concerning the uniqueness see the next remark. This finishes the proof of Theorem 1.11.2.   □

Remark 1.11.5   The extension ψ of φ from Theorem 1.11.2 can be "described" as follows: if $\alpha \in R$ is a zero of the polynomial $0 \neq f \in K[t]$, if $\alpha_1 < \cdots < \alpha_r$ are the distinct zeroes of f in R, and if $\beta_1 < \cdots < \beta_s$ are the distinct zeroes of f^{φ} (the polynomial f, transferred to S by φ) in S, then $r = s$ by Hermite, and $\psi(\alpha_i) = \beta_i$ for $i = 1, \ldots, r$. (A proof of the homomorphy of ψ based on this definition requires more effort however, cf. [4, p. 93].)

Theorem 1.11.6 *Let L/K be an algebraic field extension and $\varphi \colon K \to S$ a homomorphism into a real closed field S. Let P denote the ordering of K induced by φ. Then there exists a bijection between the extensions $\psi \colon L \to S$ of φ to L and the extensions Q of the ordering P to L; it is given by $\psi \mapsto Q := \psi^{-1}(S^2)$.*

Proof For every extension ψ of φ, $\psi^{-1}(S^2)$ is an extension of P. Conversely, let Q be an ordering of L such that $P \subseteq Q$, and let R be the real closure of (L, Q) (thus also of (K, P)). By Theorem 1.11.2 there exists an extension $R \to S$ of φ, and its restriction ψ to L satisfies $Q = \psi^{-1}(S^2)$. Finally, if $\psi_1, \psi_2 \colon L \rightrightarrows S$ are extensions of φ such that $\psi_1^{-1}(S^2) = Q = \psi_2^{-1}(S^2)$, then there exist extensions $\chi_i \colon R \to S$ of ψ_i $(i = 1, 2)$ by Theorem 1.11.2. Since $\chi_1|_K = \varphi = \chi_2|_K$, we have $\chi_1 = \chi_2$, and thus also $\psi_1 = \psi_2$. $\qquad\square$

Corollary 1.11.7 *Let (K, P) be an ordered field with real closure R, and let L/K be a finite algebraic extension. If $L = K(\alpha)$ and $f \in K[t]$ is the minimal polynomial of α over K, then the extensions of P to L correspond bijectively to the zeroes of f in R. In particular, P has at most $[L : K]$ distinct extensions to L.*

Example 1.11.8 Let $L = \mathbb{Q}(\alpha)$ be a number field and $f \in \mathbb{Q}[t]$ the minimal polynomial of α over $\mathbb{Q}$. The zeroes of f correspond to the field embeddings $L \hookrightarrow \mathbb{C}$. The real zeroes of f thus correspond to the embeddings $L \hookrightarrow \mathbb{R}$, and therefore also to the orderings of L.

The following theorem ties in with Corollary 1.6.3:

Theorem 1.11.9 (Orderings and Involutions in the Galois Group) *Let K be a field (of arbitrary characteristic) with absolute Galois group $\Gamma = \mathrm{Gal}(K_s/K)$. There exists a natural bijection Φ from the set of all conjugacy classes $[\tau]$ of involutions in Γ to the set of orderings of K. More precisely: $\Phi([\tau])$ is the ordering of K defined by the inclusion $K \hookrightarrow \mathrm{Fix}(\tau)$, i.e., $\Phi([\tau]) = \{a \in K : \tau(\sqrt{a}) = \sqrt{a}\}$. (Note that $\mathrm{Fix}(\tau)$ is real closed by Sect. 1.6.)*

Proof (Sketch) Φ is well-defined since $\mathrm{Fix}(\sigma\tau\sigma^{-1}) = \sigma(\mathrm{Fix}\,\tau)$ $(\sigma, \tau \in \Gamma)$. The injectivity of φ follows from Theorem 1.11.2, and the surjectivity from the existence of the real closure and the findings in Sect. 1.6. We leave the (easy) details to the reader. $\qquad\square$

If τ is an involution in Γ, then the centralizer of τ in Γ is equal to $\{1, \tau\}$; this is just a reformulation of $\mathrm{Aut}(\mathrm{Fix}(\tau)/K) = \{1\}$ (Proposition 1.5.6(b)). Therefore, the conjugates of τ are in bijective correspondence with the elements of the homogeneous space $\Gamma/\{1, \tau\}$.

To conclude this section, we express several of our previous results in the language of Witt rings and signatures. Let K be a field of char $K \neq 2$. By

Sect. 1.2 the orderings of K correspond bijectively to the signatures of K. Every homomorphism $\varphi: K \to R$ into a real closed field R gives rise to a signature

$$W(\varphi) := i_{R/K}: W(K) \to W(R) = \mathbb{Z}$$

of K (Exercise 1.2.17), and in this way one obtains all signatures in case $R/\varphi(K)$ is algebraic (existence of the real closure). If $\varphi: K \to R$, $\varphi': K \to R'$ are homomorphisms into real closed fields R, R', and if $R/\varphi(K)$ is algebraic, then $W(\varphi) = W(\varphi')$ if and only if there exists a K-homomorphism $R \to R'$ (uniqueness of the real closure).

Theorem 1.11.6 can be formulated as follows: If L/K is an algebraic field extension, $\varphi: K \to R$ a homomorphism into a real closed field R, and $\sigma = W(\varphi)$ the induced signature of K, then the extensions $\psi: L \to R$ of φ correspond bijectively to the signatures τ of L that extends σ (i.e., that satisfy $\sigma = \tau \circ i_{L/K}$) via $\tau = W(\psi)$.

1.12 Transfer of Quadratic Forms

In this section all field have characteristic $\neq 2$.

If L/K is an arbitrary field extension, then the functoriality of the Witt ring yields a ring homomorphism $i_{L/K}: W(K) \to W(L)$. If $[L: K]$ is finite, then every K-linear form s on L defines a map s_* in the reverse direction, the transfer via s. This is only a homomorphism of additive groups. The special case $s = \mathrm{tr}_{L/K}$ leads to a trace formula for the extension of signatures.

In the remainder of this section let L/K be a finite field extension and $s: L \to K$ a nonzero linear form on the K-vector space L.

Definition 1.12.1 Let $\varphi = (V, b)$ be a bilinear space over L. The *transfer* $s_*\varphi$ of φ (to K, via s) is the bilinear space $s_*\varphi = (V, s \circ b)$ over K (where V is considered as K-vector space).

Example 1.12.2 Let $L = \mathbb{Q}(\sqrt{2})$, $\mathrm{tr} := \mathrm{tr}_{L/\mathbb{Q}}$ and $\varphi = \langle 1 \rangle$, i.e., $\varphi(x, y) = xy$. The matrix of $\mathrm{tr}_* \varphi$ with respect to the $\mathbb{Q}$-basis $(1, \sqrt{2})$ of L is $\left(\begin{smallmatrix} 2 & 0 \\ 0 & 4 \end{smallmatrix}\right)$. Hence, $\mathrm{tr}_* \varphi \cong \langle 2, 1 \rangle$.

Lemma 1.12.3 *Let $W \subseteq V$ be an L-subspace. Then, for all $v \in V$,*

$$b(v, W) = 0 \Leftrightarrow (s \circ b)(v, W) = 0.$$

Proof One direction is trivial. For the other direction we note that the K-bilinear form

$$\beta: L \times L \to K, \quad \beta(a, a') := s(aa')$$

is nondegenerate. (By assumption there exists $a_0 \in L$ such that $s(a_0) \neq 0$; thus if $0 \neq a \in L$, then $\beta(a, a^{-1}a_0) \neq 0$.) Hence, if $(s \circ b)(v, W) = 0$ and $w \in W$, then

$$0 = s(b(v, aw)) = s(a \cdot b(v, w)) = \beta(a, b(v, w))$$

for all $a \in L$, and it follows that $b(v, w) = 0$.   □

Lemma 1.12.4 (Elementary Properties of the Transfer) *Let* φ, φ' *be bilinear spaces over* L, *and let* s *be as above.*

(a) $\dim_K s_*\varphi = [L : K] \cdot \dim_L \varphi$;
(b) $s_*(\varphi \perp \varphi') \cong_K s_*\varphi \perp s_*\varphi'$;
(c) $\mathrm{Rad}(\varphi) = \mathrm{Rad}(s_*\varphi)$; *in particular,* $s_*\varphi$ *is nondegenerate if and only if* φ *is nondegenerate;*
(d) *if* φ *is hyperbolic, then* $s_*\varphi$ *is also hyperbolic.*

Proof (a) and (b) are clear, and (c) follows from Lemma 1.12.3.

(d): If $\varphi = (V, b)$ is hyperbolic, then φ is nondegenerate and there exists a subspace $U \subseteq V$ with $U = U^\perp$ (cf. Sect. 1.2). By Lemma 1.12.3 both properties transfer to $s_*\varphi$.   □

From this we immediately obtain:

Proposition 1.12.5 *Let* L/K *be a finite field extension and* $s\colon L \to K$ *a nonzero* K-*linear form. Then* s_* *induces a homomorphism* $s_*\colon W(L) \to W(K)$ *of additive groups (which is not multiplicative in general).*

Remark 1.12.6 The additive subgroup $s_*W(L)$ of $W(K)$ does not depend on s. Indeed, if $s'\colon L \to K$ is another K-linear map, $s' \neq 0$, then there exists $a \in L^*$ with $s'(b) = s(ab)$ for all $b \in L$. It follows that $s'_*(\eta) = s_*(\langle a \rangle \cdot \eta)$ for $\eta \in W(L)$. The following theorem shows in particular that $s_*W(L)$ is an ideal of $W(K)$.

Proposition 1.12.7 (Projection Formula) *For all* $\xi \in W(K)$ *and* $\eta \in W(L)$ *we have*

$$s_*\big(i_{L/K}(\xi) \cdot \eta\big) = \xi \cdot s_*(\eta).$$

In particular, the composition $s_* \circ i_{L/K}$ *is multiplication by* $s_*(\langle 1 \rangle)$ *in* $W(K)$.

Proof Let ξ, resp. η, be represented by bilinear spaces (V, b) over K, resp. (W, c) over L. Then $\xi \cdot s_*(\eta)$ is represented by $(V \otimes_K W, b \otimes (s \circ c))$, and $s_*\big(i_{L/K}(\xi) \cdot \eta\big)$ is represented by $((V \otimes_K L) \otimes_L W, s \circ ((b \otimes 1) \otimes c)))$. For simplicity we write $b' := b \otimes (s \circ c)$, $c' := (b \otimes 1) \otimes c$; so b' is K-bilinear on $V \otimes_K W$ and c' is L-bilinear on $(V \otimes_K L) \otimes_L W$.

There exists a canonical K-vector space isomorphism $\varphi \colon (V \otimes_K L) \otimes_L W \xrightarrow{\sim} V \otimes_K W$, given by $\varphi((v \otimes \lambda) \otimes w) = v \otimes \lambda w$. This φ is an isometry between $s \circ c'$ and b'; indeed for $x := (v \otimes \lambda) \otimes w$ and $x' := (v' \otimes \lambda') \otimes w'$ we have

$$
\begin{aligned}
(s \circ c')(x, x') &= s\left((b \otimes 1)(v \otimes \lambda, v' \otimes \lambda') \cdot c(w, w')\right) \\
&= s\left(b(v, v') \cdot \lambda \lambda' \cdot c(w, w')\right) \\
&= b(v, v') \cdot s \circ c(\lambda w, \lambda' w') \\
&= b'(v \otimes \lambda w, v' \otimes \lambda' w') = b'(\varphi x, \varphi x'). \qquad \square
\end{aligned}
$$

The projection formula shows that s_* is a homomorphism of $W(K)$-modules, where the $W(K)$-module structure of $W(L)$ is defined via $i_{L/K}$.

Corollary 1.12.8 *Let $W(L/K)$ denote the kernel of $i_{L/K} \colon W(K) \to W(L)$. Then*

$$
s_*(W(L)) \cdot W(L/K) = 0. \qquad \square
$$

Thus, if σ is a signature of K and if there exists $\eta \in W(L)$ such that $\sigma(s_* \eta) \neq 0$, then σ has an extension to L! (Proposition 1.4.4).

Theorem 1.12.9 (Trace Formula) *Let L/K be a finite field extension, $\mathrm{tr} := \mathrm{tr}_{L/K}$ the trace of L over K, and $\sigma \colon W(K) \to \mathbb{Z}$ a signature of K. Then, for all $\eta \in W(L)$,*

$$
\sigma(\mathrm{tr}_* \eta) = \sum_{\tau \mid \sigma} \tau(\eta),
$$

where the sum is over all signatures τ of L that extend σ.

Proof Since K is real, $\mathrm{char}\, K = 0$, hence $\mathrm{tr} \neq 0$ and tr_* is defined. Let R be the real closure of K with respect to σ. By the Primitive Element Theorem ([55, Vol. I, §4.14], [73, V, Theorem 4.6]) there exists $\alpha \in L$ such that $L = K(\alpha)$. Let $f \in K[t]$ be the minimal polynomial of α over K, $\alpha_1, \ldots, \alpha_r$ the zeroes of f in R, and $\varphi_i \colon L \to R$ the embeddings over K defined by $\varphi_i(\alpha) = \alpha_i$ for $i = 1, \ldots, r$. By Corollary 1.11.7, the $\tau_i := W(\varphi_i)$ $(i = 1, \ldots, r)$ are precisely the signatures of L that extend σ.

Since tr_* is additive, it suffices to prove the statement for $\eta = \langle \beta \rangle$, $\beta \in L^*$. Let $g \in K[t]$ with $\beta = g(\alpha)$. We note that for $i = 1, \ldots, r$,

$$
\tau_i(\eta) = \mathrm{sign}_R\, g(\alpha_i). \qquad (\star)
$$

On the other hand, $\mathrm{tr}_*(\eta)$ is precisely the Sylvester form $\mathrm{Syl}_K(f; g)$ (Sect. 1.8). Then, by Lemma 1.8.5(a),

$$\sigma\left(\mathrm{tr}_*(\eta)\right) = \sigma\left(\mathrm{Syl}_K(f; g)\right) = \mathrm{sign}_R\,\mathrm{Syl}_R(f; g)$$

$$\underset{(\star\star)}{=} \sum_{i=1}^{r} \mathrm{sign}_R\, g(\alpha_i) \underset{(\star)}{=} \sum_{i=1}^{r} \tau_i(\eta) = \sum_{\tau|\sigma} \tau(\eta),$$

where the identity $(\star\star)$ follows from the explicit computation of the Sylvester form over real closed fields (see the end of Sect. 1.8). □

Corollary 1.12.10 *Under the assumptions of Theorem 1.12.9, $\sigma\left(\mathrm{tr}_*(\langle 1\rangle_L)\right)$ is precisely the number of extensions of σ to L.* □

At this point the circle is complete: Since $\mathrm{tr}_*(\langle 1\rangle_L) = \mathrm{Syl}_K(f; 1)$ we have with Proposition 1.8.7 that

$$\sigma\left(\mathrm{tr}_*(\langle 1\rangle_L)\right) = \mathrm{sign}_R\,\mathrm{Syl}_R(f; 1) = \mathrm{sign}_R\, S(f)$$

is the signature of the Sylvester form of f. By Corollary 1.11.7 we thus obtain Hermite's Theorem (proved in Sect. 1.7 and again in Sect. 1.8) in this special case.

Chapter 2
Convex Valuation Rings and Real Places

This chapter gives a detailed introduction to valuation theory of fields. Valuations are presented from three different points of view: valuation maps, valuation rings, and places. Valuations are naturally associated with orderings, since every convex subring of an ordered field is a valuation ring. The precise relationship between the orderings of a field that are compatible with a valuation ring, and the orderings of the residue field of the latter, is exhibited by the Baer–Krull Theorem. Finally, the Artin–Lang Theorem is presented in the form of a *Stellensatz*. Other than the Artin–Schreier concept of orderings and real closures, this is our essential input for the solution of Hilbert's 17th Problem, which we present at the end of the chapter.

2.1 Convex Subrings of Ordered Fields

Definition 2.1.1 Let $(M, \leq)$ be a partially ordered set. A subset $X \subseteq M$ is called *convex in M* if for all $x, y, z \in M$,

$$x \leq z \leq y \text{ and } x, y \in X \Rightarrow z \in X.$$

Arbitrary intersections and upward directed unions of convex subsets are again convex. In particular, for every subset $Y \subseteq M$ there exists a smallest convex superset X of Y in M, the *convex hull* of Y in M.

Definition 2.1.2

(a) An *ordered abelian group* is a pair $(\Gamma, \leq)$, where Γ is an abelian group (usually written additively) and $\leq$ is a total order relation on the set Γ such that for all $\alpha, \beta, \gamma \in \Gamma$,

$$\alpha \leq \beta \Rightarrow \alpha + \gamma \leq \beta + \gamma.$$

If there is no ambiguity, we usually simply write Γ instead of $(\Gamma, \leq)$.

© Springer Nature Switzerland AG 2022
M. Knebusch, C. Scheiderer, *Real Algebra*, Universitext,
https://doi.org/10.1007/978-3-031-09800-0_2

(b) A homomorphism $\varphi\colon \Gamma \to \Gamma'$ between ordered abelian groups is called *order preserving* if for all $\alpha \in \Gamma$,

$$\alpha \geq 0 \Rightarrow \varphi(\alpha) \geq 0.$$

Definition 2.1.3 Let Γ be an ordered abelian group and $\Delta \leq \Gamma$ a subgroup. For $\alpha \in \Gamma$ we let $|\alpha| := \alpha$ or $-\alpha$, depending on whether $\alpha \geq 0$ or $\alpha \leq 0$.

(a) An element $\gamma \in \Gamma$ is called *infinitely small* with respect to Δ if $n\gamma < |\delta|$ for all $0 \neq \delta \in \Delta$ and all $n \in \mathbb{Z}$. Conversely, γ is called *infinitely large* with respect to Δ if $\delta < |\gamma|$ for all $\delta \in \Delta$.
(b) The group Γ is called *archimedean over* Δ if Γ does not contain any infinitely large elements with respect to Δ. The group Γ is called *archimedean* if for all $\alpha, \beta \in \Gamma$ with $\beta \neq 0$, there exists $n \in \mathbb{Z}$ such that $\alpha < n|\beta|$ (i.e., if Γ is archimedean over every subgroup different from $\{0\}$).

Remarks 2.1.4

(1) An ordered abelian group can also be defined as a pair (Γ, Π), where Γ is an abelian group and $\Pi \subseteq \Gamma$ is a subset such that

$$\Pi + \Pi \subseteq \Pi, \quad \Pi \cap (-\Pi) = \{0\}, \text{ and } \Pi \cup (-\Pi) = \Gamma.$$

(2) Let $(K, \leq)$ be an ordered field and denote the set of positive elements of K by K_+^*. Then $(K, +, \leq)$ and $(K_+^*, \cdot, \leq)$ are ordered abelian groups.
(3) Ordered abelian groups are *torsion free*: $n\alpha \neq 0$ for all $0 \neq \alpha \in \Gamma$ and $0 \neq n \in \mathbb{Z}$.
(4) Let Γ be an ordered abelian group. The convex hulls of subgroups of Γ are subgroups again. A subgroup Δ is convex if and only if $\gamma \in \Gamma, \delta \in \Delta$ and $0 \leq \gamma \leq \delta$ imply that $\gamma \in \Delta$. The convex subgroups of Γ form a *chain*: every two convex subgroups are comparable with respect to inclusion.
(5) If $\varphi\colon \Gamma \to \Gamma'$ is an order preserving homomorphism of ordered abelian groups, then $\mathrm{Ker}(\varphi)$ is convex in Γ. Conversely, for every convex subgroup Δ of Γ there is precisely one ordering $\overline{\Pi}$ of $\overline{\Gamma} = \Gamma/\Delta$ that makes $\overline{\Gamma}$ into an ordered abelian group and $\pi\colon \Gamma \to \overline{\Gamma}$ order preserving, namely $\overline{\Pi} = \pi(\Pi)$, where $\Pi := \{\alpha \in \Gamma\colon \alpha \geq 0\}$ (Remark (1)). The quotient group $\overline{\Gamma} = \Gamma/\Delta$ will always be equipped with this ordering. There exists an obvious homomorphism theorem for ordered abelian groups.
(6) Convex subgroups are sometimes called *isolated* subgroups in the literature (see for instance [13, Chap. VI]).

Examples 2.1.5

(1) Let $\Gamma_1, \ldots, \Gamma_r$ be ordered abelian groups. We equip $\Gamma := \Gamma_1 \times \cdots \times \Gamma_r$ with
the *lexicographic order*: $(\gamma_1, \ldots, \gamma_r) > 0$ if and only if

there exists $i \in \{0, \ldots, r - 1\}$ such that $\gamma_1 = \cdots = \gamma_i = 0$ and $\gamma_{i+1} > 0$.

Then $(\Gamma, >)$ is an ordered abelian group which we denote by $(\Gamma_1 \times \cdots \times \Gamma_r)_{\mathrm{lex}}$.
The projections $\Gamma \rightarrow (\Gamma_1 \times \cdots \times \Gamma_i)_{\mathrm{lex}}$ where $i = 0, \ldots, r$, are order
preserving, and so their kernels Δ_i are convex subgroups of Γ. Clearly we have
$\Gamma = \Delta_0 \supseteq \Delta_1 \supseteq \cdots \supseteq \Delta_r = 0$. If the Γ_i are archimedean, then $\Delta_0, \ldots, \Delta_r$
are the only convex subgroups of Γ.
(2) In the special case $\Gamma_1 = \cdots = \Gamma_r = \mathbb{Z}$ we call $(e_1, \ldots, e_r)$ the *lexicographic
basis* of $\mathbb{Z}^r_{\mathrm{lex}}$, where e_i denotes i-th unit vector.

Exercise 2.1.6 Let G be a torsion free abelian group and $H \subseteq G$ a subgroup. Let
an ordering $\leq$ of H be given. Show that $\leq$ can be extended to an ordering of G, and
that this extension is uniquely determined if G/H is a torsion group.

In the rest of this section (K, P) will be an ordered field. Whenever it is clear
that convexity, the $\leq$ sign, etc. refer to P we will not mention this explicitly.

Proposition 2.1.7 *Let A be a subring of K.*

(a) *The convex hull of A in K is a subring of K.*
(b) *A is convex in K if and only if $[0, 1] \subseteq A$. In particular, if A is convex in K,
then every overring of A is also convex in K.*

Proof (a) This is clear. (b) Obviously $[0, 1] \subseteq A$ is necessary for the convexity of
A. Conversely, if $[0, 1] \subseteq A$ and if $a \in A$, $b \in K$ with $0 < b < a$, then $ba^{-1} \in A$
and so $b = ba^{-1} \cdot a \in A$. □

Definition 2.1.8 Let (K, P) be an ordered field and $A \subseteq K$ a subring.

(a) We denote the convex hull of A in K (with respect to P) by $o_P(K/A)$ or simply
$o(K/A)$. The smallest convex subring of K with respect to P is $o_P(K/\mathbb{Z}) =: o_P(K)$.
(b) When discussing *infinitely small* or *infinitely large* elements of K in relation to
A (and P), we always refer to the underlying additive groups. In particular,
(K, P) is called *archimedean over* A if $o_P(K/A) = K$; the ordered field
(K, P) is called *archimedean* if $o_P(K) = K$, i.e., if K does not contain any
proper convex subrings. The last condition is equivalent to the well-known
"Axiom of Archimedes", which says that for all $a \in K$ there exists an $n \in \mathbb{N}$
such that $a < n$.

One should note that a given (real) field in general possesses archimedean as well
as non-archimedean orderings. For the simple transcendental extension $\mathbb{Q}(t)$ of $\mathbb{Q}$,
examples of both cases were given in Sect. 1.1.

In a (non-trivial) example we determine the convex hull of a subfield explicitly:

Example 2.1.9 Let (F, P) be an ordered field, $K = F(t)$ the rational function field in one variable over F, and $Q := P_{0,+}$ the ordering of K that extends P (where t is positive and infinitely small compared to F) as presented in Example 1.1.13(3). We will show that $o_Q(K/F) = F[t]_{(t)}$. In (K, Q) we have $|at^r| < |b|$ for all $r \in \mathbb{N}$ and all $a, b \in F^*$. Thus, if $g(t) = a_0 t^d + \cdots + a_d \in F[t]$ with $g(0) = a_d \neq 0$, then

$$\tfrac{1}{2}|g(0)| < |g(t)| < 2|g(0)|$$

in (K, Q). Hence,

$$\tfrac{1}{4}|f(0)| < |f(t)| < 4|f(0)|$$

for all $f \in F(t)$ with $f(0) \notin \{0, \infty\}$. Now let $h(t) = t^r f(t) \in F(t)^*$, where $r \in \mathbb{Z}$ and $f \in F(t)$ with $f(0) \notin \{0, \infty\}$. By what we have just established, there exist $a, b \in F$ with $0 < a < b$ and $at^r < |h(t)| < bt^r$. It follows that

$$h(t) \in o_Q(K/F) \iff r \geq 0 \iff h(t) \in F[t]_{(t)}.$$

The following classical theorem of Hölder gives (in principle) a summary of all archimedean ordered abelian groups and fields:

Theorem 2.1.10 (O. Hölder [52], 1901)

(a) *Let Γ be an archimedean ordered abelian group and let $0 < \gamma \in \Gamma$. Then there exists precisely one order preserving injective group homomorphism $\varphi \colon \Gamma \hookrightarrow \mathbb{R}$ such that $\varphi(\gamma) = 1$.*

(b) *Let (K, P) be an archimedean ordered field with underlying archimedean ordered abelian group $(K, +, P)$. The order preserving embedding $\varphi \colon K \hookrightarrow \mathbb{R}$ such that $\varphi(1) = 1$, furnished by (a), is a ring homomorphism.*

Corollary 2.1.11 *Up to order compatible isomorphism the archimedean ordered abelian groups are precisely the subgroups of $(\mathbb{R}, +)$, and the archimedean ordered fields are precisely the subfields of $\mathbb{R}$ (each with the ordering induced by $\mathbb{R}$).*   □

Corollary 2.1.12 (von Staudt's Theorem) *The identity is the only endomorphism of the field $\mathbb{R}$ of real numbers.*   □

Corollary 2.1.13 *A proper overfield of $\mathbb{R}$ does not have any archimedean orderings.*   □

Proof of Theorem 2.1.10 With some heuristics one sees quickly how the proof should proceed. Namely, if there exists a map φ as in (a), then for all $\alpha \in \Gamma$, $m \in \mathbb{Z}$, and $n \in \mathbb{N}$ we have

$$\tfrac{m}{n} \leq \varphi(\alpha) \iff m\varphi(\gamma) \leq n\varphi(\alpha) \iff m\gamma \leq n\alpha.$$

Therefore we define for $\alpha \in \Gamma$:

$$U(\alpha) := \{\tfrac{m}{n} : m \in \mathbb{Z}, n \in \mathbb{N}, m\gamma \leq n\alpha\}, \quad O(\alpha) := \{\tfrac{m}{n} : m \in \mathbb{Z}, n \in \mathbb{N}, m\gamma > n\alpha\}.$$

The pair $\big(U(\alpha), O(\alpha)\big)$ constitutes a proper Dedekind cut of $\mathbb{Q}$. In other words (cf. Definition 2.9.3):

(1) $U(\alpha) \cup O(\alpha) = \mathbb{Q}$;
(2) $U(\alpha), O(\alpha)$ are nonempty;
(3) for all $a \in U(\alpha)$ and $b \in O(\alpha)$ we have $a < b$.

Here (1) is trivial, (2) uses the archimedean property, and (3) uses the torsion freeness of Γ.

As is well-known, the real numbers are axiomatically defined via such cuts. Thus, for every $\alpha \in \Gamma$ there exists precisely one $\varphi(\alpha) \in \mathbb{R}$ such that

$$U(\alpha) =]{-\infty}, \varphi(\alpha)]_{\mathbb{R}} \cap \mathbb{Q} \quad \text{and} \quad O(\alpha) =]\varphi(\alpha), \infty[_{\mathbb{R}} \cap \mathbb{Q}.$$

One verifies that for all $\alpha, \beta \in \Gamma$, $U(\alpha) + U(\beta) \subseteq U(\alpha + \beta)$ and $O(\alpha) + O(\beta) \subseteq O(\alpha + \beta)$, from which it follows that $\varphi(\alpha + \beta) = \varphi(\alpha) + \varphi(\beta)$. Then φ is order preserving since $\gamma > 0$ and Γ is archimedean. Also, φ is injective since $\mathrm{Ker}(\varphi) = \{\alpha \in \Gamma : -\gamma \leq n\alpha \leq \gamma \text{ for all } n \in \mathbb{Z}\}$.

For (b), let $\Gamma = K$ and $\gamma = 1$. The map φ is then also multiplicative: since $\mathbb{Q} \subseteq K$ we have

$$U(\alpha) =]{-\infty}, \alpha]_{K} \cap \mathbb{Q} \quad \text{and} \quad O(\alpha) =]\alpha, \infty[_{K} \cap \mathbb{Q}$$

for all $\alpha \in K$, from which the homomorphy of φ follows. □

Warning: By Hölder's Theorem, $\mathbb{Q}$ is dense in every archimedean ordered field (K, P) (with respect to the topology induced by P on K, cf. Sect. 2.6). The obvious guess that this is always so for archimedean extensions K/L of ordered fields is false, as the following exercise shows.

Exercise 2.1.14 Let $L = \mathbb{R}(t)$ and $K = \mathbb{R}(t^2)$, a subfield of L, and let Q be the positive cone of L with $0 <_Q nt <_Q 1$ for all $n \in \mathbb{N}$. Show that K fails to be dense in L with respect to the order topology of Q.

We mention the book [93] as a very extensive source on ordered groups, rings and fields.

2.2 Valuation Rings

The theory of Krull valuations revolves around three fundamental concepts: valuations, valuation rings, and places. These are more or less equivalent to each other, but it may be more convenient to work with one concept and not the other two, depending on the situation. Therefore it is important to be familiar with all three of them and to be able to translate from one to the other, as and when required. In this section we investigate the first part of this triptych: valuation rings. In Sect. 2.4 we will consider valuations. Finally, places will be investigated in Sect. 2.8.

Definition 2.2.1 A subring A of a field K is called a *valuation ring of K* if for every $a \in K^*$ we have $a \in A$ or $a^{-1} \in A$. An arbitrary ring A is called a *valuation ring* if it has no zero divisors and is a valuation ring of its field of fractions.

Definition 2.2.2 A ring A is called *local* if $A \neq 0$ and A has just one maximal ideal, which is usually denoted by $\mathfrak{m}_A$ or simply $\mathfrak{m}$. We call the field $\kappa(A) := A/\mathfrak{m}_A$ the *residue field* of A.

Proposition 2.2.3 *Every valuation ring is a local ring.*

Proof Let A be a valuation ring. We must show that $\mathfrak{m} := A \setminus A^*$ is an ideal in A. It is clear that $a\mathfrak{m} \subseteq \mathfrak{m}$ for $a \in A$. Let $a, b \in \mathfrak{m} \setminus \{0\}$ and assume without loss of generality that $ab^{-1} \in A$. Then $a - b = (ab^{-1} - 1)b \in \mathfrak{m}$. We conclude that $\mathfrak{m}$ is an ideal.   □

The following fact, although almost trivial, is of great importance for real algebra and geometry:

Proposition 2.2.4 *If (K, P) is an ordered field, then every subring of K that is convex with respect to P is a valuation ring of K.*

Proof Let $A \subseteq K$ be a convex subring and $a \in K^*$. If $|a| \leq 1$, then $a \in A$, otherwise $a^{-1} \in A$.   □

Definition 2.2.5 A valuation ring A is called *residually real* if the field $\kappa(A) = A/\mathfrak{m}_A$ is real.

Remarks 2.2.6

(1) Let (K, P) be an ordered field, $A \subseteq K$ a convex subring of K and $\pi : A \to \kappa(A) = A/\mathfrak{m}_A$ the residue homomorphism. We see immediately that

$$\overline{P} := \pi(A \cap P)$$

is an ordering of the residue field $\kappa(A)$. In particular, A is residually real. We call $\overline{P}$ the ordering of $\kappa(A)$, *induced* by P.

(2) Let (K, P) be an ordered field and $L \subseteq K$ a subfield. The maximal ideal of the valuation ring $\mathfrak{o}_P(K/L)$ consists precisely of the elements of K that are infinitely small with respect to L (i.e., the elements $a \in K$ with $n|a| < |b|$ for all $n \in \mathbb{N}$ and $0 \neq b \in L$).

Proposition 2.2.7 *Let (K, P) be an ordered field and $A \subseteq K$ a valuation ring of K. The following statements are equivalent:*

(i) *A is convex in K;*
(ii) *$\mathfrak{m}_A$ is convex in K;*
(iii) *$\mathfrak{m}_A$ is convex in A;*
(iv) *$\mathfrak{m}_A \subseteq]-1, 1[$;*
(v) *$1 + \mathfrak{m}_A > 0$;*
(vi) *$\mathfrak{o}_P(K) \subseteq A$.*

Proof (i) $\Rightarrow$ (ii): Let $a, b \in K$ with $0 < b < a$ and $a \in \mathfrak{m}_A$. From $a^{-1} \notin A$ and $0 < a^{-1} < b^{-1}$ it follows that $b^{-1} \notin A$, hence $b \in \mathfrak{m}_A$.

The implications (ii) $\Rightarrow$ (iii) $\Rightarrow$ (iv) $\Rightarrow$ (v) are trivial.

(v) $\Rightarrow$ (vi): Let $0 < a \in \mathfrak{o}_P(K)$ and let $n \in \mathbb{N}$ with $a < n$. If $a \notin A$, then $a^{-1} \in \mathfrak{m}_A$ and $1 - na^{-1} < 0$, which contradicts (v).

(vi) $\Rightarrow$ (i): This follows from Proposition 2.1.7. □

Further examples of valuation rings can be obtained from:

Proposition 2.2.8 *If A is a unique factorization domain and $\pi \in A$ a prime element, then the localization $A_{\pi A}$ is a valuation ring (more precisely: a discrete valuation ring of rank 1, cf. Sect. 2.4).*

Proof Every $0 \neq b \in \mathrm{Quot}\, A$ can be written as $b = \pi^e \cdot a/a'$ with $e \in \mathbb{Z}$ and $a, a' \in A$ not divisible by π. If $e \geq 0$, then $b \in A_{\pi A}$; otherwise $b^{-1} \in A_{\pi A}$. □

Next we will discuss an important connection between the prime ideals of a valuation ring A and its overrings in $K := \mathrm{Quot}\, A$. Note that every overring of A in K is itself a valuation ring of K (this follows immediately from the definition).

Proposition 2.2.9 *Let A be a valuation ring of K.*

(a) *For every prime ideal $\mathfrak{p}$ of A we have $\mathfrak{p} = \mathfrak{p}A_{\mathfrak{p}}$. Thus $\mathfrak{p}$ is the maximal ideal of the valuation ring $B := A_{\mathfrak{p}}$.*
(b) *For every overring B of A in K, $\mathfrak{p} := \mathfrak{m}_B$ is contained in A, hence is a prime ideal of A, and $B = A_{\mathfrak{p}}$.*
(c) *The set of overrings of A in K is totally ordered with respect to inclusion. The same is true for the set of prime ideals of A.*

Corollary 2.2.10 *There exists an inclusion reversing bijection between the set of prime ideals $\mathfrak{p}$ of A and the set of overrings B of A in K, given by $\mathfrak{p} \mapsto A_{\mathfrak{p}} =: B$. The inverse is given by $B \mapsto \mathfrak{m}_B =: \mathfrak{p}$. Both sets are totally ordered (by inclusion).* □

Corollary 2.2.11 *For every two valuation rings A and B of K we have*

$$A \subseteq B \Leftrightarrow \mathfrak{m}_A \supseteq \mathfrak{m}_B.$$

□

Proof of Proposition 2.2.9 (a) Let $a \in \mathfrak{p}$ and $b \in A \setminus \mathfrak{p}$. If $ab^{-1} \notin A$, then $(ab^{-1})^{-1} = ba^{-1} \in A$, and thus $b = (ba^{-1})a \in \mathfrak{p}$, contradiction. Therefore $ab^{-1} \in A$, and it follows from $a = (ab^{-1})b$ and $b \notin \mathfrak{p}$ that $ab^{-1} \in \mathfrak{p}$.

(b) For all $0 \neq b \in \mathfrak{m}_B =: \mathfrak{p}$ we have $b^{-1} \notin B$, thus also $b^{-1} \notin A$, and hence $b \in A$. This shows $\mathfrak{p} \subseteq A$. The inclusion $A_\mathfrak{p} \subseteq B$ follows from $A \setminus \mathfrak{p} \subseteq B \setminus \mathfrak{p} = B^*$. Conversely, assume that $b \in B$. If $b \notin A$, then $b^{-1} \in A$ and $b^{-1} \notin \mathfrak{p} = \mathfrak{m}_B$. Therefore, $b = 1/b^{-1} \in A_\mathfrak{p}$.

(c) Let B and B' be overrings of A in K. Assume for the sake of contradiction that there exist elements $b \in B \setminus B'$ and $b' \in B' \setminus B$. Consider $a := b^{-1}b' \in K^*$. We have $a \notin B$ (for otherwise $b' = ba \in B$) and $a^{-1} \notin B'$ (for otherwise $b = b'a^{-1} \in B'$). In particular, $a \notin A$ and $a^{-1} \notin A$, contradiction. Hence, $B \subseteq B'$ or $B' \subseteq B$. $\qquad\square$

We will see in Sect. 2.4 that every two *arbitrary* ideals of a valuation ring are comparable.

2.3 Integral Elements

Definition 2.3.1 Let B be a ring and $A \subseteq B$ a subring.

(a) An element $b \in B$ is called *integral over A* if there exist $n \in \mathbb{N}$ and $a_1, \ldots, a_n \in A$ such that

$$b^n + a_1 b^{n-1} + \cdots + a_n = 0.$$

(It is essential that b^n has coefficient 1!)
(b) If every $b \in B$ is integral over A, then B is called *integral over A* and $A \subseteq B$ is called an *integral ring extension*.

Proposition 2.3.2 *If B is a ring and $A \subseteq B$ a subring, then the elements of B that are integral over A form a subring of B.*

Recall that an A-module M is called *faithful* if for every $0 \neq a \in A$ we have $aM \neq 0$. We require the following

Lemma 2.3.3 *An element $b \in B$ is integral over A if and only if there exists a faithful $A[b]$-module M that is finitely generated as an A-module.*

Proof If b is integral over A, then $A[b]$ is a finitely generated A-module. Conversely, assume that there exists a module M as in the statement of the lemma and write $M = Au_1 + \cdots + Au_n$. There exist elements $a_{ij} \in A$ (with $1 \le i, j \le n$) such that $bu_i = \sum_j a_{ij} u_j$ for $i = 1, \ldots, n$. Thus we have a system of equations

$$\sum_j (a_{ij} - b\delta_{ij}) u_j = 0 \quad (i = 1, \ldots, n)$$

where δ_{ij} denotes the Kronecker symbol. Let $f(t) \in A[t]$ be the characteristic polynomial of the matrix (a_{ij}). Since for every $n \times n$-matrix S we have $S\widehat{S} = \widehat{S}S = \det(S) \cdot 1$ (where $\widehat{S}$ denotes the adjugate of S), it follows that $f(b)u_i = 0$ for $i = 1, \ldots, n$, and so $f(b)M = 0$. Since M is faithful, we have $f(b) = 0$, which shows that b is integral over A by Definition 2.3.1.   □

Proof of Proposition 2.3.2 If $b, b' \in B$ are integral over A, then $A[b, b']$ is a finitely generated $A[b]$-module and $A[b]$ is a finitely generated A-module. Hence, $A[b, b']$ is also a finitely generated A-module. It is a faithful $A[c]$-module for every $c \in A[b, b']$, and the statement follows from Lemma 2.3.3.   □

Definition 2.3.4

(a) Let B be a ring and $A \subseteq B$ a subring. The subring of elements of B that are integral over A is called the *integral closure* of A in B. If the integral closure of A in B is equal to A, then A is called *integrally closed in B*.
(b) A ring A is called *integrally closed* if it has no zero divisors and coincides with its integral closure in Quot A.

If $\{A_\alpha\}$ is a family of subrings of a ring B and if all A_α are integrally closed in B, then $\bigcap_\alpha A_\alpha$ is also integrally closed in B.

Theorem 2.3.5 (Cohen–Seidenberg) *Let $A \subseteq B$ be an integral ring extension.*

(a) *For every prime ideal $\mathfrak{p}$ of A there exists a prime ideal $\mathfrak{q}$ of B such that $\mathfrak{p} = A \cap \mathfrak{q}$.*
(b) *For every prime ideal $\mathfrak{q}$ of B we have: $\mathfrak{q}$ is maximal in B $\Leftrightarrow$ $A \cap \mathfrak{q}$ is maximal in A.*

Proof We first prove (b). It suffices to show that for every integral extension $K \subseteq L$ of rings K and L without zero divisors, K is a field if and only if L is a field (consider the integral extension $A/A \cap \mathfrak{q} \subseteq B/\mathfrak{q}$). If K is a field and $0 \ne b \in L$, then there exists a monic *irreducible* polynomial $f \in K[t]$ such that $f(b) = 0$, and it follows that $b^{-1} \in L$. If L is a field and $0 \ne a \in K$, then there exist $a_1, \ldots, a_n \in K$ such that $a^{-n} + a_1 a^{1-n} + \cdots + a_n = 0$, and multiplying by a^{n-1} shows that $a^{-1} \in K$.

Next we prove (a). Let $\mathfrak{p}$ be a prime ideal of A and let $S := A \setminus \mathfrak{p}$. The natural homomorphism $A_\mathfrak{p} = S^{-1}A \to S^{-1}B$ is injective, and $A_\mathfrak{p} \subseteq S^{-1}B$ is also an integral ring extension. Choosing a maximal ideal $\mathfrak{q}'$ of $S^{-1}B$ gives $\mathfrak{q}' \cap A_\mathfrak{p} = \mathfrak{p}A_\mathfrak{p}$ by (b) ($A_\mathfrak{p}$ is local), and it follows that $\mathfrak{q} \cap A = \mathfrak{p}$ for the preimage $\mathfrak{q}$ of $\mathfrak{q}'$ in B.   □

Proposition 2.3.6 *Valuation rings are integrally closed.*

Proof Let A be a valuation ring of K, $b \in K$, and $a_1, \ldots, a_n \in A$ with

$$b^n = a_1 b^{n-1} + \cdots + a_n.$$

If $b \notin A$, then $b^{-1} \in \mathfrak{m}_A$, and so $1 = a_1 b^{-1} + \cdots + a_n b^{-n} \in \mathfrak{m}_A$, a contradiction.
□

It follows that every intersection of valuation rings of a field is integrally closed. The converse is also true:

Theorem 2.3.7 (Krull [67]) *Let K be a field, $A \subseteq K$ a subring, and $\widetilde{A}$ the integral closure of A in K.*

(a) *$\widetilde{A}$ is the intersection of all valuation rings B of K with $A \subseteq B$.*
(b) *In (a) we may restrict ourselves to those valuation rings B of K with $A \subseteq B$ for which $\mathfrak{p} := A \cap \mathfrak{m}_B$ is a maximal ideal of A and $A/\mathfrak{p} \subseteq \kappa(B)$ is an algebraic field extension.*

We consider another definition before turning to the proof of this theorem:

Definition 2.3.8 If A and B are local subrings of a field K, then we say that A *is dominated by* B if $A \subseteq B$ and $\mathfrak{m}_A \subseteq \mathfrak{m}_B$ (and so, $\mathfrak{m}_A = A \cap \mathfrak{m}_B$). When this happens, $A \subseteq B$ induces an embedding $\kappa(A) \subseteq \kappa(B)$.

Note that domination is a transitive relation and hence determines a partial ordering on the set of local subrings of K.

Proof of Theorem 2.3.7 The inclusion $\widetilde{A} \subseteq \bigcap B$ follows from Proposition 2.3.6. Conversely, assume that $x \in K$ is not integral over A, i.e., $x \notin \widetilde{A}$. Then $x \notin A[x^{-1}]$, for otherwise there would exist an equation $x^n = a_1 x^{n-1} + \cdots + a_n$ with $a_i \in A$. Therefore, $A[x^{-1}]$ has a maximal ideal $\mathfrak{q}$ such that $x^{-1} \in \mathfrak{q}$. Let $\mathfrak{p} := A \cap \mathfrak{q}$. Then $A/\mathfrak{p} \to A[x^{-1}]/\mathfrak{q}$ is an isomorphism, and $\mathfrak{p}$ is in particular a maximal ideal of A.

Let $A' := A[x^{-1}]_\mathfrak{q}$. By Zorn's Lemma there exists a local ring $B \subseteq K$ which is maximal with respect to the properties

(1) B dominates A', and
(2) $\kappa(B)$ is algebraic over $\kappa(A')$.

Since $x^{-1} \in \mathfrak{q} \subseteq \mathfrak{m}_{A'} \subseteq \mathfrak{m}_B$ we have $x \notin B$. The proof is complete once we have established that B is a valuation ring of K. Thus, let $y \in K^*$. We must show that $y \in B$ or $y^{-1} \in B$.

Assume first that y is not integral over B. Upon replacing A by B and x by y in the arguments above, we obtain a local ring $C \subseteq K$ that dominates B, and for which $y^{-1} \in C$ and $\kappa(C)$ is algebraic over $\kappa(B)$. Then C also satisfies (1) and (2), and it follows from the maximality of B that $B = C$, and thus $y^{-1} \in B$.

On the other hand, if y is integral over B, then $B[y]$ contains a maximal ideal $\mathfrak{n}$ with $\mathfrak{m}_B = B \cap \mathfrak{n}$ by Theorem 2.3.5. Hence, the local ring $C := B[y]_\mathfrak{n}$ dominates B. Furthermore, the extension $\kappa(B) \subseteq \kappa(C)$ is algebraic since $\kappa(C)$ is generated over $\kappa(B)$ by the residue class of y and y satisfies an integrality equation. It follows again that $B = C$, and so $y \in B$. □

Corollary 2.3.9 *The valuation rings of K are precisely those local subrings that are not properly dominated by any other local subring.*

Proof It follows from Theorem 2.3.7 that the integral closure of a local subring $A \subseteq K$ is equal to the intersection of all valuation rings of K that dominate A. Thus, if A is not properly dominated by any local subring, then A must be a valuation ring. The converse direction follows from Proposition 2.2.9. □

Corollary 2.3.10 *For every subring A of a field K and every prime ideal $\mathfrak{p}$ of A there exists a valuation ring B of K such that $A \subseteq B$, $\mathfrak{p} = A \cap \mathfrak{m}_B$, and $\kappa(B)$ is algebraic over $\kappa(\mathfrak{p}) = \operatorname{Quot} A/\mathfrak{p}$.*

Proof Choose B such that B dominates the local ring $A_\mathfrak{p}$ and $\kappa(B)$ is algebraic over $\kappa(A_\mathfrak{p}) = \kappa(\mathfrak{p})$ (Theorem 2.3.7). Then $A \cap \mathfrak{m}_B = A \cap A_\mathfrak{p} \cap \mathfrak{m}_B = A \cap \mathfrak{p} A_\mathfrak{p} = \mathfrak{p}$. □

Corollary 2.3.11 *The fields K that have no proper (i.e., different from K) valuation rings are precisely the algebraic field extensions of finite fields.*

Proof Let K be algebraic over $\mathbb{F}_p$. Since every subring of K contains the prime field $\mathbb{F}_p$, K is the only subring of K that is integrally closed in K. Conversely, let K be a field without any proper valuation rings. Then char $K =: p > 0$, for otherwise K would be integral over $\mathbb{Z}$ by Theorem 2.3.7, contradicting that $\mathbb{Z}$ is integrally closed in $\mathbb{Q}$. If there was a $t \in K$, transcendental over $\mathbb{F}_p$, then t^{-1} would not be integral over $\mathbb{F}_p[t]$, a contradiction. □

2.4 Valuations, Ideals of Valuation Rings

Throughout this section K denotes an arbitrary field.

Let Γ be an ordered abelian group and ∞ an element not in Γ. We write the disjoint union $\Gamma \cup \{\infty\}$ simply as $\Gamma \cup \infty$. Declaring

$$\alpha < \infty \quad \text{and} \quad \alpha + \infty = \infty + \alpha = \infty + \infty = \infty$$

for all $\alpha \in \Gamma$ gives $\Gamma \cup \infty$ the structure of a totally ordered semigroup.

Definition 2.4.1 Let K be a field and Γ an ordered abelian group. A *valuation* of K with values in Γ is a map $v \colon K \to \Gamma \cup \infty$ such that

(V1) $v(a) = \infty \Leftrightarrow a = 0$;
(V2) $v(ab) = v(a) + v(b)$;
(V3) $v(a + b) > \min\{v(a), v(b)\}$

for all $a, b \in K$. If there exists an $a \in K^*$ with $v(a) \neq 0$, then v is called *non-trivial*. Otherwise, v is called *trivial*. The ordered abelian group $v(K^*)$ is called the *value group* of v and is denoted Γ_v. We will sometimes call valuations *Krull valuations* in order to distinguish them from absolute values, cf. Sect. 2.6.

Remarks 2.4.2 Let $v: K \to \Gamma \cup \infty$ be a valuation of K.

(1) v is determined by its restriction to K^*. Valuations of K can also be defined as those group homomorphisms $K^* \to \Gamma$ that satisfy (V3) for all $a, b \in K^*$ with $a + b \neq 0$.
(2) For all roots of unity $a \in K^*$, we have $v(a) = 0$ since Γ is torsion free.

Exercise 2.4.3 Let $a, b \in K$ with $v(a) \neq v(b)$. Show that equality holds in (V3), i.e., that v satisfies the following stronger property:

$$v(a + b) = \min\{v(a), v(b)\}.$$

Example 2.4.4 Let A be a unique factorization domain and $K := \operatorname{Quot} A$. Every prime element π of A defines a valuation $v_\pi : K^* \to \mathbb{Z}$ of K, the *order at* π, as follows: $v_\pi(\pi^e ab^{-1}) := e$ for $e \in \mathbb{Z}$ and $a, b \in A$ not divisible by π. The valuation ring $A_{\pi A}$ of K (cf. Proposition 2.2.8) is the "valuation ring belonging to v_π" in the sense of:

Proposition 2.4.5 *Let* $v: K \to \Gamma \cup \infty$ *be a valuation of* K. *Then*

$$\mathfrak{o}_v := \{a \in K : v(a) \geq 0\}$$

is a valuation ring of K *with maximal ideal*

$$\mathfrak{m}_v := \{a \in K : v(a) > 0\}.$$

We call $\mathfrak{o}_v$ *the* valuation ring of K associated to v.

Proof If $a \in K$ and $a \notin \mathfrak{o}_v$, then $v(a) < 0$ and so $a^{-1} \in \mathfrak{m}_v$ since $v(a^{-1}) = -v(a) > 0$. □

Conversely, we will show that every valuation ring comes from a valuation (which is essentially unique), hence the name.

Observe that if $v: K^* \to \Gamma$ is a valuation, assumed to be surjective, then v and Γ can be obtained from the valuation ring $\mathfrak{o}_v$ as follows: since $\operatorname{Ker} v = \mathfrak{o}_v^*$, the valuation v induces a group homomorphism $\bar{v}: K^*/\mathfrak{o}_v^* \xrightarrow{\sim} \Gamma$. For all $a, b \in K^*$ we have

$$v(a) \leq v(b) \Leftrightarrow v(a^{-1}b) \geq 0 \Leftrightarrow a^{-1}b \in \mathfrak{o}_v,$$

and so $\bar{v}$ is an order preserving isomorphism if we order $K^*/\mathfrak{o}_v^*$ via

$$a\,\mathfrak{o}_v^* \leq b\,\mathfrak{o}_v^* :\Leftrightarrow a^{-1}b \in \mathfrak{o}_v.$$

(The multiplicative notation for this ordered abelian group should not be a source of confusion for the reader!)

Now, let A be an arbitrary valuation ring of K. For all $a, b \in K^*$ we define analogously,

$$aA^* \leq bA^* :\Leftrightarrow a^{-1}b \in A.$$

It is immediately clear that this definition turns K^*/A^* into an ordered abelian group.

Definition 2.4.6 Let A be a valuation ring of K. The *value group of A* is the ordered abelian group $\Gamma_A := K^*/A^*$ with ordering as above. The canonical epimorphism $v_A: K^* \to \Gamma_A$ is called the *canonical valuation of K associated to A.*

The second definition is justified by:

Proposition 2.4.7 *Let A be a valuation ring of K. Then $v_A: K \to \Gamma_A \cup \infty$ is a valuation of K and the associated valuation ring satisfies $\mathfrak{o}_{v_A} = A$.*

Proof The map $v_A|_{K^*}$ is a homomorphism, so it remains to show that (V3) holds. Let $a, b \in K^*$ with $a + b \neq 0$ and $v_A(a) \leq v_A(b)$. Then $a^{-1}b \in A$ and it follows that $v_A(a+b) = (a+b)A^* = a(1 + a^{-1}b)A^* \geq aA^* = v_A(a)$. Clearly, $\mathfrak{o}_{v_A} = A$. $\square$

Conversely, if $v: K \to \Gamma \cup \infty$ is a valuation with valuation ring $A := \mathfrak{o}_v$, then there exists precisely one order preserving embedding $i: \Gamma_A \hookrightarrow \Gamma$ such that $v = i \circ v_A$. Therefore, a valuation of K can be interpreted as a pair (A, i), where A is a valuation ring of K and $i: \Gamma_A \hookrightarrow \Gamma$ an order preserving embedding of ordered abelian groups.

Definition 2.4.8 Let v and v' be valuations of K. Then v is a *coarsening* of v' if $\mathfrak{o}_v \subseteq \mathfrak{o}_{v'}$. If $\mathfrak{o}_v = \mathfrak{o}_{v'}$, then v and v' are *equivalent*.

Lemma 2.4.9 *Let $v: K^* \to \Gamma$ and $v': K^* \to \Gamma'$ be valuations of K, and assume that v is surjective. The following statements are equivalent:*

(i) *v' is a coarsening of v;*
(ii) *there exists an order preserving homomorphism $\varphi: \Gamma \to \Gamma'$ such that $v' = \varphi \circ v$.*

Proof (i) $\Rightarrow$ (ii): Let $A := \mathfrak{o}_v$ and $A' := \mathfrak{o}_{v'}$. We have a commutative diagram

$$
\begin{array}{ccccc}
\Gamma & \xleftarrow{\;\;v\;\;} & K^* & \xrightarrow{\;\;v'\;\;} & \Gamma' \\
{\scriptstyle i}\big\uparrow{\scriptstyle \cong} & \;\swarrow{\scriptstyle v_A} & & {\scriptstyle v_{A'}}\searrow\; & \big\uparrow{\scriptstyle i'} \\
K^*/A^* & & \xrightarrow{\;\;\pi\;\;} & & K^*/A'^*
\end{array}
$$

where $\pi\colon K^*/A^* \to K^*/A'^*$ denotes the canonical epimorphism, i and i' are order preserving, and i is an isomorphism. Let $\varphi := i' \circ \pi \circ i^{-1}$.

(ii) $\Rightarrow$ (i): This is trivial. $\square$

Definition 2.4.10 Let (K, P) be an ordered field and $v\colon K \twoheadrightarrow \Gamma \cup \infty$ a valuation of K. Then v and P are *compatible* if $\mathfrak{o}_v$ is a convex subring of K with respect to P. A number of equivalent characterizations is given in Proposition 2.2.7.

Example 2.4.11 Every ordered abelian group Γ occurs as the value group of a surjective valuation. Indeed, let k be a field and $A := k[\Gamma_+]$ the *semigroup algebra* of $\Gamma_+ := \{\alpha \in \Gamma\colon \alpha \geq 0\}$. This means that A has a k-vector space basis $\{x_\alpha\colon \alpha \in \Gamma_+\}$ with $x_\alpha \neq x_\beta$ for $\alpha \neq \beta$, and multiplication in A is given by $x_\alpha x_\beta = x_{\alpha+\beta}$ for all $\alpha, \beta \in \Gamma_+$.

For $0 \neq a = \sum_{\alpha\in\Gamma_+} a_\alpha x_\alpha \in A$ (with $a_\alpha \in k$ and almost all $a_\alpha = 0$) let

$$\widetilde{v}(a) := \min\{\alpha \in \Gamma_+\colon a_\alpha \neq 0\}.$$

Let $K := \operatorname{Quot} A$ (note that A has no zero divisors). Then $\widetilde{v}$ extends uniquely to a surjective homomorphism $v\colon K^* \to \Gamma$, and v is a valuation of K.

Exercise 2.4.12 Verify that:

(1) $\widetilde{v}(ab) = \widetilde{v}(a) + \widetilde{v}(b)$ for all $a, b \in A \setminus \{0\}$;
(2) $\widetilde{v}(a + b) \geq \min\{\widetilde{v}(a), \widetilde{v}(b)\}$ if $a + b \neq 0$.

Example 2.4.13 We continue with the set-up of the previous example and assume in addition that k has an ordering P. Then P induces an ordering Q of K as follows: for $0 \neq a = \sum_\alpha a_\alpha x_\alpha \in A$ we define the "leading coefficient" of a as

$$L(a) := a_{v(a)} \;\; (\in k^*),$$

and we let $Q := \{0\} \cup \{\frac{a}{b}\colon 0 \neq a, b \in A \text{ and } L(ab) \in P\}$. Now Q is indeed an ordering of K since $L(ab) = L(a)L(b)$ for all $a, b \in A \setminus \{0\}$, and if $a + b \neq 0$, then

$$
L(a + b) = \begin{cases} L(a) + L(b) & \text{if } v(a) = v(b) \text{ and } L(a) + L(b) \neq 0 \\ L(a) & \text{if } v(a) < v(b) \end{cases}.
$$

We claim that the valuation ring $\mathfrak{o}_v$ is convex in K with respect to Q. To see this, let $0 \neq x \in \mathfrak{m}_v$. There exist $a, b \in A$ with $v(a) = v(b) = 0$, and $0 < \alpha \in \Gamma$ such that $x = x_\alpha \cdot a/b$. It follows that $1 + x = b^{-2}(b^2 + abx_\alpha) \in Q$ since $L(b^2 + abx_\alpha) = L(b)^2 \in P$. This shows that $1 + \mathfrak{m}_v \subseteq Q$, from which the claim follows by Proposition 2.2.7.

Example 2.4.14 Let (K, P) be an ordered field, $F \subseteq K$ a subfield, and $A = \mathfrak{o}_P(K/F)$ its convex hull in K. Then Γ_A and v_A can be described as follows: we call $a, b \in K^*$ *archimedean equivalent over F* if there exist $\lambda, \mu \in F^*$ such that $|a| \leq \lambda |b|$ and $|b| \leq \mu |a|$. Then Γ_A is the (multiplicative) group of the resulting equivalence classes, ordered by the inverse absolute value, and v_A is the residue map.

Recall that an abelian group G is called *divisible* if $nG = G$ for all $n \geq 1$. The torsion free divisible abelian groups are precisely the $\mathbb{Q}$-vector spaces.

Example 2.4.15 The value group Γ_A of any valuation ring A of a real closed or algebraically closed field is divisible (and thus a $\mathbb{Q}$-vector space), since a real closed field contains all roots of all positive elements and an algebraically closed field contains all roots of all elements.

We will now investigate the ideals of valuation rings.

Definition 2.4.16 Let $(M, \leq)$ be an ordered set and $X \subseteq M$ a subset (where $X = \varnothing$ is allowed). We call X an *upper subset* of M if for all $x \in X$ and $y \in M$,

$$x \leq y \Rightarrow y \in X.$$

Proposition 2.4.17 (Ideals of Valuation Rings) *Let A be a valuation ring of K with surjective valuation $v = v_A \colon K \to \Gamma \cup \infty$. Let $\Gamma_+ := \{\alpha \in \Gamma : \alpha \geq 0\}$.*

(a) *The map*

$$\mathfrak{a} \mapsto v(\mathfrak{a} \setminus \{0\})$$

defines an inclusion preserving bijection from the ideals of A to the upper subsets of Γ_+. The inverse map is given by

$$M \mapsto v^{-1}(M \cup \{\infty\}).$$

(b) *The map*

$$\mathfrak{p} \mapsto v(A_\mathfrak{p}^*) = v(A \setminus \mathfrak{p}) \cup \left(-v(A \setminus \mathfrak{p})\right)$$

defines an inclusion reversing bijection from the prime ideals of A to the convex subgroups of Γ. The inverse map is given by

$$\Delta \mapsto \{0\} \cup v^{-1}(\Gamma_+ \setminus \Delta).$$

(c) *The ideals of A form a chain (i.e., are totally ordered by inclusion).*
(d) *A is a Bézout ring, i.e., A has no zero divisors and every finitely generated ideal is principal. More precisely: for $a_1, \ldots, a_n \in A$ and $\mathfrak{a} = Aa_1 + \cdots + Aa_n$, if $v(a_1) \leq v(a_i)$ for $i = 1, \ldots, n$, then $\mathfrak{a} = Aa_1$.*
(e) *If $\mathfrak{a} \neq A$ is a radical ideal (i.e., $\mathfrak{a} = \sqrt{\mathfrak{a}}$), then $\mathfrak{a}$ is a prime ideal.*

Proof The proofs of (a) and (b) are elementary, (c) follows immediately from (a), and (d) is clear.

(e) Let $a, b \in A$ with $ab \in \mathfrak{a}$. We may assume that $a \mid b$ in A, i.e., $b = ac$ for some $c \in A$. Then $b^2 = abc \in \mathfrak{a}$, and so $b \in \mathfrak{a}$.   □

Recall that a sequence

$$\cdots \to G_{i-1} \xrightarrow{\varphi_{i-1}} G_i \xrightarrow{\varphi_i} G_{i+1} \to \cdots$$

of abelian groups and homomorphisms is called *exact* if $\mathrm{Im}(\varphi_{i-1}) = \mathrm{Ker}(\varphi_i)$ for all i. Sequences of the form

$$0 \to G' \to G \to G'' \to 0$$

are called *short*.

Proposition 2.4.18 *Let A and B be valuation rings of K, and assume that $A \subseteq B$. Let $\pi : B \to \kappa(B)$ denote the residue map of B.*

(a) $C := \pi(A) = A/\mathfrak{m}_B$ *is a valuation ring of $\kappa(B)$.*
(b) *There exists a natural short exact sequence*

$$0 \to \Gamma_C \to \Gamma_A \to \Gamma_B \to 0$$

of order preserving homomorphisms.

Proof (a) If $c \in \kappa(B)^*$ and $b \in B^*$ with $c = \pi(b)$, then $b \in A$ or $b^{-1} \in A$, and so $c \in C$ or $c^{-1} = \pi(b^{-1}) \in C$.

(b) Since $A^* \subseteq B^*$, there is an order preserving epimorphism $\Gamma_A = K^*/A^* \twoheadrightarrow K^*/B^* = \Gamma_B$ whose kernel is the (convex) subgroup $B^*/A^* =: \Gamma_{B/A}$ of Γ_A. Furthermore, $\pi|_{B^*} : B^* \to \kappa(B)^*$ is an epimorphism, and $\pi^{-1}(C^*) = A^*$. Therefore π induces an isomorphism $\Gamma_{B/A} = B^*/A^* \xrightarrow{\sim} \kappa(B)^*/C^* = \Gamma_C$, which is clearly order preserving.   □

Corollary 2.4.19 *Let A be a valuation ring of K with natural valuation $v_A : K^* \to \Gamma_A$. Then there exists a canonical inclusion preserving bijection from the set of overrings B of A in K to the set of convex subgroups Δ of Γ_A, namely*

$$B \mapsto \Gamma_{B/A} := B^*/A^* = v_A(B^*).$$

The inverse map is

$$\Delta \mapsto \{0\} \cup v_A^{-1}(\Delta \cup \Gamma_+).$$

Furthermore, for each overring B of A there is a canonical order preserving isomorphism

$$\Gamma_A / \Gamma_{B/A} \xrightarrow{\sim} \Gamma_B.$$

Proof This follows from Corollary 2.2.10 and Propositions 2.4.17 and 2.4.18 □

Definition 2.4.20

(a) Let Γ be an ordered abelian group. The *rank* of Γ, denoted rank Γ, is the number of convex subgroups of Γ that are different from Γ (a natural number or ∞).
(b) Let A be a valuation ring. The *rank* of A, denoted rank A, is defined as the rank of the value group $\Gamma_A = K^*/A^*$, where $K = \operatorname{Quot} A$.

Remarks 2.4.21

(1) Many authors (Bourbaki, for example) refer to the rank of Γ as the height of Γ.
(2) By Corollaries 2.4.19 and 2.2.10, the rank of a valuation ring A is also equal to the number of prime ideals of A that are different from $\{0\}$ (the Krull dimension of A), as well as the number of overrings of A in K that are different from K.
(3) The valuation rings of rank 0 are the fields. The ordered abelian groups of rank 1 are precisely the nontrivial subgroups of $(\mathbb{R}, +)$ (Theorem 2.1.10). The only "discrete" ones among these are the infinite cyclic groups, which justifies the following definition:

Definition 2.4.22 A valuation ring A is called *discrete of rank one* if the value group Γ_A is infinite cyclic.

Proposition 2.4.23 *Let A be a valuation ring which is not a field. The following are equivalent:*

(i) *A is discrete of rank one;*
(ii) *A is noetherian;*
(iii) *A is a principal ideal domain.*

Proof (i) $\Rightarrow$ (ii): This follows from Proposition 2.4.17(a). (ii) $\Rightarrow$ (iii): This follows from Proposition 2.4.17(d). (iii) $\Rightarrow$ (i): Let $m_A = A\pi$. In particular, π is a prime element of A. If π' is a further prime element of A, then $\pi \mid \pi'$ or $\pi' \mid \pi$ (since A is a valuation ring), and thus π and π' are associated, i.e., $A\pi = A\pi'$. Since A is factorial, every element $0 \neq a \in A$ is of the form $a = u\pi^e$ with $u \in A^*$ and $e \geq 0$. Hence Γ_A is cyclic with positive generator $v(\pi)$. $\square$

Exercise 2.4.24 Show that a field K has a non-trivial valuation ring with real residue field if, and only if, K has a non-archimedean ordering.

Exercise 2.4.25 Let k be a field, let t be a variable, and let $\mathcal{V}$ be the set of all non-trivial valuation rings of $k(t)$ that contain k. Show that there is a natural bijection between $\mathcal{V}$ and $\mathcal{P} \cup \{\infty\}$, where $\mathcal{P}$ is the set of monic irreducible polynomials in $k[t]$ and ∞ is an additional symbol. Every $B \in \mathcal{V}$ is a discrete valuation ring of rank one.

Exercise 2.4.26 Let k be a field and $n \in \mathbb{N}$, put $x = (x_1, \ldots, x_n)$. For each polynomial $f = \sum_{\alpha \in \mathbb{Z}_+^n} c_\alpha x^\alpha$ in $k[x]$ let

$$v(f) := \inf\{|\alpha| : c_\alpha \neq 0\}$$

(multinomial notation, with $|\alpha| = \sum_i \alpha_i$ as usual). Show the following:

(a) v extends to a valuation v of $k(x) = k(x_1, \ldots, x_n)$ which is discrete of rank one. The valuation ring $\mathfrak{o}_v$ and its maximal ideal $\mathfrak{m}_v$ satisfy $k[x] \subseteq \mathfrak{o}_v$ and $\mathfrak{m}_v \cap k[x] = (x_1, \ldots, x_n)$.
(b) The residue field of v is isomorphic to the rational function field $k(y_1, \ldots, y_{n-1})$ in $n-1$ variables.

2.5   Residue Fields and Subfields of Convex Valuation Rings

Throughout this section K denotes an arbitrary field.

Theorem 2.5.1 *Let A be a valuation ring of K and let $\kappa(A) = A/\mathfrak{m}_A$.*

(a) *If K is real closed, then so is $\kappa(A)$.*
(b) *If K is real closed, then:*

$$\kappa(A) \text{ is real closed } \Leftrightarrow \sqrt{-1} \notin \kappa(A) \Leftrightarrow A \text{ is convex in } K.$$

Proof Let $f \in A[t]$ be a monic polynomial. If f has a root in K, this root is already in A since A is integrally closed. In particular, the mod $\mathfrak{m}_A$ reduced polynomial has a root in $\kappa(A)$, from which (a) follows immediately.

Assume now that K is real closed. It follows from the previous argument that $\kappa(A)$ has no proper field extensions of odd degree and that $\overline{P} := \kappa(A)^2$ satisfies the ordering axioms (O1) and (O2) (Sect. 1.1). Therefore, $\kappa(A)$ is real closed if and only if $-1 \notin \kappa(A)^2$ (Proposition 1.5.2).

If A is convex in K, then $\mathfrak{m}_A \cap [1, \infty[= \varnothing$ (Proposition 2.2.7). Hence, $1 + a^2 \notin \mathfrak{m}_A$ for $a \in A$, and so $-1 \notin \kappa(A)^2$. Conversely, assume that A is not convex. Then there exists $a \in K$ such that $1 + a^2 \in \mathfrak{m}_A$ (Proposition 2.2.7). It follows that $a \in A$ for if not, we have $a^{-1} \in \mathfrak{m}_A$ and thus $1 = a^{-2}(1 + a^2) - a^{-2} \in \mathfrak{m}_A$. We conclude that $-1 \in \kappa(A)^2$. $\qquad\square$

Let A be a valuation ring of K. If K is ordered and A is convex in K, then A contains subfields of K (e.g., $\mathbb{Q}$). More generally, one sees easily that A contains a subfield if and only if $\operatorname{char} K = \operatorname{char} \kappa(A)$ (the so-called "equal characteristic case"; the unequal characteristic case only occurs when $\operatorname{char} K = 0$ and $\operatorname{char} \kappa(A) > 0$). Every subfield F of A is contained in a maximal subfield of A (Zorn's Lemma). Also, via $F \hookrightarrow A \to \kappa(A)$ we always consider F as a subfield of $\kappa(A)$.

Proposition 2.5.2 *Let A be a valuation ring of K and F a maximal subfield of A. Then F is algebraically closed in K, and the field extension $\kappa(A)/F$ is algebraic.*

Proof Since A is integrally closed in K, it contains the algebraic closure of F in K. Let $\pi: A \to \kappa(A)$ be the residue homomorphism. If there were an $a \in A$ with $\pi(a)$ transcendental over $\pi(F)$, we would have $F[a] \cap \mathfrak{m}_A = \{0\}$, and thus $F \subsetneqq F(a) \subseteq A$, contradicting the maximality of F. $\qquad\square$

Proposition 2.5.3 *Let R be a real closed field, A a convex subring of R and F a maximal subfield of A. Then the restriction of $\pi: A \to \kappa(A)$ to F is an isomorphism from F to $\kappa(A)$.*

(In the case of algebraically closed fields, there is an analogous statement for arbitrary valuation rings of the field.)

Proof Since F is algebraically closed in R (Proposition 2.5.2), it is real closed (Corollary 1.5.5). Since $\kappa(A)$ is algebraic over $\pi(F)$ (Proposition 2.5.2) and $\kappa(A)$ is real (Theorem 2.5.1), we have $\kappa(A) = \pi(F)$. $\qquad\square$

Proposition 2.5.4 *Let (K, P) be an ordered field and A a convex subring of K. Then A is the convex hull of any of its maximal subfields.*

Proof Let F be a maximal subfield of A and let $B := \mathfrak{o}_P(K/F)$. Consider an element $a \in A$. By Proposition 2.5.2 there exist $n \in \mathbb{N}$ and elements $b_1, \ldots, b_n \in F$ such that $a^n + b_1 a^{n-1} + \cdots + b_n \in \mathfrak{m}_A \subseteq \mathfrak{m}_B$. Assume that $a \notin B$. Then $a^{-1} \in \mathfrak{m}_B$, and multiplication with a^{-n} gives $1 + (b_1 a^{-1} + \cdots + b_n a^{-n}) \in \mathfrak{m}_B$, a contradiction. $\qquad\square$

Definition 2.5.5 Let (K, P) be an ordered field. We call a subfield F of K *archimedean saturated in K* (with respect to P) if K has no proper overfields F' of F that are archimedean over F.

It follows that the subfields of K that are archimedean saturated in K are precisely the maximal subfields of convex subrings of K.

Proposition 2.5.6 *Let A be a convex subring of an ordered field K and $F \subseteq A$ a subfield, then* tr.deg.$(K/F) \geq$ rank A.

Proof Let $A = A_0 \subsetneqq A_1 \subsetneqq \cdots \subsetneqq A_r = K$ be a chain of overrings of A. Let F_0 be a maximal overfield of F in A_0 and F_i a maximal overfield of F_{i-1} in A_i, where $i = 1, \ldots, r$. Since $A_i = \mathfrak{o}(K/F_i)$ for $i = 0, \ldots, r$, it follows that all F_i are different. By Proposition 2.5.2 we have tr.deg.$(F_i/F_{i-1}) \geq 1$ for $i = 1, \ldots, r$, hence tr.deg.$(K/F) \geq r$.   □

Proposition 2.5.7 *Let K be a real closed field and $F \subseteq K$ a subfield. If*

$$F \subseteq F_0 \subsetneqq F_1 \subsetneqq \cdots \subsetneqq F_r = K$$

and

$$F \subseteq F_0' \subsetneqq F_1' \subsetneqq \cdots \subsetneqq F_s' = K$$

are two chains of archimedean saturated subfields F_i, F_i' of K that cannot be made any finer, then $r = s$ and $F_i \cong F_i'$ for $i = 0, \ldots, r$.

Proof Since the $\mathfrak{o}(K/F_i)$, resp. the $\mathfrak{o}(K/F_i')$, are precisely the distinct convex overrings of F in K (Proposition 2.5.4), we have $r = s = $ rank $\mathfrak{o}(K/F)$ and $\mathfrak{o}(K/F_i) = \mathfrak{o}(K/F_i')$ for $i = 0, \ldots, r$. Then, by Proposition 2.5.3, $F_i \cong \kappa(\mathfrak{o}(K/F_i)) \cong F_i'$.   □

Example 2.5.8 (E. Artin) It is essential that K is real closed for the isomorphism $F_i \cong F_i'$!

Let R_0 be the real closure of $\mathbb{Q}$ (i.e., the relative algebraic closure of $\mathbb{Q}$ in $\mathbb{R}$) and F the relative algebraic closure of $\mathbb{Q}(e)$ in $\mathbb{R}$ (note that $e = 2.71828\ldots$ is transcendental). Consider $K = F(t)$, equipped with the ordering $P_{0,+}$ (Example 1.1.13(3)). Then $0 < t < a$ for all $0 < a \in F$. Finally, let $F' := R_0(e + t) \subseteq K$. Since F' is not real closed, F and F' are not isomorphic. We claim that not only F, but also F', is archimedean saturated in K. This is clear for F. To see this for F', we need to cite a result from field theory (cf. e.g., [54, Vol. 3, p. 199]):

> If E/k is a field extension such that k is (relatively) algebraically closed in E, and if $E(x)/E$ is a simple transcendental extension, then $k(x)$ is also algebraically closed in $E(x)$.

In the above situation, it suffices to verify that F' is algebraically closed in K since tr.deg.$(K/F') = 1$ and $\mathfrak{o}(K/F') = \mathfrak{o}(K/\mathbb{Z}) \neq K$. This, however, is an immediate application of the cited theorem (R_0 is algebraically closed in F and $K = F(e + t)$).

In this example, one should also observe that $R_0(e)$ is a subfield of K which is order isomorphic to F', but which is not archimedean saturated in K (F is archimedean over $R_0(e)$).

2.6 The Topology of Ordered and Valued Fields

Let (K, P) be an ordered field. The open intervals

$$]a, b[_P = \{x \in K : a <_P x <_P b\}$$

form an open basis of a topology on K that we denote by $\mathcal{T}_P$. For $n \geq 1$, we also denote the induced product topology on K^n by $\mathcal{T}_P$ and remark that K^n is a Hausdorff space with respect to this topology. Moreover, $(K, \mathcal{T}_P)$ is a topological field (i.e., the maps $(x, y) \mapsto x - y$ and $(x, y) \mapsto xy$ from $K \times K$ to K are continuous, as is the map $K^* \to K^*$, $x \mapsto x^{-1}$). It follows that for every rational function $f = g/h \in K(t_1, \ldots, t_n)$ (with g, h polynomials), the evaluation map $x \mapsto f(x)$ from $\{x \in K^n : h(x) \neq 0\}$ to K is continuous. We usually call $\mathcal{T}_P$ the *strong topology* on K^n (with respect to P) in order to distinguish it from the "weaker" Zariski topology (see Sect. 3.1).

There is however really just one case where the topology $\mathcal{T}_P$ has satisfying properties, namely when $K = \mathbb{R}$. This is one of the reasons that makes *semialgebraic topology* (see [31–33] and the references in [33]) indispensable when engaging in geometry over arbitrary real closed fields:

Proposition 2.6.1 *Let (K, P) be an ordered field. If $K \neq \mathbb{R}$, then the topological space $(K, \mathcal{T}_P)$ is totally disconnected. For $a, b \in K$ with $a < b$, the interval $[a, b]$ is not compact and $(K, \mathcal{T}_P)$ is in particular not locally compact.*

Proof If (K, P) is archimedean, then there exists an order preserving embedding $K \hookrightarrow \mathbb{R}$, and $\mathcal{T}_P$ is precisely the subspace topology on K induced by $\mathbb{R}$. If $K \neq \mathbb{R}$, the statements can be verified directly.

Assume thus that (K, P) is not archimedean. Then there exists a proper convex subring A of K, and A is open and closed in K. Let $x \in K$ and $0 < \varepsilon \in K$. Choose $a \in K^*$ with $\varepsilon/a \notin A$. Then $x + aA \subseteq]x - \varepsilon, x + \varepsilon[$ and $x + aA$ is a clopen (i.e., open and closed) neighbourhood of x. This shows that K is totally disconnected. Now let $a, b \in K$ with $a < b$. Then there exists an $\varepsilon \in K$, $\varepsilon > 0$, such that $n\varepsilon < b - a$ for all $n \in \mathbb{N}$ (if $b - a \in \mathfrak{m}_A$, then one can choose $\varepsilon = (b - a)^2$, otherwise $\varepsilon \in \mathfrak{m}_A$ suffices). The family $\{]x - \varepsilon, x + \varepsilon[: x \in K\}$ contains no finite subcovering of $[a, b]$. $\square$

Assume now that K is an arbitrary field and that $v \colon K \to \Gamma \cup \infty$ is a surjective valuation of K. The sets

$$B_\alpha(a) := \{x \in K : v(x - a) > \alpha\} \quad (a \in K, \alpha \in \Gamma)$$

form an open basis of a topology $\mathcal{T}_v$ on K and $(K, \mathcal{T}_v)$ is again a topological field. The topology $\mathcal{T}_v$ is the discrete topology if and only if v is the trivial valuation (i.e., $\Gamma = 0$).

As a consequence of the *strong triangle inequality* (V3) in Definition 2.4.1, all sets $v^{-1}(\alpha)$ (where $\alpha \in \Gamma$) are open in K (for every $a \in K$ with $v(a) = \alpha$ we have $B_\alpha(a) \subseteq v^{-1}(\alpha)$). Since

$$K \setminus v^{-1}(\alpha) = B_\alpha(0) \cup \bigcup_{\beta < \alpha} v^{-1}(\beta),$$

we have that $v^{-1}(\alpha)$ is also closed. All sets $B_\alpha(a)$ in the basis that defines $\mathcal{T}_v$ are thus clopen. In particular, $(K, \mathcal{T}_v)$ is totally disconnected. Furthermore, the obvious assumption that the closure and boundary of $B_\alpha(a)$ are given by $\{x: v(x - a) \geq \alpha\}$ and $\{x: v(x - a) = \alpha\}$, respectively, is *false*.

Proposition 2.6.2 *Let (K, P) be an ordered field and $v: K \to \Gamma \cup \infty$ a nontrivial surjective valuation, compatible with P. Then the order and valuation topologies on K coincide: $\mathcal{T}_P = \mathcal{T}_v$.*

Proof For every $0 < a \in K$ there exists $\alpha \in \Gamma$ such that $B_\alpha(0) \subseteq \,]-a, a[$, for instance $\alpha = v(a)$. Indeed, if $x \in K$ and $v(x) > \alpha = v(a)$, then $x/a \in \mathfrak{m}_v \subseteq \,]-1, 1[$, hence $|x| < a$. Conversely, for every $\alpha \in \Gamma$ there exists $0 < a \in K$ such that $]-a, a[\,\subseteq B_\alpha(0)$. Indeed, choose a such that $v(a) > \alpha$. Then if $|x| < a$, it follows that $|x/a| < 1$, hence $v(x/a) \geq 0$ and so $v(x) \geq v(a) > \alpha$. $\qquad\square$

The converse of this theorem is false: there may exist valuations that are not compatible with P, yet induce the same topology as P. This follows easily from the following:

Exercise 2.6.3 Let v and v' be surjective nontrivial valuations of K. Show that if v' is a coarsening of v, then $\mathcal{T}_v = \mathcal{T}_{v'}$.

In addition to Krull valuations we introduce absolute values. In number theory especially they are indispensable.

Definition 2.6.4 Let K be a field. An *absolute value* on K is a map $f: K \to \mathbb{R}$ such that for all $a, b \in K$:

(A1) $f(a) \geq 0$, and $f(a) = 0 \Leftrightarrow a = 0$;
(A2) $f(ab) = f(a)\,f(b)$;
(A3) $f(a + b) \leq f(a) + f(b)$.

If instead of (A3) the *ultrametric inequality*

(A4) $f(a + b) \leq \max\{f(a), f(b)\}$ for all $a, b \in K$,

is satisfied, then f is called an *ultrametric absolute value*.

In the literature the term "valuation" is often used instead of absolute value. Depending on whether it is ultrametric or not, this "valuation" is then called "non-archimedean" or "archimedean". Clearly this easily causes confusion.

Remarks 2.6.5

(1) If $\varphi\colon K \hookrightarrow \mathbb{C}$ is a field embedding, then $a \mapsto |\varphi(a)|$ is a non-ultrametric absolute value on K. The converse is essentially also true, as the following classical theorem of A. Ostrowski ([86], 1916) shows:

> If $f\colon K \to \mathbb{R}$ is a non-ultrametric absolute value on a field K, then there exists an embedding $\varphi\colon K \hookrightarrow \mathbb{C}$ and a real number s with $0 < s \le 1$, such that $f(a) = |\varphi(a)|^s$ for all $a \in K$. (See for instance also [55, Vol. II, §9.5].)

(2) If $v\colon K \to \Gamma \cup \infty$ is a surjective valuation with *archimedean* value group Γ, then one obtains an ultrametric absolute value f_v on K as follows: choose an order preserving embedding $i\colon \Gamma \hookrightarrow \mathbb{R}$ (cf. Sect. 2.1) as well as a real number c with $0 < c < 1$ and let $f_v(a) := c^{i(v(a))}$ $(a \ne 0)$, $f_v(0) := 0$. Conversely, every ultrametric absolute value can be "logarithmified" to a valuation of rank ≤ 1. Upon defining an equivalence of absolute values in the obvious way, one then obtains a bijective correspondence between the equivalence classes of valuations of rank ≤ 1 and the equivalence classes of ultrametric absolute values. Together with Remark (1) we thus obtain a description of all absolute values on a field.

Every absolute value f on K defines a metric on K via $(a, b) \mapsto f(a - b)$ and makes K into a metric topological space. We denote the associated topology by $\mathcal{T}_f$.

Proposition 2.6.6 *Let (K, P) be an ordered field and assume that K is either archimedean, or that K contains a proper convex subring of finite rank. Then there exists an absolute value f on K such that $\mathcal{T}_P = \mathcal{T}_f$. (In particular, $\mathcal{T}_P$ is metrizable).*

Proof If (K, P) is archimedean with order preserving embedding $\varphi\colon K \hookrightarrow \mathbb{R}$, then one obtains f as in Remark 2.6.5(1). In the other case, K contains a maximal proper convex subring A. The valuation $v_A = v$ has rank 1, and one obtains f as in Remark 2.6.5(2). It is clear that $\mathcal{T}_f = \mathcal{T}_v$, and Proposition 2.6.2 gives $\mathcal{T}_f = \mathcal{T}_P$. $\qquad\square$

2.7 The Baer–Krull Theorem

We have seen that for every convex subring A of an ordered field (K, P) there is an induced ordering $\overline{P}$ of the residue field $\kappa(A)$ (Remark 2.2.6(1)). Conversely, given a valuation ring A of a field K and an ordering Q of $\kappa(A)$, one can wonder if there exists an ordering P of K that makes A convex and for which $\overline{P} = Q$.

The Baer–Krull Theorem provides a (positive) answer to this question. In fact, the theorem is considerably more precise in that it determines all such orderings Q.

Consider a surjective valuation $v\colon K^* \to \Gamma$ and let $A := \mathfrak{o}_v$. Then v induces a surjective homomorphism $v_2\colon K^*/K^{*2} \twoheadrightarrow \Gamma/2\Gamma$. Let $\{\gamma_i\colon i \in I\} \subseteq \Gamma$ be a subset such that $\{\gamma_i + 2\Gamma\colon i \in I\}$ is a basis of the $\mathbb{F}_2$-vector space $\Gamma/2\Gamma$ and $\gamma_i > 0$ for all

$i \in I$. We choose elements $\pi_i \in K^*$ such that $v(\pi_i) = \gamma_i$ for each $i \in I$. All π_i are in $\mathfrak{m}_A$.

The system $\{\pi_i : i \in I\}$ will be kept fixed in what follows and is called a *quadratic system of representatives* of K with respect to A. Every $a \in K^*$ has a representation of the form

$$a = uc^2 \prod_{j \in J} \pi_j \tag{2.1}$$

with $J \subseteq I$ finite, $c \in K^*$ and $u \in A^*$, where J is uniquely determined by a. (The set J is characterized by $v_2(aK^{*2}) = \sum_{j \in J} (\gamma_j + 2\Gamma)$; since $v\left(a \prod_{j \in J} \pi_j^{-1}\right) \in 2\Gamma$ one then obtains the form (2.1).)

Theorem 2.7.1 (Baer–Krull) *Let A be a valuation ring of K, Q an ordering of $\kappa(A)$ and $\{\pi_i : i \in I\}$ a quadratic system of representatives as above. Let Y_Q be the set of all orderings P of K such that A is convex with respect to P and $Q = \overline{P}$. Then the map $P \mapsto \left(\mathrm{sign}_P(\pi_i)\right)_{i \in I}$ is a bijection from the set Y_Q to the set $\{\pm 1\}^I$.*

The theorem says in particular that Y_Q contains precisely one element when $I = \varnothing$, i.e., when $\Gamma = 2\Gamma$.

Proof We denote the homomorphism $A \to \kappa(A)$ by $a \mapsto \overline{a}$. Let $\varepsilon = (\varepsilon_i)_{i \in I} \in \{\pm 1\}^I$. Then there exists at most one $P \in Y_Q$ such that $\mathrm{sign}_P(\pi_i) = \varepsilon_i$ for all $i \in I$ since for every $a \in K^*$ the sign of a with respect to P is determined by (2.1). In order to prove the existence of such a $P \in Y_Q$ we may assume $\varepsilon_i = 1$ for all i (replace π_i by $\varepsilon_i \pi_i$ for all $i \in I$).

Let $a \in K^*$ and let

$$a = u\,c^2 \prod_{j \in J} \pi_j = u'c'^{\,2} \prod_{j \in J} \pi_j$$

be two representations of the form (2.1). Then $\frac{c}{c'} \in A^*$ and, since $u' = \left(\frac{c}{c'}\right)^2 u$, it follows that $\mathrm{sign}_Q(\overline{u'}) = \mathrm{sign}_Q(\overline{u})$. Hence $\sigma : a \mapsto \mathrm{sign}_Q(\overline{u})$ is a well-defined map $\sigma : K^* \to \{\pm 1\}$. This is a homomorphism since, if

$$a = u\,c^2 \prod_{j \in J} \pi_j \quad \text{and} \quad b = v\,d^2 \prod_{\ell \in L} \pi_\ell$$

are representations of the form (2.1), then there is a representation

$$ab = w\,e^2 \prod_{m \in M} \pi_m$$

of the form (2.1) with $w = uv$.

We will now show that $P := \{0\} \cup \mathrm{Ker}(\sigma)$ is an ordering of K. Since $\sigma(-1) = -1$ it is clear that $PP \subseteq P$, $P \cup (-P) = K$ and $P \cap (-P) = \{0\}$. It remains to show that $P + P \subseteq P$. Let $0 \neq a, b \in P$ with $a + b \neq 0$ and with representations

$$a = u\,c^2 \prod_{j \in J} \pi_j, \quad b = v\,d^2 \prod_{\ell \in L} \pi_\ell$$

of the form (2.1). Then $\bar{u}, \bar{v} \in Q$. We consider two cases.

Case 1: $J = L$. After exchanging a and b if needed, we obtain $x := \frac{c}{d} \in A$, and so

$$a + b = (uc^2 + vd^2) \prod_{j \in J} \pi_j = (ux^2 + v) \cdot d^2 \prod_{j \in J} \pi_j.$$

Since $\bar{u}\,\bar{x}^2 + \bar{v} \in Q$ we conclude that $\sigma(a + b) = 1$, i.e., $a + b \in P$.

Case 2: $J \neq L$. We have $v(a) < v(b)$ after a possible exchange of a and b. Hence there exists $x \in \mathfrak{m}_A$ with $b = ax$, and it follows that

$$a + b = (1 + x)a = (1 + x)u \cdot c^2 \prod_{j \in J} \pi_j,$$

i.e., $a + b \in P$.

Thus we have shown that P is an ordering of K. From the definition of P the convexity of A with respect to P follows immediately (since $1 + \mathfrak{m}_A \subseteq P$), as does $\overline{P} = Q$. $\qquad\square$

Corollary 2.7.2 *Let A be a residually real valuation ring of K (Definition 2.2.5). Then there exists an ordering of K that makes A convex.* $\qquad\square$

A special case of this result is contained in Theorem 2.5.1.

Remark 2.7.3 Our formulation of the Baer–Krull Theorem in this section exhibits a certain arbitrariness due to the choice of a system of representatives. We explain briefly how the statement of the theorem can be presented in a more "invariant" manner.

Let $v\colon K^* \twoheadrightarrow \Gamma$ be a surjective valuation with valuation ring A. Let $k := \kappa(A)$, Y the set of orderings P of K that make A convex, and X_k the set of all orderings of k. For an abelian group G we denote by $\widehat{G} := \mathrm{Hom}(G, S^1)$ its character group, where $S^1 = \{\zeta \in \mathbb{C}\colon |\zeta| = 1\}$. Upon identifying an ordering P of K with the character $\sigma_P\colon aK^{*2} \mapsto \mathrm{sign}_P(a)$ of K^*/K^{*2}, we consider P as an element of $\widehat{K^*/K^{*2}}$. In particular, Y becomes a subset of $\widehat{K^*/K^{*2}}$ in this way.

A quadratic system of representatives $\Pi = \{\pi_i : i \in I\}$ is just a homomorphism $s\colon \Gamma/2\Gamma \to K^*/K^{*2}$ such that $v_2 \circ s = \mathrm{id}_{\Gamma/2\Gamma}$ (Π corresponds to $\gamma_i + 2\Gamma \mapsto \pi_i K^{*2}$). If we keep such a section s fixed, we obtain a map $Y \to \widehat{\Gamma/2\Gamma} \times X_k$, namely $P \mapsto (\sigma_P \circ s, \overline{P})$. The assertion of Baer–Krull is that this map is a bijection.

In order to avoid an arbitrary choice of the section s in the formulation, we observe that the epimorphism $v_2 \colon K^*/K^{*2} \twoheadrightarrow \Gamma/2\Gamma$ induces an embedding $\widehat{v_2} \colon \widehat{\Gamma/2\Gamma} \hookrightarrow \widehat{K^*/K^{*2}}$ via dualization (for $\chi \in \widehat{\Gamma/2\Gamma}$ we have $\widehat{v_2}(\chi) = \chi \circ v_2$). Interpreting $\widehat{\Gamma/2\Gamma}$ as a subgroup of $\widehat{K^*/K^{*2}}$ via $\widehat{v_2}$ (and Y as a subset of $\widehat{K^*/K^{*2}}$, as explained above), we obtain $\widehat{\Gamma/2\Gamma} \cdot Y \subseteq Y$, i.e., $\widehat{\Gamma/2\Gamma}$ acts on Y via translation. This is precisely what we showed in the essential part of the proof of the theorem. For $\chi \in \widehat{\Gamma/2\Gamma}$ and $P \in Y$, the ordering $Q := \chi \cdot P$ of K is determined by the following signature:

$$a \mapsto \chi\big(v(a) + 2\Gamma\big) \cdot \operatorname{sign}_P(a) \quad \text{(for all } a \in K^*).$$

As a result, the map $P \mapsto \overline{P}$ from Y to X_k is invariant under this action of $\widehat{\Gamma/2\Gamma}$ and takes different values on different orbits. Thus we obtain:

Theorem 2.7.4 (Baer–Krull Theorem, "Invariant" Formulation) *Let $v \colon K^* \to \Gamma$ be a surjective valuation with valuation ring A and residue field $k = A/\mathfrak{m}_A$. Let Y be the set of all orderings of K compatible with v, and X_k the set of all orderings of k. Then there is a natural free action of $\widehat{\Gamma/2\Gamma}$ on Y and a canonical bijection $Y/\widehat{\Gamma/2\Gamma} \to X_k$.*

2.8   Places

Let K and L be fields. We denote the disjoint union $K \cup \{\infty\}$ by $\widetilde{K}$. The operations $+$ and $\cdot$ are partially extended to $\widetilde{K}$ as follows:

$$a + \infty = \infty + a = \infty \quad \text{(for all } a \in K)$$

$$a \cdot \infty = \infty \cdot a = \infty \quad \text{(for all } a \in \widetilde{K}, \ a \neq 0).$$

The expressions $\infty + \infty$, $0 \cdot \infty$ and $\infty \cdot 0$ are not defined. The construction of inverses with respect to $+$ and $\cdot$ is extended via

$$-\infty = \infty \quad \text{and} \quad 0^{-1} = \infty, \ \infty^{-1} = 0.$$

As usual we write $x - y = x + (-y)$ and $\frac{x}{y} = xy^{-1}$ whenever the right hand side is defined. One can verify that the associative and distributive laws are satisfied whenever "everything" is defined.

Definition 2.8.1 A *place of K with values in L* is a map $\lambda \colon \widetilde{K} \to \widetilde{L}$ such that for all $a, b \in \widetilde{K}$:

(P1)   if $\lambda(a) + \lambda(b)$ is defined, then $a + b$ is defined and $\lambda(a + b) = \lambda(a) + \lambda(b)$;
(P2)   if $\lambda(a)\lambda(b)$ is defined, then ab is defined and $\lambda(ab) = \lambda(a)\lambda(b)$;
(P3) $\lambda(1) = 1$.

Let $k \subseteq K$ be a subfield and $j: k \hookrightarrow L$ a fixed embedding. If $\lambda(a) = j(a)$ for all $a \in k$, then λ is called a *place over k* (with respect to j) and we write $\widetilde{K} \xrightarrow{k} \widetilde{L}$ (there is no confusion about j).

Exercise 2.8.2 Let A be a valuation ring of K, $L := \kappa(A)$ and $a \mapsto \bar{a}$ the residue map $A \to L$. Show that the map $\lambda: \widetilde{K} \to \widetilde{L}$ defined by $\lambda(a) = \bar{a}$ if $a \in A$ and $\lambda(a) = \infty$ if $a \in \widetilde{K} \setminus A$ is a place. We often write λ_A instead of λ and call λ_A the *canonical place of K associated with the valuation ring A*.

Exercise 2.8.3 Show that every field homomorphism $\varphi: K \to L$ defines a place $\widetilde{\varphi}: \widetilde{K} \to \widetilde{L}$ via $\widetilde{\varphi}|_K = \varphi$ and $\widetilde{\varphi}(\infty) = \infty$. Such places are called *trivial*.

We will shortly see that canonical places and trivial places are essentially the only examples of places.

Proposition 2.8.4 (Properties of Places) *Let* $\lambda: \widetilde{K} \to \widetilde{L}$ *be a place. Then:*

(a) $\lambda(0) = 0$ *and* $\lambda(\infty) = \infty$.
(b) $\lambda(-a) = -\lambda(a)$ *and* $\lambda(a^{-1}) = \lambda(a)^{-1}$ *for all* $a \in \widetilde{K}$.
(c) *If* $\mu: \widetilde{L} \to \widetilde{M}$ *is a further place, then* $\mu \circ \lambda: \widetilde{K} \to \widetilde{M}$ *is also a place.*

Proof (a) If $\lambda(\infty) \neq \infty$, then $\lambda(\infty) + \lambda(\infty)$ is defined, and thus also $\infty + \infty$, contradicting (P1). If $\lambda(0) \neq 0$, then $\lambda(0) \cdot \lambda(\infty)$ is defined, and thus also $0 \cdot \infty$, contradicting (P2).

(b) This is clear for $a \in \{0, \infty\}$ by (a). Thus let $a \in K^*$. If $\lambda(a) = \lambda(-a) = \infty$, then $\lambda(-a) = -\lambda(a)$. Otherwise $\lambda(a) + \lambda(-a)$ is defined and it follows from (a) that $\lambda(a) + \lambda(-a) = \lambda(0) = 0$, hence $\lambda(-a) = -\lambda(a)$. Similarly, $\lambda(a^{-1}) = \lambda(a)^{-1}$.

(c) This is trivial.   □

Proposition 2.8.5 (Relationship Between Places and Valuation Rings)

(a) *If* $\lambda: \widetilde{K} \to \widetilde{L}$ *is a place, then*

$$A := \lambda^{-1}(L) = \{a \in K : \lambda(a) \neq \infty\}$$

is a valuation ring of K with maximal ideal $\mathfrak{m} = \lambda^{-1}(0)$. *There is exactly one homomorphism* $\varphi: \kappa(A) \to L$ *such that* $\lambda = \widetilde{\varphi} \circ \lambda_A$ *(cf. Exercises 2.8.2 and 2.8.3).*
(b) *Conversely, if A is a valuation ring of K and* $\varphi: \kappa(A) \to L$ *a homomorphism, then* $A = \lambda^{-1}(L)$ *for the place* $\lambda := \widetilde{\varphi} \circ \lambda_A: \widetilde{K} \to \widetilde{L}$.

In other words, a place of K with values in L is "the same" as a valuation ring A of K together with a homomorphism $\kappa(A) \to L$.

Proof (a) It is clear that A is a subring of K and that $\lambda|_A : A \to L$ is a homomorphism. If $a \in K$ and $a \notin A$, then $\lambda(a) = \infty$ and so $\lambda(a^{-1}) = 0$ by Proposition 2.8.4(b). It follows that A is a valuation ring of K with maximal ideal $\lambda^{-1}(0)$ and that $\varphi : \kappa(A) \to L$ is the homomorphism induced by $\lambda|_A$.

(b) This is clear.   □

Notation 2.8.6

(a) A place $\lambda : \widetilde{K} \to \widetilde{L}$ will from now on be written as $\lambda : K \to L \cup \infty$. If $\varphi : K \to L$ is a homomorphism, we denote its associated trivial place $K \to L \cup \infty$ by φ as well (instead of by $\widetilde{\varphi}$).
(b) Let $\lambda : K \to L \cup \infty$ be a place. We denote the valuation ring $\lambda^{-1}(L)$ of K by $\mathfrak{o}_\lambda$, its maximal ideal by $\mathfrak{m}_\lambda$, its residue field $\mathfrak{o}_\lambda/\mathfrak{m}_\lambda$ by κ_λ, and the induced homomorphism $\kappa_\lambda \to L$ by $\overline{\lambda}$. We also denote the value group of $\mathfrak{o}_\lambda$ by $\Gamma_\lambda :=$ $K^*/\mathfrak{o}_\lambda^*$ and the associated canonical valuation by $v_\lambda : K^* \to \Gamma_\lambda$.

Places provide an alternative way of looking at valuations and valuation rings. We illustrate this in the results below.

Proposition 2.8.7 *Let $A \subseteq K$ be a subring. An element $a \in K$ is integral over A if and only if $\lambda(a) \neq \infty$ for every place λ of K that is finite on A.*

Proof By Theorem 2.3.7(a).   □

Theorem 2.8.8 (Chevalley; Extension of Places) *Let $A \subseteq K$ be a subring and assume that L is algebraically closed. Any homomorphism $\varphi : A \to L$ can be extended to an L-valued place of K. In other words, there exists a place $\lambda : K \to L \cup \infty$ such that $\lambda|_A = \varphi$.*

Proof Let $\mathfrak{p} = \operatorname{Ker} \varphi$ and let $\varphi' : \kappa(\mathfrak{p}) = A_\mathfrak{p}/\mathfrak{p} A_\mathfrak{p} \to L$ be the homomorphism induced by φ. By Theorem 2.3.7 there exists a valuation ring B of K which dominates $A_\mathfrak{p}$ and such that $\kappa(B)$ is algebraic over $\kappa(\mathfrak{p})$. It follows that φ' factors through a homomorphism $\psi' : \kappa(B) \to L$. Letting $\lambda := \psi' \circ \lambda_B$ finishes the proof.
                                                                                                      □

Next we turn our attention to places of ordered fields.

Definition 2.8.9

(a) A place $\lambda : K \to L \cup \infty$ is called *real* if L is a real field.
(b) Let P and Q be orderings of K and L, respectively. A place $\lambda : K \to L \cup \infty$ is called *compatible* with P and Q (or *order preserving* with respect to P and Q) if $\lambda(P) \subseteq Q \cup \{\infty\}$.

Proposition 2.8.10 *Let* (K, P) *and* (L, Q) *be ordered fields and let* $\lambda\colon K \to L\cup\infty$ *be a place. The following statements are equivalent:*

(i) λ *is compatible with* P *and* Q;
(ii) *the subring* $\mathfrak{o}_\lambda$ *of* K *is convex with respect to* P *and the homomorphism* $\overline{\lambda}\colon \kappa_\lambda \to L$ *is order preserving with respect to* $\overline{P}$ *and* Q.

Proof (i) $\Rightarrow$ (ii): It follows from $\lambda(1 + \mathfrak{m}_\lambda) = \{1\}$ that $1 + \mathfrak{m}_\lambda \subseteq P$, proving the convexity of $\mathfrak{o}_\lambda$ with respect to P. It is clear that $\overline{\lambda}$ is order preserving.

(ii) $\Rightarrow$ (i): Let $a \in P \cap \mathfrak{o}_\lambda$ and let $\overline{a}$ be the image of a in κ_λ. Then $\overline{a} \in \overline{P}$ and so $\lambda(a) = \overline{\lambda}(\overline{a}) \in Q$. □

Corollary 2.8.11 *Let* $\lambda\colon K \to L \cup \infty$ *be a place. For every ordering* Q *of* L *there exists an ordering* P *of* K *such that* λ *is compatible with* P *and* Q.

Proof By the Baer–Krull Theorem (Sect. 2.7) and Proposition 2.8.10. □

Example 2.8.12 If K is a real closed field and (L, Q) an ordered field, then every place $K \to L \cup \infty$ is order preserving with respect to Q.

We finish this section with an extension theorem for *real* places.

Lemma 2.8.13 *Let* $K \subseteq L$ *be a field extension and* $B \subseteq L$ *a valuation ring of* L. *Then* $A := K \cap B$ *is a valuation ring of* K *and* $\mathfrak{m}_A = K \cap \mathfrak{m}_B$. *If* L *is algebraic over* K, *then* $\kappa(B)$ *is algebraic over* $\kappa(A)$.

Proof It is clear that A is a valuation ring of K. If $a \in K^*$, then

$$a \in \mathfrak{m}_A \Leftrightarrow a^{-1} \notin A \Leftrightarrow a^{-1} \notin B \Leftrightarrow a \in \mathfrak{m}_B.$$

Assume now that L is algebraic over K and let $b \in B$. There exist $a_0, \ldots, a_n \in K$ such that $a_0 \neq 0$ and $a_0 b^n + \cdots + a_n = 0$. Let m be the smallest index i such that $v_A(a_i) \leq v_A(a_j)$ for $j = 0, \ldots, n$. Division by a_m gives

$$\overline{b}^{n-m} + c_1 \overline{b}^{n-m-1} + \cdots + c_{n-m} = 0$$

in $\kappa(B)$, where the $c_j = \overline{a_{j+m}/a_m}$ are in $\kappa(A)$. It follows that $\overline{b}$ is algebraic over $\kappa(A)$. □

Lemma 2.8.14 *Let* $K \subseteq L$ *be an algebraic field extension, and let* B *and* C *be valuation rings of* L *such that* $K \cap B = K \cap C$ *and* $B \subseteq C$. *Then* $B = C$.

Proof Let $A := K \cap B = K \cap C$. Then there are inclusions $\kappa(A) = A/\mathfrak{m}_A \hookrightarrow B/\mathfrak{m}_C \hookrightarrow C/\mathfrak{m}_C = \kappa(C)$. Since $B/\mathfrak{m}_C$ is a valuation ring of $\kappa(C)$ by Proposition 2.4.18(a) and $\kappa(C)$ is algebraic over $\kappa(A)$ by Lemma 2.8.13, it follows that $B/\mathfrak{m}_C = \kappa(C)$ and so $B = C$ since valuation rings are integrally closed. □

Theorem 2.8.15 (Extension of Real Places) *Let* (K, P) *be an ordered field, R a real closed field and* $\lambda\colon K \to R \cup \infty$ *a P-compatible place. Let $L \supseteq K$ be an algebraic field extension and Q an extension of the ordering P to L. Then there exists a unique Q-compatible place* $\mu\colon L \to R \cup \infty$ *that extends λ.*

Proof The subring $A := \mathfrak{o}_\lambda$ of K is convex with respect to P and $\bar{\lambda}\colon \kappa(A) \to R$ is order preserving with respect to $\bar{P}$ by Proposition 2.8.10. Let B be the Q-convex hull of A in L. By Lemma 2.8.13 $\kappa(B)$ is algebraic over $\kappa(A)$ and $\bar{Q}$ is an extension of $\bar{P}$. Hence there exists a (unique) order preserving homomorphism $\psi\colon \kappa(B) \to R$ such that $\psi|_{\kappa(A)} = \bar{\lambda}$ (by uniqueness of the real closure; Theorem 1.11.2). Therefore the place $\mu := \psi \circ \lambda_B\colon L \to R \cup \infty$ satisfies $\mu|_K = \lambda$ and is compatible with Q.

To show the uniqueness of μ, assume that $\mu'\colon L \to R\cup\infty$ is also a Q-compatible extension of λ and let $B' := \mathfrak{o}_{\mu'}$. It suffices to show that $B = B'$. Since B' is convex with respect to Q and $K \cap B' = K \cap B = A$ it follows that $B \subseteq B'$. Hence $B = B'$ by Lemma 2.8.14. □

Corollary 2.8.16 *Let R be a real closed field and* $\lambda\colon K \to R \cup \infty$ *a place. There exists a real closure S of K that admits an extension* $\lambda_S\colon S \to R \cup \infty$ *of λ. More precisely: let S be a real closure of K, then λ has an extension* $\lambda_S\colon S \to R \cup \infty$ *if and only if λ is compatible with the ordering of K induced by S.*

Proof By Theorem 2.8.15 and Corollary 2.8.11. □

2.9 The Orderings of $R(t)$, $R((t))$ and Quot $\mathbb{R}\{t\}$

Let R be a real closed field. In this section we will explicitly determine the orderings of $R(t)$, $R((t))$ and Quot $\mathbb{R}\{t\}$.

We start with the rational function field in one variable $R(t)$. Let $c \in R$. Then c determines a place $\lambda_c\colon R(t) \to R \cup \infty$, namely $\lambda_c(f) = f(c)$ (setting $f(c) = \infty$ if f has a pole at c, as usual). The valuation ring associated to λ_c is $R[t]_{(t-c)}$. It is discrete of rank one and $t - c$ is a generator of its maximal ideal. By the Baer–Krull Theorem, $R(t)$ has precisely two λ_c-compatible orderings: $P_{c,+}$ and $P_{c,-}$. These are characterized by

$$t - c \in P_{c,+} \quad \text{and} \quad -(t - c) = c - t \in P_{c,-}$$

and are of course the orderings of Example 1.1.13(3).

There is a further R-valued place of $R(t)$, namely $\lambda_\infty\colon R(t) \to R \cup \infty$ which is defined by $\lambda_\infty := \lambda_0 \circ \sigma$ where σ is the R-automorphism of $R(t)$ that satisfies $\sigma(t) = t^{-1}$. For $f \in R(t)$ we again have $\lambda_\infty(f) = f(\infty)$ (correctly interpreted). The valuation ring of λ_∞ is

$$R[t^{-1}]_{(t^{-1})} = \left\{ \tfrac{f}{g}\colon\ f, g \in R[t],\ g \neq 0,\ \deg f \leq \deg g \right\}.$$

Its maximal ideal is generated by t^{-1}, and as above there are precisely two λ_∞-compatible orderings $P_{\infty,+}$ and $P_{\infty,-}$, characterized by $t^{-1} \in P_{\infty,+}$ and $-t^{-1} \in P_{\infty,-}$. (In a certain sense this notation would be more consistent the other way around. See the following paragraph however.)

The following geometric interpretation of these orderings may be helpful. If $0 \neq f \in R(t)$, then f has only finitely many zeroes and poles. Given any $c \in R$ it therefore makes sense to speak of the sign of f immediately to the right or to the left of c. Here "$f > 0$ immediately to the right of c" means that there exists an $\varepsilon > 0$ such that $f(a) > 0$ for all $a \in \,]c, c + \varepsilon[$. Similarly, $f(c)$ has a well-defined sign for $c \to +\infty$ and for $c \to -\infty$. Thus we obtain

$$P_{c,+(-)} = \{0\} \cup \{f \in R(t)^* : f > 0 \text{ immediately to the right (left) of } c\},$$

$$P_{\infty,+(-)} = \{0\} \cup \{f \in R(t)^* : f(c) > 0 \text{ for } c \gg 0 \ (c \ll 0)\}.$$

Proposition 2.9.1 *The orderings $P_{c,+}$ and $P_{c,-}$, for all $c \in R \cup \{\infty\}$, are precisely the orderings of $R(t)$ that are non-archimedean over R. If $R = \mathbb{R}$, these are all the orderings of $\mathbb{R}(t)$.*

Proof Let P be an ordering of $R(t)$ that is non-archimedean over R and let $A := \mathfrak{o}_P(R(t)/R)$ be the convex hull of R.

Let $t \in A$. Then $\mathfrak{p} := \mathfrak{m}_A \cap R[t]$ is a prime ideal of $R[t]$ (the *centre* of A in $R[t]$). Since $R[t]_{\mathfrak{p}} \subseteq A$ and $A \neq R(t)$ we have $\mathfrak{p} \neq (0)$. Thus $\mathfrak{p} = R[t] \cdot f$ for some monic irreducible polynomial $f \in R[t]$. Since $\kappa(A)$ is a real field and $R[t]/(f)$ is a subfield of $\kappa(A)$ there exists an element $c \in R$ such that $f = t - c$. Since $R[t]_{(t-c)}$ is a valuation ring of $R(t)$ (by Proposition 2.2.8) and is dominated by A, we have $A = R[t]_{(t-c)}$ (by Corollary 2.3.9) and it follows that $P = P_{c,+}$ or $P = P_{c,-}$.

If $t \notin A$, we let $u := t^{-1}$ and obtain as above that $A = R[u]_{(u)}$. Hence $P = P_{\infty,+}$ or $P = P_{\infty,-}$.

The statement about $\mathbb{R}$ follows from Corollary 2.1.13. $\square$

As a consequence of this proof we obtain:

Corollary 2.9.2 *Let k be an arbitrary field. Then the nontrivial valuation rings of $k(t)$ that contain k are precisely the following:*

(1) $A = k[t]_{(f)}$ *with $f \in k[t]$ monic and irreducible;*
(2) $A = k[t^{-1}]_{(t^{-1})}$.

In each case the value group is $\mathbb{Z}$ and the residue field is a finite simple field extension of k. $\square$

In order to determine the remaining orderings of $R(t)$ we consider for every ordering P of $R(t)$ the sets

$$U_P := \{a \in R : t - a \in P\} \quad \text{and} \quad O_P := \{a \in R : a - t \in P\}.$$

The pair $\eta_P = (U_P, O_P)$ is a generalized Dedekind cut of R in the following sense:

Definition 2.9.3 Let $(M, \leq)$ be a totally ordered set. A *(generalized) Dedekind cut* of M is a pair (U, O) of subsets of M such that $U \cup O = M$ and $u < v$ for all $u \in U$ and all $v \in O$. If U and O are not empty, the Dedekind cut is called *proper*.

The orderings of $R(t)$ determined above thus yield

for $P = P_{\infty,+}$, resp. $P_{\infty,-}$: $\eta_P = (R, \varnothing)$, resp. $(\varnothing, R)$;

for $P = P_{c,+}$, resp. $P_{c,-}$: $\eta_P = (]-\infty, c],]c, \infty[)$, resp. $(]-\infty, c[, [c, \infty[)$.

These Dedekind cuts are not free in the following sense:

Definition 2.9.4 A Dedekind cut (U, O) of M is called *free* if it is proper, U contains no largest element and O contains no smallest element.

It is clear that the non-free Dedekind cuts of R are precisely those η_P where P runs through all orderings of $R(t)$ that are non-archimedean over R.

Consider thus a free Dedekind cut $\xi = (U, O)$ of R. We construct an ordering $P = P_\xi$ of $R(t)$ with $\xi = \eta_P$ that is archimedean over R as follows: Let $0 \neq f \in R(t)$ and let $-\infty < c_1 < \cdots < c_r < \infty$ be the distinct zeroes and poles of f in R. Then f has constant sign $\varepsilon_0, \varepsilon_1, \ldots, \varepsilon_r$, respectively, on the interval

$$I_0 :=]-\infty, c_1[, \quad I_i :=]c_i, c_{i+1}[\text{ (for } i = 1, \ldots, r-1), \quad I_r :=]c_r, \infty[,$$

respectively, cf. Proposition 1.7.2. There is precisely one $i \in \{0, \ldots, r\}$ such that $I_i \cap U \neq \varnothing \neq I_i \cap O$ and so we say that f has sign ε_i at ξ. Thus, if we let

$$P_\xi := \{0\} \cup \{f \in R(t)^* : f \text{ is positive at } \xi\}$$

$$= \{f \in R(t): \exists u \in U, \, v \in O \text{ such that } f \text{ has no pole in }]u, v[\text{ and } f \geq 0\},$$

then it is immediately clear that $P := P_\xi$ is an ordering of $R(t)$ with $\xi = \eta_P$.

In fact we have obtained all orderings of $R(t)$ since the converse correspondence also holds:

Proposition 2.9.5 *For every ordering P of $R(t)$ that is archimedean over R, $\eta := \eta_P$ is a free Dedekind cut of R and $P = P_\eta$. Thus there is a canonical bijection*

$$\{P: P \text{ is an ordering of } R(t) \text{ that is archimedean over } R\}$$

$$\updownarrow$$

$$\{\eta: \eta \text{ is a free Dedekind cut of } R\},$$

given by $P \mapsto \eta_P$ and $\eta \mapsto P_\eta$.

Proof Let P be archimedean over R and $\eta := \eta_P = (U_P, O_P)$. Then $U_P \neq \varnothing \neq O_P$ and for every $a \in U_P$ there exists $0 < b \in R$ such that $b < t - a$ (otherwise

$t - a$ would be infinitely small compared to R), i.e., such that $a + b \in U_P$. Hence U_P does not contain a largest element. In a similar manner we see that O_P does not contain a smallest element. It follows that η_P is free. In order to show that $P = P_\eta$ it suffices to prove that $P \cap R[t] = P_\eta \cap R[t]$. Let $0 \neq f \in R[t]$ and write

$$f(t) = \alpha \prod_{i=1}^{r} (t - a_i) \prod_{j=1}^{s} ((t - b_j)^2 + c_j^2)$$

with $\alpha \in R^*$, $a_i, b_j, c_j \in R$ and $c_j \neq 0$ (for $i = 1, \ldots, r$, $j = 1, \ldots, s$). The quadratic factors are positive at each ordering of $R(t)$ and for $i = 1, \ldots, r$ we have by definition, $t - a_i \in P \Leftrightarrow a_i \in U_P \Leftrightarrow t - a_i \in P_\eta$. We conclude that $f \in P \Leftrightarrow f \in P_\eta$.   □

Corollary 2.9.6 *There exists a natural bijection between the orderings of $R(t)$ and the generalized Dedekind cuts of R.*   □

Next on our list is the field $R((t))$ of formal Laurent series $\sum_{i \geq n} a_i t^i$ (with $n \in \mathbb{Z}$ and $a_i \in R$). Recall that $R((t))$ is the field of fractions of the ring $A = R[[t]]$ of formal power series in one variable over R. Let $0 \neq f(t) = \sum_{i \geq n} a_i t^i$ with $a_n \neq 0$. Then $v(f) := n$ defines a valuation v of $R((t))$ with valuation ring A, maximal ideal $\mathfrak{m}_A = tA$, value group $\mathbb{Z}$ and residue field $A/\mathfrak{m}_A = R$. Since for every $f \in \mathfrak{m}_A$ the element

$$1 + f = \left(\sum_{i \geq 0} \binom{1/2}{i} f^i \right)^2$$

is a square in A (binomial sequence!), A is convex with respect to every ordering of $R((t))$. Thus, by the Baer–Krull Theorem (Sect. 2.7), $R((t))$ has precisely two orderings:

$$P_+ = \{0\} \cup \{t^n f(t) : n \in \mathbb{Z}, \ f \in A, \ f(0) > 0\}$$

and

$$P_- = \{0\} \cup \{(-t)^n f(t) : n \in \mathbb{Z}, \ f \in A, \ f(0) > 0\}.$$

Finally, let $B := \mathbb{R}\{t\}$ be the ring of convergent power series in one variable over $\mathbb{R}$, i.e., the ring of real function germs that are real-analytic at zero. Consider its field of fractions $K = \mathrm{Quot}\, B \subseteq R((t))$. Restricting the valuation v of $R((t))$ to K gives a valuation w of K with valuation ring B and value group $\mathbb{Z}$. The same argument as above then shows that K has precisely two orderings, namely the restrictions of the two orderings of $R((t))$.

2.10   Composition and Decomposition of Places

Let K, L and M be fields.

Proposition 2.10.1 (Factorization of Places) *Let* $\lambda\colon K \to L \cup \infty$ *and* $\mu\colon K \to U\infty$ *be places and let* μ *be a coarsening of* λ *(i.e.,* $\mathfrak{o}_\mu \subseteq \mathfrak{o}_\lambda$*). If* λ *is surjective, then there exists a unique place* $\nu\colon L \to M \cup \infty$ *such that* $\mu = \nu \circ \lambda$. *Furthermore,* $\mathfrak{o}_\nu = \lambda(\mathfrak{o}_\mu)$.

Proof Let $A := \mathfrak{o}_\lambda$, $B := \mathfrak{o}_\mu$ and $C := \lambda(B)$. Then C is a valuation ring of L by Proposition 2.4.18. The restriction $\lambda|_B$ induces an isomorphism $B/\mathfrak{m}_A \stackrel{\sim}{\to} C$ and thus an isomorphism $\alpha\colon \kappa(B) \stackrel{\sim}{\to} \kappa(C)$ such that the diagram

$$
\begin{array}{ccc}
B & \xrightarrow{\;\lambda|_B\;} & C \\
{\scriptstyle \pi_B}\downarrow & & \downarrow{\scriptstyle \pi_C} \\
\kappa(B) & \xrightarrow[\;\alpha\;]{\sim} & \kappa(C)
\end{array}
$$

commutes. Consider the embedding $\overline{\mu}\colon \kappa(B) \to M$ induced by μ, the map $\psi := \overline{\mu} \circ \alpha^{-1}\colon \kappa(C) \to M$ and the place $\nu := \psi \circ \lambda_C\colon L \to M \cup \infty$ induced by ψ. Then $\mathfrak{o}_\nu = C$ and $\mu = \nu \circ \lambda$. The uniqueness of ν is clear (since λ is surjective).
                                                                                                    □

The following result is Proposition 2.4.18 for places:

Proposition 2.10.2 *Let* $\lambda\colon K \to L \cup \infty$ *and* $\nu\colon L \to M \cup \infty$ *be places. If* λ *is surjective, there exists a canonical exact sequence*

$$0 \to \Gamma_\nu \to \Gamma_{\nu\circ\lambda} \to \Gamma_\lambda \to 0 \qquad\qquad (2.2)$$

of order preserving homomorphisms.
                                                                                                    □

Recall that a short exact sequence

$$0 \to G' \to G \stackrel{\pi}{\to} G'' \to 0$$

of abelian groups *splits* if there exists a *section* of π, i.e., a homomorphism $s\colon G'' \to G$ such that $\pi \circ s = \mathrm{id}_{G''}$. In this case $G \to G'' \times G'$, $x \mapsto (\pi x, x - s\pi x)$ is an isomorphism.

If the exact sequence (2.2) splits we can compute $\Gamma_{\nu\circ\lambda}$ from Γ_ν and Γ_λ. Indeed:

Lemma 2.10.3 *Let* Γ, Γ' *and* Γ'' *be ordered abelian groups and let*

$$0 \to \Gamma' \to \Gamma \stackrel{\pi}{\to} \Gamma'' \to 0$$

*be a short exact sequence of order preserving homomorphisms. If π has a section s
(as a group homomorphism), then*

$$\varphi\colon \alpha \mapsto (\pi\alpha, \alpha - s\pi\alpha)$$

is an order preserving *isomorphism from* Γ *to* $(\Gamma'' \times \Gamma')_{\text{lex}}$ *(where we identify* Γ'
with its image in Γ.)

Proof The map φ is an isomorphism of abelian groups. Let $0 < \alpha \in \Gamma$. If $\alpha \in \Gamma'$,
then $\varphi(\alpha) = (0, \alpha) > 0$. If $\alpha \notin \Gamma'$, then $\varphi(\alpha) > 0$ since $\pi(\alpha) > 0$.   □

For example, if Γ'' is a free abelian group, then π always has a section. The same
is true if Γ is divisible, i.e., a vector space over $\mathbb{Q}$ (since Γ'' is then also a vector
space over $\mathbb{Q}$ and π is $\mathbb{Q}$-linear).

Assume again that $\lambda\colon K \to L \cup \infty$ and $\nu\colon L \to M \cup \infty$ are places, and that λ
is surjective. Assume furthermore that ν_λ is discrete of rank one, i.e., $\Gamma_\lambda \cong \mathbb{Z}$. By
Proposition 2.10.2 and Lemma 2.10.3 there exists an order preserving isomorphism
(in general not unique)

$$\Gamma_{\nu\circ\lambda} \overset{\sim}{\to} (\Gamma_\lambda \times \Gamma_\nu)_{\text{lex}} = (\mathbb{Z} \times \Gamma_\nu)_{\text{lex}}$$

that we will describe explicitly in order to express $\nu_{\nu\circ\lambda}$ in terms of ν_λ and ν_ν.

To this end, let $h \in \mathfrak{m}_\lambda$ be a generator of $\mathfrak{m}_\lambda$. Then $\alpha_0 := \nu_\lambda(h)$ is the positive
generator of Γ_λ. The choice of h yields a section $s\colon \Gamma_\lambda \to \Gamma_{\nu\circ\lambda}$ of $\pi\colon \Gamma_{\nu\circ\lambda} \to \Gamma_\lambda$
via $s(\alpha_0) = \nu_{\nu\circ\lambda}(h)$. Let $\varphi\colon \Gamma_{\nu\circ\lambda} \overset{\sim}{\to} (\Gamma_\lambda \times \Gamma_\nu)_{\text{lex}}$ be the isomorphism induced
by s as in Lemma 2.10.3. Identifying $\text{Ker}(\pi) = \mathfrak{o}_\lambda^* / \mathfrak{o}_{\nu\circ\lambda}^*$ with Γ_ν (via λ, cf.
Proposition 2.4.18) gives

$$\varphi \circ \nu_{\nu\circ\lambda}(f) = \left(\nu_\lambda(f), \nu_\nu \circ \lambda(fh^{-\nu_\lambda(f)})\right)$$

for all $f \in K^*$. This solves the problem of expressing the valuation $\nu_{\nu\circ\lambda}$ in terms
of ν_λ, ν_ν and λ. (The argument can be repeated without any assumptions on Γ_λ, but
then the description of the section s will be more complicated in general.)

Notation 2.10.4 We denote by $\mathbb{Z}_{\text{antilex}}^r$ the group $\mathbb{Z}^r$ equipped with the *antilexicographic* order, i.e., the positive elements are the tuples $(a_1, \ldots, a_s, 0, \ldots, 0)$
$(1 \leq s \leq r)$ with $a_s > 0$. Of course, $\mathbb{Z}_{\text{antilex}}^r \cong \mathbb{Z}_{\text{lex}}^r$ as ordered abelian groups;
we use the antilexicographic order just to simplify our notation below.

Let us consider the case of rational function fields (in several variables). Let
k be an arbitrary field, $t_1, \ldots, t_r$ algebraically independent variables over k and
$K_i := k(t_1, \ldots, t_i)$ for $0 \leq i \leq r$. Given a tuple $c = (c_1, \ldots, c_r) \in k^r$ we construct
a place $\lambda\colon K_r \to k \cup \infty$ such that $\lambda(f) = f(c)$ for all $f \in k[t_1, \ldots, t_r]$ as follows:

For every $i = 1, \ldots, r$ there is a unique place $\lambda_i\colon K_i \to K_{i-1} \cup \infty$ over K_{i-1}
such that $\lambda_i(t_i) = c_i$ (cf. Corollary 2.9.2). Let $\lambda := \lambda_1 \circ \cdots \circ \lambda_r$. Since $\Gamma_{\lambda_i} = \mathbb{Z}$
for $i = 1, \ldots, r$, it follows by induction from Proposition 2.10.2 and Lemma 2.10.3
that $\Gamma_\lambda \cong \mathbb{Z}_{\text{lex}}^r$. We wish to determine ν_λ explicitly. For $\alpha = (\alpha_1, \ldots, \alpha_r) \in \mathbb{Z}^r$ we

write $(t - c)^\alpha := (t_1 - c_1)^{\alpha_1} \cdots (t_r - c_r)^{\alpha_r} \in K_r$, as usual. Let $f \in k[t_1, \ldots, t_r]$. Then f can be expressed uniquely in the form

$$f = \sum_{\alpha \in \mathbb{Z}_+^r} a_\alpha (t - c)^\alpha \quad \text{(with } a_\alpha \in k \text{ and almost all } a_\alpha = 0\text{).}$$

After an inductive application of the considerations above we may conclude that $v_\lambda(f)$ is the smallest index α in $\mathbb{Z}_{\text{antilex}}^r$ such that $a_\alpha \neq 0$. In particular we obtain:

Proposition 2.10.5 *The valuation* $v \colon k(t_1, \ldots, t_r)^* \to \mathbb{Z}_{\text{antilex}}^r$, *defined by*

$$v\big((t_1 - c_1)^{\alpha_1} \cdots (t_r - c_r)^{\alpha_r}\big) = (\alpha_1, \ldots, \alpha_r)$$

is associated to the place $\lambda \colon k(t_1, \ldots, t_r) \to k \cup \infty$ *with* $\lambda(f) = f(c)$ *for all* $f \in k[t_1, \ldots, t_r]$, *described above. Furthermore,* $v(t_r - c_r), \ldots, v(t_1 - c_1)$ *is the lexicographic basis of* Γ_λ. $\qquad\square$

Note that there are of course many other places $\lambda' \colon k(t) \to k \cup \infty$ over k with $\lambda'(t) = c$. For instance, by choosing a different transcendence basis $u = (u_1, \ldots, u_r)$ such that $k[t] = k[u]$ and replacing c by $u(c)$.

Assume now that $k = R$ is real closed. Let $K := R(t_1, \ldots, t_r)$ and $\lambda \colon K \to R \cup \infty$ the place described above. Let P be an ordering of K, compatible with λ (by the Baer–Krull Theorem, there are precisely 2^r such orderings) and S the real closure of (K, P). By Theorem 2.8.15, λ has a unique extension $\mu \colon S \to R \cup \infty$.

Proposition 2.10.6 Γ_μ *is isomorphic to* $\mathbb{Q}_{\text{lex}}^r$.

Indeed, more generally we have:

Proposition 2.10.7 *Let* $L \supseteq K$ *be an algebraic field extension with* L *real closed or algebraically closed. If* $\mu \colon L \to F \cup \infty$ *is a place and* $\lambda := \mu|_K$, *then there exists an order preserving isomorphism* $\Gamma_\lambda \otimes \mathbb{Q} \xrightarrow{\sim} \Gamma_\mu$.

(Let Γ be an ordered abelian group. Then $\Gamma \otimes \mathbb{Q}$ has a unique ordering that makes the embedding $\Gamma \hookrightarrow \Gamma \otimes \mathbb{Q}$ order preserving. In the statement of the theorem, $\Gamma_\lambda \otimes \mathbb{Q}$ is equipped with this ordering.)

Proof The inclusion $K^* \subseteq L^*$ induces an order preserving embedding $\Gamma_\lambda \subseteq \Gamma_\mu$. We claim that the group $\Gamma_\mu / \Gamma_\lambda$ is torsion. Indeed, let $b \in L^*$ and let $a_0, \ldots, a_n \in K$ be such that $a_n b^n + \cdots + a_1 b + a_0 = 0$ and $a_n \neq 0$. Then there exist indices i and j such that $0 \leq i < j \leq n$ and $v_\mu(a_i b^i) = v_\mu(a_j b^j) < \infty$ (for otherwise, $v_\mu(\sum a_k b^k) = \min_k v(a_k b^k) < \infty$)), and it follows that $(j - i) v_\mu(b) = v_\mu\big(\frac{a_i}{a_j}\big) \in \Gamma_\lambda$.

Since L is real closed or algebraically closed, Γ_μ is divisible (every positive element, resp. every element of L has arbitrary roots) and so $\Gamma_\mu = \mathbb{Q} \cdot \Gamma_\lambda = \Gamma_\lambda \otimes \mathbb{Q}$. $\qquad\square$

Let k be a field, $K_0 := k$ and $K_i := K_{i-1}((t_i)) = \operatorname{Quot} K_{i-1}[\![t_i]\!]$ for $i = 1, 2, \ldots, r$. (Note that for $r \geq 2$, the field $k((t_1, \ldots, t_r))$ is a *proper* subfield of the field $K_r = k((t_1)) \cdots ((t_r))$!) Let $\lambda_i : K_i \to K_{i-1} \cup \infty$ be the place over K_{i-1}, associated to $K_{i-1}[\![t_i]\!]$, and let $\lambda = \lambda_1 \circ \cdots \circ \lambda_r : K_r \to k \cup \infty$. Then $\Gamma_\lambda \cong \mathbb{Z}^r_{\text{lex}}$ with lexicographic basis $v_\lambda(t_r), \ldots, v_\lambda(t_1)$, as in Proposition 2.10.5. If $k \subseteq K \subseteq K_r$ is an intermediate field with $t_1, \ldots, t_r \in K$, then we also have $\Gamma_{\lambda|K} = \mathbb{Z}^r_{\text{lex}}$.

Using some more commutative algebra we obtain an interesting application (which will play no further role however):

Let A be a regular local k-algebra with $\kappa(A) = k$, $\widehat{A} = \varprojlim A/\mathfrak{m}_A^n$ its completion and $(f_1, \ldots, f_r)$ a minimal set of generators of $\mathfrak{m}_A$. It is well-known that there exists a k-isomorphism $\widehat{A} \xrightarrow{\sim} k[\![t_1, \ldots, t_r]\!]$ such that $f_i \mapsto t_i$ for all $i = 1, \ldots, r$, cf. [81, p. 206] or [13, Ch. IX, § 3, no. 3]. Let $K := \text{Quot } A$. It follows from $K \subseteq \text{Quot } \widehat{A} \cong k((t_1, \ldots, t_r)) \subseteq K_r$ that the place $\lambda|_K : K \xrightarrow[k]{} k \cup \infty$ also has value group $\mathbb{Z}^r_{\text{lex}}$, with lexicographic basis $(v(f_r), \ldots, v(f_1))$ (where $v := v_{\lambda|_K}$). Furthermore, its valuation ring dominates A.

2.11 Existence of Real Places of Function Fields

We briefly recall some facts from field theory. Let $L \supseteq K$ be a field extension. Then all maximal families of elements of L that are algebraically independent over K have the same cardinality. Any such family is called a *transcendence basis* of L over K and its cardinality is called the *transcendence degree* of L over K, denoted $\text{tr.deg.}(L/K)$. The field L is called a d-dimensional *function field* over K if L is finitely generated over K and $\text{tr.deg.}(L/K) = d$. (The terminology comes from the fact that up to K-isomorphism such fields are precisely the fields of rational functions on irreducible algebraic K-varieties of dimension d, cf. [97, I, §3] or [45].)

The purpose of this section is to provide a proof of:

Theorem 2.11.1 *Let R be a real closed field, K an r-dimensional real function field over R and $(t_1, \ldots, t_r)$ a fixed transcendence basis of K over R. Then there exist elements $a_i, b_i \in R$ with $a_i < b_i$ for $i = 1, \ldots, r$ such that for every real closed overfield $S \supseteq R$ the following property is satisfied:*

(T_r) *For every tuple $c = (c_1, \ldots, c_r) \in S^r$ such that $a_i < c_i < b_i$ for $i = 1, \ldots, r$, there exists a place $\lambda : K \to S \cup \infty$ over R that satisfies $\lambda(t_i) = c_i$ for $i = 1, \ldots, r$.*

Note that such a place λ must be trivial (i.e., a homomorphism) if $c_1, \ldots, c_r$ are algebraically independent over R. (Indeed, in this case we have $R[t_1, \ldots, t_r] \cap \lambda^{-1}(0) = \{0\}$, which implies $R(t_1, \ldots, t_r) \subseteq \mathfrak{o}_\lambda$. It follows that $\mathfrak{o}_\lambda = K$ since $\mathfrak{o}_\lambda$ is integrally closed.)

Corollary 2.11.2 (Lang's Embedding Theorem) *Let R be a real closed field and K a real function field over R. For every real closed overfield S of R such that $\text{tr.deg.}(S/R) \geq \text{tr.deg.}(K/R)$ there exists an R-homomorphism $K \to S$.*

Proof The statement follows from Theorem 2.11.1 and the remark following it. Indeed, there are elements $c_i \in S$ with $a_i < c_i < b_i$ for $i = 1, \ldots, r$ such that $c_1, \ldots, c_r$ are algebraically independent over R; this is clear for $r = 1$ and follows by induction for all r.   □

We first establish two auxiliary results.

Lemma 2.11.3 *Let $L \supseteq K$ be a finite field extension, B a valuation ring of L and $A = K \cap B$. Then the* ramification index $e = [\Gamma_B : \Gamma_A]$ *and the* residue degree $f = [\kappa(B) : \kappa(A)]$ *are finite and $ef \leq [L : K]$.*

Proof Write $v = v_B$. Consider elements $b_1, \ldots, b_r \in B \setminus \{0\}$ such that the images of the $v(b_i)$ in Γ_B / Γ_A are distinct, and elements $c_1, \ldots, c_s \in B$ such that the images $\overline{c_j}$ of the c_j in $\kappa(B)$ are linearly independent over $\kappa(A)$. We will show that for all $a_{ij} \in K$ ($1 \leq i \leq r, 1 \leq j \leq s$),

$$v\left(\sum_{i,j} a_{ij} b_i c_j\right) = \min_{i,j} v(a_{ij} b_i).$$

This implies in particular that the $b_i c_j$ are linearly independent over K.

We may assume that for every $i \in \{1, \ldots, r\}$ there exists $j \in \{1, \ldots, s\}$ such that $a_{ij} \neq 0$ (for otherwise b_i can be omitted). For every $i \in \{1, \ldots, r\}$ we choose $k(i) \in \{1, \ldots, s\}$ such that $v(a_{ik(i)}) = \min_j v(a_{ij})$. Then $a_{ik(i)} \neq 0$ and it follows that for all i,

$$\sum_j a_{ij} b_i c_j = a_{ik(i)} b_i \sum_j a'_{ij} c_j$$

where $a'_{ij} := a_{ij}/a_{ik(i)} \in A$. Since the $\overline{c_j}$ are linearly independent over $\kappa(A)$ (and $a'_{ik(i)} = 1$) we have $\sum_j a'_{ij} c_j \in B^*$ and thus

$$v\left(\sum_j a_{ij} b_i c_j\right) = v(a_{ik(i)}) + v(b_i) \quad (i = 1, \ldots, r).$$

These values are all distinct by assumption and we conclude that

$$v\left(\sum_i \sum_j a_{ij} b_i c_j\right) = \min_i \left(v(a_{ik(i)}) + v(b_i)\right) = \min_{i,j} v(a_{ij} b_i).\qquad □$$

Lemma 2.11.4 *Let K be a field, L a function field over K and $n := \mathrm{tr.deg.}(L/K)$. Let $B \neq L$ be a proper valuation ring of L such that $K \subseteq B$. Then $\mathrm{tr.deg.}(\kappa(B)/K) \leq n - 1$, and if $\mathrm{tr.deg.}(\kappa(B)/K) = n - 1$, then B is discrete of rank one and $\kappa(B)$ is a function field (i.e., finitely generated) over K.*

Proof Let $u_1, \ldots, u_r \in B$ be such that their images $\overline{u_1}, \ldots, \overline{u_r}$ in $\kappa(B)$ are algebraically independent over K. Then $u_1, \ldots, u_r$ (in L) are also algebraically independent over K and so $r \leq n$. Furthermore, $K[u_1, \ldots, u_r] \cap m_B = \{0\}$. It follows that $K(u_1, \ldots, u_r) \subseteq B$. If $r = n$, then L is algebraic and hence integral over $K(u_1, \ldots, u_r)$ and thus $B = L$, a contradiction. Assume that $r = \text{tr.deg.}(\kappa(B)/K) = n - 1$. Let $F := K(u_1, \ldots, u_{n-1})$ and choose $t \in L$ such that $(u_1, \ldots, u_{n-1}, t)$ is a transcendence basis of L over K. Then $A := B \cap F(t)$ is a proper valuation ring of $F(t)$ and $F \subseteq A$. It follows from Corollary 2.9.2 that $[\kappa(A) : F] < \infty$ and $\Gamma_A \cong \mathbb{Z}$, and from Lemma 2.11.3 that $[\kappa(B) : \kappa(A)]$ and Γ_B / Γ_A are finite. We conclude that $\kappa(B)$ is finitely generated over K and $\Gamma_B \cong \mathbb{Z}$. $\qquad \square$

Exercise 2.11.5 Let K be a field, let v be a valuation of K with value group Γ and residue field κ, and assume $\text{char}(\kappa) \neq 2$. Let $a \in K^*$ with $\sqrt{a} \notin K$, and let $L = K(\sqrt{a})$. Show the following (extension always means "up to equivalence"):

(a) If $v(a) \notin 2\Gamma$, then v has a unique extension to L, and $e = 2$, $f = 1$.
(b) If $v(a) = 0$ and $\sqrt{\overline{a}} \notin \kappa$, then v has a unique extension to L, and $e = 1$, $f = 2$.
(c) If $v(a) = 0$ and $\sqrt{\overline{a}} \in \kappa$, then v has exactly two extensions to L, and both satisfy $e = f = 1$.

We will now show that in order to prove Theorem 2.11.1 it suffices to prove the following weaker property for all $r \geq 1$, under the same assumptions:

($\widetilde{T}_r$) *There exist $a_i, b_i \in R$ with $a_i < b_i$ for $i = 1, \ldots, r$ such that for every real closed overfield $S \supseteq R$ and every tuple $c = (c_1, \ldots, c_r) \in S^r$ with $a_i < c_i < b_i$ for $i = 1, \ldots, r$ and $c_1, \ldots, c_r$ algebraically independent over R, there exists a homomorphism $\varphi \colon K \to S$ such that $\varphi(t_i) = c_i$ for $i = 1, \ldots, r$.*

Proposition 2.11.6 ($\widetilde{T}_r$) *implies* (T_r) *for all $r \geq 1$.*

Proof Let $c \in S^r$ with $a_i < c_i < b_i$ and $c_1, \ldots, c_r$ not necessarily algebraically independent over R. Let $u_1, \ldots, u_r$ be independent variables over S and $T' := S(u_1, \ldots, u_r)$ the r-dimensional rational function field over S. Consider the place $\mu' \colon T' \to S \cup \infty$ over S with $\mu'(u_i) = 0$ for $i = 1, \ldots, r$, described in Sect. 2.10. By Corollary 2.8.16 there exists a real closure T of T' and an extension $\mu \colon T \to S \cup \infty$ of μ'. Since the u_i are infinitely small with respect to S we have $a_i < c_i + u_i < b_i$ for $i = 1, \ldots, r$. Since the $c_i + u_i$ are algebraically independent over R, there exists a homomorphism $\varphi \colon K \to T$ such that $\varphi(t_i) = c_i + u_i$ by property ($\widetilde{T}_r$). The place $\lambda := \mu \circ \varphi \colon K \to S \cup \infty$ then satisfies $\lambda(t_i) = c_i$ for $i = 1, \ldots, r$. $\qquad \square$

Let us now prove ($\widetilde{T}_1$). Assume thus that $r = 1$, that R and K are as in Theorem 2.11.1 and that $F := R(t)$, where $t := t_1$. Let $\text{tr} \colon K \to F$ be the trace of the finite extension $K \supseteq F$ and let $g_1, \ldots, g_n \in F^*$ be such that $\text{tr}_*(\langle 1 \rangle_K) \cong \langle g_1, \ldots, g_n \rangle$. After multiplying the g_i with the square of their denominators we may assume without loss of generality that all g_i are in $R[t]$. Let $d_1 < \cdots < d_N$ be the distinct zeroes of $g_1 \cdots g_n$ in R, $d_0 := -\infty$ and $d_{N+1} := +\infty$. Every g_i has a fixed sign $\varepsilon_{ij} \in \{\pm 1\}$ on $]d_{j-1}, d_j[$ (interpreted as an interval in R or in any ordered overfield of R) for $j = 1, \ldots, N + 1, i = 1, \ldots, n$.

Since K is real there exists an ordering Q of K. Let $P := F \cap Q$ and let $j \in \{1, \ldots, N\}$ be the index for which $d_{j-1} <_P t <_P d_j$. Then $\mathrm{sign}_P(g_i) = \varepsilon_{ij}$ for $i = 1, \ldots, n$. Since

$$\sum_{i=1}^{n} \varepsilon_{ij} = \mathrm{sign}_P \, \mathrm{tr}_*(\langle 1 \rangle_K)$$

is the number of extensions of P to K (cf. Corollary 1.12.10), this number is ≥ 1. We will show that $(\widetilde{T}_1)$ is satisfied for every choice of a_1, b_1 such that $]a_1, b_1[\subseteq]d_{j-1}, d_j[$.

Let S be a real closed overfield of R, $c \in S \setminus R$ with $d_{j-1} < c < d_j$ and $\psi : F \to S$ the R-homomorphism that satisfies $\psi(t) = c$. By Theorem 1.11.6 the extensions $\varphi : K \to S$ of ψ are in bijective correspondence with the by ψ induced extensions to F of the orderings P_c of K. Their number is equal to

$$\mathrm{sign}_{P_c} \, \mathrm{tr}_*(\langle 1 \rangle_K) = \sum_{i=1}^{n} \mathrm{sign}_{P_c}(g_i) = \sum_{i=1}^{n} \varepsilon_{ij} \geq 1,$$

which shows in particular that such an extension φ exists.

Before proving Theorem 2.11.1 in general we consider some consequences of (T_1).

Proposition 2.11.7 *Let* $\lambda : K \to L \cup \infty$ *be a real place and* $z_1, \ldots, z_n \in K$. *If* $\lambda(\sum_i z_i^2) \neq \infty$, *then* $\lambda(z_i) \neq \infty$ *for* $i = 1, \ldots, n$.

Proof By Corollary 2.8.11 there exists a λ-compatible ordering P of K. Since $\mathfrak{o}_\lambda$ is convex with respect to P and $0 \leq z_i^2 \leq z_1^2 + \cdots + z_n^2 \in \mathfrak{o}_\lambda$, we have $z_i^2 \in \mathfrak{o}_\lambda$ and conclude that $z_i \in \mathfrak{o}_\lambda$ for $i = 1, \ldots, n$. $\qquad\square$

Theorem 2.11.8 (Artin–Lang Stellensatz) *Let* R *be a real closed field and* K *a real function field over* R. *For all finitely many* $z_1, \ldots, z_n \in K$ *there exists a place* $\lambda : K \to R \cup \infty$ *over* R *such that* $\lambda(z_i) \neq \infty$ *for* $i = 1, \ldots, n$. *Furthermore, if* $r = \mathrm{tr.deg.}(K/R)$, *then there exists such a place with* $\Gamma_\lambda \cong \mathbb{Z}^r_{\mathrm{lex}}$.

Proof By Proposition 2.11.7 we may assume that $n = 1$. We proceed by induction on r. If $r = 0$, the statement is trivially true since $K = R$. If $r = 1$, the existence of λ follows from (T_1). Furthermore, $\Gamma_\lambda \cong \mathbb{Z}$ for every such λ by Lemma 2.11.4. Thus, let $r > 1$ and consider a transcendence basis $t_1, \ldots, t_r$ of K over R with $z \in R(t_1, \ldots, t_{r-1}) =: F$. Let T be a real closure of K, S the algebraic closure of F in T and $L := KS$ (the subfield of T generated by K and S). Since L is a one-dimensional real function field over S, there exists a place $L \xrightarrow{S} S \cup \infty$ (by the $r = 1$ case) and via restriction we obtain a place $\mu : K \xrightarrow{F} S \cup \infty$. Since tr.deg.$(K/F) = 1$ (and S/F is algebraic) this place is nontrivial and satisfies $\mu(z) \neq \infty$. Let $A := \mathfrak{o}_\mu$ and $\lambda_A \colon K \to \kappa(A) \cup \infty$ the canonical place associated to A. By Lemma 2.11.4 we have $[\kappa(A) \colon F] < \infty$ and $\Gamma_A \cong \mathbb{Z}$. Furthermore, since $\kappa(A)$ is real and tr.deg.$\bigl(\kappa(A)/R\bigr) = $ tr.deg.$(F/R) = r - 1$ it follows by induction that there exists a place $v \colon \kappa(A) \to R \cup \infty$ over R such that $v\bigl(\lambda_A(z)\bigr) \neq \infty$ and $\Gamma_v \cong \mathbb{Z}_{\mathrm{lex}}^{r-1}$. Let $\lambda := v \circ \lambda_A \colon K \underset{R}{\to} R \cup \infty$. Then $\lambda(z) \neq \infty$ and $\Gamma_\lambda \cong \mathbb{Z}_{\mathrm{lex}}^r$ by Sect. 2.10. $\quad\square$

Theorem 2.11.9 (Artin–Lang Stellensatz, Refined Version) *In the situation of Theorem 2.11.8, consider also an ordering Q of K. Then there exists a place $\lambda \colon K \underset{R}{\to} R \cup \infty$ as in Theorem 2.11.8 that satisfies the additional property* $\mathrm{sign}_Q(z_i) = \mathrm{sign}\,\lambda(z_i)$ *for* $i = 1, \ldots, n$.

Proof After multiplying z_i with $\mathrm{sign}_Q(z_i)$ we may assume without loss of generality that $z_i >_Q 0$ for $i = 1, \ldots, n$. Applying Theorem 2.11.8 to $K' := K(\sqrt{z_1}, \ldots, \sqrt{z_n})$ and the elements $z_1, \ldots, z_n, \frac{1}{z_1 \cdots z_n} \in K'$ we obtain a place $\lambda' \colon K' \underset{R}{\to} R \cup \infty$ such that $\lambda'(z_i) \notin \{0, \infty\}$, and thus such that $\lambda'(z_i) = \lambda'(\sqrt{z_i})^2 > 0$ for $i = 1, \ldots, n$. Let $\lambda := \lambda'|_K$. Since $\Gamma_{\lambda'}/\Gamma_\lambda$ is finite (by Lemma 2.11.3), we have rank $\Gamma_\lambda = $ rank $\Gamma_{\lambda'} = r$ and it follows easily that $\Gamma_\lambda \cong \mathbb{Z}_{\mathrm{lex}}^r$. $\quad\square$

We will now finish the proof of Theorem 2.11.1 by showing that property $(\widetilde{\mathrm{T}}_r)$ holds for all $r \geq 2$. Thus, let $t_1, \ldots, t_r$ be a transcendence basis of K over R and let $F := R(t_1, \ldots, t_r)$. Assume again that $g_1, \ldots, g_n \in R[t_1, \ldots, t_r]$ are such that $\mathrm{tr}_*(\langle 1 \rangle_K) \cong \langle g_1, \ldots, g_n \rangle$ (where $n = [K \colon F]$). There exists an ordering Q of K. By Theorem 2.11.9 there exists a place $\mu \colon K \underset{R}{\to} R \cup \infty$ such that $\mu(t_i) \neq \infty$ for $i = 1, \ldots, r$, and $\mu(g_j) \neq \infty$ and $\mathrm{sign}\,\mu(g_j) = \mathrm{sign}_Q(g_j)$ for $j = 1, \ldots, n$. Let $p := \bigl(\mu(t_1), \ldots, \mu(t_r)\bigr) \in R^r$ and $P := F \cap Q$. As before, it follows that

$$1 \leq \mathrm{sign}_P \,\mathrm{tr}_*\bigl(\langle 1 \rangle_K\bigr) = \sum_{j=1}^n \mathrm{sign}_P(g_j) = \sum_{j=1}^n \mathrm{sign}\,\mu(g_j) = \sum_{j=1}^n \mathrm{sign}\,g_j(p),$$

where the final equality follows from the fact that $\mu(f) = f(p)$ for all $f \in R[t_1, \ldots, t_r]$.

In order to finish the proof of Theorem 2.11.1 we need the following proposition that we will prove at the end of this section:

Proposition 2.11.10 *Let* $g \in R[x_1, \ldots, x_r]$ *be a polynomial and* $p \in R^r$ *such that* $g(p) \neq 0$. *Then there exists an element* $\varepsilon \in R$, $\varepsilon > 0$, *such that for every real closed overfield* S *of* R *and every* $q \in S^r$ *with* $p_i - \varepsilon < q_i < p_i + \varepsilon$ *for* $i = 1, \ldots, r$ *we have* $\mathrm{sign}\,g(p) = \mathrm{sign}\,g(q)$.

For $g_1, \ldots, g_n$ and p as above, the lemma provides us with an $0 < \varepsilon \in R$. Let $S \supseteq R$ be a real closed overfield and consider a tuple $c = (c_1, \ldots, c_r) \in S^r$ such that $c_1, \ldots, c_r$ are algebraically independent over R and $p_i - \varepsilon < c_i < p_i + \varepsilon$ for $i = 1, \ldots, r$. Let $\psi \colon F \to S$ be the embedding determined by $\psi(t_i) = c_i$ and P_c

the ordering of F pulled back from S via ψ. As in the proof of property $(\widetilde{T}_1)$ we obtain

$$\text{sign}_{P_c}\,\text{tr}_*(\langle 1 \rangle_K) = \sum_{j=1}^{n} \text{sign}_{P_c}(g_j) = \sum_{j=1}^{n} \text{sign}\,g_j(c) = \sum_{j=1}^{n} \text{sign}\,g_j(p) \geq 1,$$

where the third equality follows from Proposition 2.11.10. It follows from Sects. 1.11 and 1.12 that ψ has an extension $\varphi\colon K \to S$ as required. This concludes the proof of Theorem 2.11.1

Proof of Proposition 2.11.10 Let $x = (x_1, \ldots, x_r)$ and let

$$g(p + x) - g(p) = \sum_{\alpha} c_\alpha x^\alpha$$

be the Taylor expansion of g at p (where the sum is over all $0 \neq \alpha \in \mathbb{N}_0^r$, and almost all c_α are 0). If $a \in S^r$ is such that $|a_i| < \varepsilon \leq 1$ for $i = 1, \ldots, r$, then $|g(p+a)-g(p)| \leq \varepsilon \sum_{\alpha} |c_\alpha|$. Hence, it suffices to choose $\varepsilon > 0$ such that $\varepsilon \sum_{\alpha} |c_\alpha| < |g(p)|$ and $\varepsilon \leq 1$. $\qquad\square$

2.12   Artin's Solution of Hilbert's 17th Problem and the Sign Change Criterion

Let k be a field. Throughout this section $t_1, \ldots, t_n$ denote variables that are algebraically independent over k. We often write t instead of $(t_1, \ldots, t_n)$. If (k, P) is an ordered field, we endow k^n with the strong topology $\mathcal{T}_P$ (cf. Sect. 2.6).

Proposition 2.12.1 *Let (k, P) be an ordered field, $U \subseteq k^n$ a nonempty open subset and $f \in k[t_1, \ldots, t_n]$ a polynomial that vanishes identically on U. Then $f = 0$.*

Proof By induction on n. If $n = 1$ this is clear. Assume the statement holds for $n - 1$. Let $t' := (t_1, \ldots, t_{n-1})$ and

$$f = f(t', t_n) = f_0(t') \cdot t_n^d + \cdots + f_d(t'),$$

where $d \geq 0$ and $f_0, \ldots, f_d \in k[t']$. We may assume without loss of generality that U is of the form $U = \prod_{i=1}^{n} \,]a_i, b_i[$ with $a_i, b_i \in k$. Let $c' \in U' := \prod_{i=1}^{n-1} \,]a_i, b_i[$ be arbitrary. Then $f(c', t_n) = f_0(c')t_n^d + \cdots + f_d(c')$ vanishes identically on $]a_n, b_n[$. Thus $f_0(c') = \cdots = f_d(c') = 0$. It follows that the f_j vanish identically on U' and so the induction hypothesis implies $f_0 = \cdots = f_d = 0$. $\qquad\square$

The zero set of a nonzero polynomial in k^n is thus *thin* in the sense that its interior is empty.

Definition 2.12.2 Let k be a field and $t = (t_1, \ldots, t_n)$.

(a) Consider a polynomial $f \in k[t]$. We denote the *zero set* of f in k^n by $Z_k(f)$ (or simply $Z(f)$), i.e.,

$$Z_k(f) = \{a \in k^n : f(a) = 0\}.$$

(b) Consider a rational function $f \in k(t)$ and write $f = g/h$ with $g, h \in k[t]$ relatively prime and $h \neq 0$. We denote the *domain* of f in k^n by $D_k(f)$ (or simply $D(f)$), i.e.,

$$D_k(f) = \{a \in k^n : h(a) \neq 0\}.$$

We also let $Z_k(f) := Z_k(g) \cap D_k(f)$.

If k is ordered, the domain of every rational function is open and dense in k^n with respect to the strong topology (Proposition 2.12.1).

Definition 2.12.3 Let (k, P) be an ordered field, $f \in k(t)$ and $M \subseteq D_k(f)$.

(a) f is *positive definite* (resp. *positive semidefinite*) on M if $f(a) > 0$ (resp. $f(a) \geq 0$) for all $a \in M$.
(b) f is *negative definite* (resp. *negative semidefinite*) on M if $-f$ is positive definite (resp. positive semidefinite) on M.
(c) f is *indefinite on M* if f is neither positive nor negative definite on M.

If $M = D_k(f)$, we say that f is *positive (semi)definite*, resp. *negative (semi)definite*, resp. *indefinite over k*.

Proposition 2.12.4 *Let (k, P) be an ordered field.*

(a) *If $f \in k(t)$ is positive semidefinite on an open dense subset of $D_k(f)$, then f is positive semidefinite over k.*
(b) *The set of rational functions that are positive semidefinite over k is a preordering of $k(t)$.*

Proof (a) This follows from the continuity of the evaluation map $D_k(f) \to k$, $a \mapsto f(a)$.

(b) Let $f, g \in k(t)^*$ be positive semidefinite over k. Since $f + g$ and fg are positive semidefinite on the dense open subset $D_k(f) \cap D_k(g)$ of k^n, they are positive semidefinite over k by (a). A similar argument shows that f^2 is positive semidefinite for every $f \in k(t)$.   □

Remark 2.12.5 Let (k, P) be an ordered field with real closure R. It follows from Proposition 2.12.4(b) that every function of the form $a_1 f_1^2 + \cdots + a_r f_r^2$ with $a_i \in P$ and $f_i \in k(t)$ is positive semidefinite over R. A rational function which is positive semidefinite over k is in general no longer positive semidefinite over R. This is because k is usually not dense in R, and f (considered over R) may become negative in the "gaps". (The reader may want to construct an example, cf. Sect. 2.1.)

In 1900 David Hilbert gave his famous address to the International Congress of Mathematicians in Paris in which he presented his list of 23 unsolved mathematics problems. Since then many of these problems have been solved, often involving the development of completely new methods. This is especially true of the 17th problem about the representation of forms by sums of squares and its solution by Emil Artin in 1927, cf. [3, 4]. Artin discovered the notion of ordering of a field, putting it to elegant and successful use. His ideas sparked new directions in the development of real algebra in the twentieth century (see the Preface).

Hilbert's 17th Problem can be formulated as follows:

Hilbert's 17th Problem *Given a positive semidefinite polynomial* $f = f(t_1, \ldots, t_n) \in \mathbb{R}[t_1, \ldots, t_n]$, *do there exist an integer* $r \geq 1$ *and rational functions* $g_1, \ldots, g_r \in \mathbb{R}(t_1, \ldots, t_n)$ *such that* $f = g_1^2 + \cdots + g_r^2$?

Supplementary question: *If* $k \subseteq \mathbb{R}$ *is a subfield that contains the coefficients of* f, *can the* g_i *already be found in* $k(t_1, \ldots, t_n)$?

Remarks 2.12.6

(1) In the formulation of the supplementary question (which is also due to Hilbert) it would be better to consider representations in the sense of Remark 2.12.5 since otherwise it is immediately clear that there is no solution in general.
(2) An obvious follow-on question is whether there exists a representation $f = g_1^2 + \cdots + g_r^2$ where $g_1, \ldots, g_r$ are actually *polynomials*. Hilbert already showed in 1888 that this is indeed the case if $n = 1$, but that for $n \geq 2$ this is false in general. (Hilbert's proof actually provided more precise details.) The first example of a positive semidefinite polynomial that cannot be written as a sum of squares of polynomials was not published until 1967. Indeed, in [84] Motzkin considered the polynomial (here presented in homogeneous form)

$$f(x, y, z) = x^4 y^2 + x^2 y^4 + z^6 - 3x^2 y^2 z^2.$$

That f is positive semidefinite follows from the arithmetic-geometric inequality:

$$\frac{1}{3}(x^4 y^2 + x^2 y^4 + z^6) \geq \sqrt[3]{x^6 y^6 z^6}.$$

A further example is given by

$$g(x, y, z) = x^4 y^2 + y^4 z^2 + z^4 x^2 - 3x^2 y^2 z^2.$$

In [26] one can find a simple method for proving that f and g are not sums of squares of polynomials, as well as further examples.

Let us now consider the solution of Hilbert's 17th Problem:

Theorem 2.12.7 (E. Artin) *Let (k, P) be an ordered field with real closure R, and let $f \in k(t_1, \ldots, t_n)$ be a rational function that is positive semidefinite over R. Then there exist an integer $r \geq 1$, functions $g_1, \ldots, g_r \in k(t_1, \ldots, t_n)$ and positive elements (with respect to P) $a_1, \ldots, a_r \in k$ such that*

$$f = a_1 g_1^2 + \cdots + a_r g_r^2.$$

Proof The proof is not constructive. Consider the set

$$T := \{a_1 g_1^2 + \cdots + a_r g_r^2 : r \geq 1, \, a_i \in P, \, g_i \in k(t)\},$$

which is a preordering of $k(t)$ and therefore equal to the intersection of all orderings Q of $k(t)$ such that $P \subseteq Q$, cf. Theorem 1.1.10. Assume for the sake of contradiction that the conclusion of the theorem is false. Then there exists an ordering Q of $k(t)$ such that $P \subseteq Q$ and $f \notin Q$. Let S denote the real closure of $k(t)$ with respect to Q. Then we can think of R as a subfield of S (cf. Theorem 1.11.2) and in particular consider the subfield $R(t)$ of S with the induced ordering Q'. Since $f \notin Q'$, there exists a place $\lambda \colon R(t) \xrightarrow{R} R \cup \infty$ such that $\lambda(t_1) \neq \infty, \ldots, \lambda(t_n) \neq \infty$, $\lambda(f) \neq \infty$ and $\lambda(f) < 0$ (cf. Theorem 2.11.9). Letting $a := \big(\lambda(t_1), \ldots, \lambda(t_n)\big) \in R^n$, we obtain $f(a) = \lambda(f) < 0$, which is impossible since f is positive semidefinite. $\square$

Hilbert's question is thus fully answered, but there are still many related questions for which answers remain elusive, at least in part. For example, having fixed k, P and n, does there exist a positive integer $r \geq 1$ with the property that every rational function $f \in k(t_1, \ldots, t_n)$ which is positive semidefinite over R has a representation $f = a_1 g_1^2 + \cdots + a_r g_r^2$ of length r, as in the theorem? And if so, what is the smallest such r? Let us write $r(k, n)$ for this integer. (To be precise we should write $r_P(k, n)$, but k has a unique ordering in the examples that follow.) When $k = R$ is real closed, it is known that $r(R, 1) = 2$ (elementary), $r(R, 2) = 4$ (Cassels, Ellison, Pfister [24], difficult) and $n + 2 \leq r(R, n) \leq 2^n$ for all $n \geq 2$ (this follows from Pfister [88] and [24]). For the field of rational numbers $k = \mathbb{Q}$ one has $r(\mathbb{Q}, 1) = 5$ (Pourchet [89]) and $r(\mathbb{Q}, 2) \leq 8$ (Kato, Colliot-Thélène [57]).

As an application of Theorem 2.12.7, we prove the following "sign change criterion":

Theorem 2.12.8 (D. Dubois, G. Efroymson) *Let (k, P) be an ordered field with real closure R, and let $f \in k[t_1, \ldots, t_n]$ be a non-constant irreducible polynomial. The following statements are equivalent:*

(i) *The ordering P extends to the function field $K = \mathrm{Quot}\, k[t_1, \ldots, t_n]/(f)$;*
(ii) *f is indefinite over R.*

The use of the word "criterion" is motivated by (ii) which signifies that f changes sign "along" the hypersurface $H := \{f = 0\}$. Algebraic geometry tells us that K is the field of rational functions on H.

The proof makes use of the following proposition, which we establish first:

Proposition 2.12.9 *Let A be a unique factorization domain, $E = \operatorname{Quot} A$ and $f \in A[u] \backslash A$ a polynomial in one variable u, irreducible in $A[u]$. Then $A[u]/(f) \to E[u]/(f)$ is injective and $E[u]/(f)$ is a field (and thus the quotient field of $A[u]/(f)$).*

Proof The statement follows immediately from the fact that since A is a UFD, $A[u]$ is also a UFD (a consequence of Gauss's Lemma, cf. [73, IV, Theorem 2.1] or [55, Vol. I, §2.16]): if $g \in A[u]$ and $h \in E[u]$ such that $g = fh$ in $E[u]$, then there exist $0 \neq a \in A$ and $h' \in A[u]$ such that $ag = fh'$ (in $A[u]$). Since f is not a divisor of a, it follows that $f \mid g$ in $A[u]$. A similar straightforward argument shows that f remains irreducible in $E[u]$. ☐

Proof of Theorem 2.12.8 (i) $\Rightarrow$ (ii): Assume that f is semidefinite. Without loss of generality we may assume that $f \geq 0$ on R^n. By Artin's theorem there exists an equation of the form

$$fh^2 = \sum_{i=1}^{r} a_i g_i^2,$$

where $0 \neq a_i \in P$, $g_i, h \in k[t]$ and (without loss of generality) $\gcd(g_1, \ldots, g_r, h) = 1$. It follows that f does not divide all g_i. In other words, not all summands on the right hand side disappear modulo (f). Hence P does not extend to K by Proposition 1.3.2.

(ii) $\Rightarrow$ (i): Assume that f is indefinite over R. Then f may no longer be irreducible over R, but still must have at least one indefinite irreducible divisor f_1 in $R[t]$. Without loss of generality we may assume that the variable t_n occurs in f_1 (and thus also in f). Let $t' := (t_1, \ldots, t_{n-1})$, $A := k[t']$, $E := k(t')$, $B := R[t']$, $F := R(t')$, and consider the commutative diagram

$$
\begin{array}{ccc}
k[t]/(f) = A[t_n]/(f) & \xrightarrow{\;\varphi\;} & B[t_n]/(f_1) = R[t]/(f_1) \\
\downarrow & & \downarrow \\
E[t_n]/(f) & \xrightarrow{\;\;\;\;\psi\;\;\;\;} & F[t_n]/(f_1)
\end{array}
\tag{2.3}
$$

where the arrows depict the canonical homomorphisms.

It follows from the lemma that the map ψ in (2.3) is a field embedding from $K = \operatorname{Quot} k[t]/(f) = E[t_n]/(f)$ into $L := \operatorname{Quot} R[t]/(f_1) = F[t_n]/(f_1)$. It suffices to show that L is a real field since every ordering of L then induces an extension of P to K. Thus we may assume without loss of generality that $k = R$ and $f = f_1$ and must show that K is a real field.

If $n = 1$, then f is linear and hence $K = R$. Thus we let $n \geq 2$ and choose $p, q \in R^n$ such that $f(p) < 0 < f(q)$. After a linear change of coordinates we may assume that $p = (a_1, \ldots, a_{n-1}, b)$ and $q = (a_1, \ldots, a_{n-1}, c)$. Let us also write $t' = (t_1, \ldots, t_{n-1})$ and $F := R(t')$. Let Q be an ordering of F such that $t_i - a_i$ is infinitely small compared to R for $i = 1, \ldots, n - 1$, cf. Sect. 2.10. Then $f = f(t', t_n) \in F[t_n]$ is indefinite over F with respect to Q. Indeed, writing $\mathfrak{o} := \mathfrak{o}_Q(F/R)$, we get $t_i \equiv a_i \bmod \mathfrak{m}_\mathfrak{o}$ for $i = 1, \ldots, n - 1$. Hence, $f(t', b) \equiv f(a_1, \ldots, a_n, b) = f(p) \bmod \mathfrak{m}_\mathfrak{o}$. In other words, $f(t', b) < 0$. Similarly, $f(t', c) > 0$. Then f is irreducible in $F[t_n]$ by the lemma. From the $n = 1$ case it follows that $F[t_n]/(f)$ possesses an ordering that extends Q. Finally, the embedding $K \hookrightarrow F[t_n]/(f)$ shows that K is a real field. $\square$

Chapter 3
The Real Spectrum

The focus of this chapter is on the real spectrum of a commutative ring—discovered in 1979 by M.F. Roy and M. Coste—and its elementary properties. We will usually be as general as possible and consider arbitrary commutative rings. However, special features arise in the "geometric setting" where coordinate rings of affine varieties over real closed base fields are considered.

The real spectrum and the Zariski spectrum exhibit many similarities (in essence one considers homomorphisms into real closed fields instead of algebraically closed fields). Since we will be using the most elementary properties of the Zariski spectrum, we recall some fundamental facts in the first section of this chapter.

We treat the Zariski spectrum and the real spectrum exclusively as topological spaces and ignore their structure sheaves. The concept of affine variety will be treated in the sense of A. Weil since the issue of reducibility will not play any role and, as such fewer technicalities will need to be considered. Nevertheless we strongly recommend gaining some familiarity with the modern language of algebraic geometry (as developed in Hartshorne's monograph [44], for example).

Throughout this chapter, all rings are assumed commutative and unitary.

3.1 The Zariski Spectrum. Affine Varieties

We start by recalling some definitions. Let A be a ring and $\mathfrak{a}$ an ideal of A. The set

$$\sqrt{\mathfrak{a}} := \{a \in A : a^n \in \mathfrak{a} \text{ for some } n \in \mathbb{N}\}$$

is an ideal of A, called the *radical* of $\mathfrak{a}$. If $\mathfrak{a} = \sqrt{\mathfrak{a}}$, then $\mathfrak{a}$ is called a *radical ideal*. The ideal $\operatorname{Nil} A := \sqrt{(0)}$ is the *nilradical* of A, and A is *reduced* if $\operatorname{Nil} A = (0)$. As usual, we also let $A_{\mathrm{red}} := A/\operatorname{Nil} A$.

© Springer Nature Switzerland AG 2022
M. Knebusch, C. Scheiderer, *Real Algebra*, Universitext,
https://doi.org/10.1007/978-3-031-09800-0_3

Definition 3.1.1 The *(Zariski) spectrum* Spec A of A is the set of all prime ideals of A. Every $\mathfrak{p} \in \operatorname{Spec} A$ has an associated *residue field*

$$\kappa_A(\mathfrak{p}) = \kappa(\mathfrak{p}) = \operatorname{Quot} A/\mathfrak{p}$$

and *evaluation homomorphism*

$$\rho_{A,\mathfrak{p}} = \rho_{\mathfrak{p}} : A \to \kappa(\mathfrak{p}), \quad f \mapsto f \bmod \mathfrak{p}.$$

We also write $f(\mathfrak{p})$ instead of $\rho_{\mathfrak{p}}(f) = f \bmod \mathfrak{p}$.

Remark 3.1.2 If A is the trivial ring, then Spec $A = \varnothing$. Otherwise, Spec $A \neq \varnothing$.

Remark 3.1.3 It is helpful to keep the following interpretation of Spec A in mind: We think of the elements $f \in A$ as "functions" on the set Spec A (which is a topological space, cf. Definition 3.1.4). To each point $\mathfrak{p} \in \operatorname{Spec} A$ we associate its function value under f, $f(\mathfrak{p}) = f \bmod \mathfrak{p} \in \kappa(\mathfrak{p})$. (Note that the function value of each point lives in a different field!) From a formal point of view we are thus considering the ring homomorphism

$$\rho : A \to \prod_{\mathfrak{p} \in \operatorname{Spec} A} \kappa(\mathfrak{p}), \quad \rho(f) = \big(f(\mathfrak{p})\big)_{\mathfrak{p} \in \operatorname{Spec} A}.$$

The kernel of ρ is equal to

$$\bigcap_{\mathfrak{p} \in \operatorname{Spec} A} \mathfrak{p} = \operatorname{Nil} A$$

(cf. Proposition 3.1.11). Hence, ρ is injective if and only if A is reduced. In this case we can view A as a subring of $\prod_{\mathfrak{p}} \kappa(\mathfrak{p})$ via ρ.

   In general the function values of $f \in A$ at distinct points $\mathfrak{p}$ cannot be compared since they live in different fields. However, there is a meaningful concept of zero set of f, namely $\{\mathfrak{p} \in \operatorname{Spec} A : f \in \mathfrak{p}\}$. Taking these zero sets as generators of closed sets gives rise to the Zariski topology on Spec A:

Definition 3.1.4

(a) For any $f \in A$ we let

$$D_A(f) := \{\mathfrak{p} \in \operatorname{Spec} A : f(\mathfrak{p}) \neq 0\} = \{\mathfrak{p} \in \operatorname{Spec} A : f \notin \mathfrak{p}\},$$
$$\mathcal{V}_A(f) := \{\mathfrak{p} \in \operatorname{Spec} A : f(\mathfrak{p}) = 0\} = \{\mathfrak{p} \in \operatorname{Spec} A : f \in \mathfrak{p}\}.$$

More generally, for any subset T of A we let

$$D_A(T) := \bigcap_{f \in T} D_A(f), \quad V_A(T) := \bigcap_{f \in T} V_A(f).$$

(We usually drop the indices A.)

(b) The *Zariski topology* on Spec A is the topology that has $\{D_A(f): f \in A\}$ as a subbasis of open sets.

We will always think of Spec A as a topological space in this way. Note that the subbasis $\{D(f): f \in A\}$ is actually a basis of the topology since for any $f, g \in A$ we have $D(f) \cap D(g) = D(fg)$. By definition, all sets $V(T)$ (with $T \subseteq A$) are closed.

Proposition 3.1.5

(a) *Let $T \subseteq A$. Then $V(T) = V(\sqrt{\mathfrak{a}})$, where $\mathfrak{a} := \sum_{t \in T} At$ is the ideal generated by T.*

(b) *Every closed subset of Spec A is of the form $V(\mathfrak{a})$ for some (radical) ideal $\mathfrak{a}$ of A.*

Proof (a) The inclusion $\supseteq$ is trivial since $T \subseteq \sqrt{\mathfrak{a}}$. For the reverse inclusion, let $\mathfrak{p} \in V(T)$ and $f \in \sqrt{\mathfrak{a}}$. Then there exist $r, n \in \mathbb{N}$ and $t_i \in T, a_i \in A$ for $i = 1, \ldots, r$, such that $f^n = a_1 t_1 + \cdots + a_r t_r$. Since $t_i \in \mathfrak{p}$ for $i = 1, \ldots, r$, it follows that $f \in \mathfrak{p}$, i.e., $\mathfrak{p} \in V(f)$.

(b) This follows from (a) and the fact that every open subset of Spec A is the union of subsets $D(f)$ with $f \in A$. □

Definition 3.1.6 Let Y be a subset of Spec A. Then

$$\mathfrak{J}(Y) := \{f \in A: f|_Y \equiv 0\} = \bigcap_{\mathfrak{p} \in Y} \mathfrak{p}$$

denotes the (radical) ideal of functions that vanish on Y.

Proposition 3.1.7 *Let Y be a subset of Spec A. Then $\overline{Y} = V(\mathfrak{J}(Y))$.*

Proof The inclusion $\subseteq$ follows from $Y \subseteq V(\mathfrak{J}(Y))$ and the fact that $V(\mathfrak{J}(Y))$ is closed. For the reverse inclusion, let $\mathfrak{a}$ be an ideal of A such that $\overline{Y} = V(\mathfrak{a})$ (cf. Proposition 3.1.5). Then $\mathfrak{a} \subseteq \mathfrak{J}(\overline{Y}) \subseteq \mathfrak{J}(Y)$, and so $\overline{Y} = V(\mathfrak{a}) \supseteq V(\mathfrak{J}(Y))$. □

Corollary 3.1.8 *The closed points of Spec A are precisely the maximal ideals of A.* □

Recall that a topological space X is called a T_0 *space* if for every two distinct points $x, y \in X$, we have $x \notin \overline{\{y\}}$ or $y \notin \overline{\{x\}}$. If we replace "or" by "and", then X is a T_1 *space*.

The Zariski topology is thus very weak as a rule and not even a T_1 topology. However:

Exercise 3.1.9 Show that the Zariski topology is a T_0 topology.

Exercise 3.1.10 Let S be a multiplicative subset of A (i.e., $SS \subseteq S$ and $1 \in S$). Show that every ideal $\mathfrak{a}$ of A that is maximal for the property $\mathfrak{a} \cap S = \varnothing$ is a prime ideal.

Proposition 3.1.11 (Abstract or Weak Nullstellensatz) *If $\mathfrak{a}$ is an ideal of A, then*

$$\mathfrak{I}(\mathcal{V}(\mathfrak{a})) = \sqrt{\mathfrak{a}}.$$

Corollary 3.1.12 *For every two ideals $\mathfrak{a}, \mathfrak{b}$ of A we have $\mathcal{V}(\mathfrak{a}) \subseteq \mathcal{V}(\mathfrak{b}) \Leftrightarrow \sqrt{\mathfrak{a}} \supseteq \sqrt{\mathfrak{b}}$.*

Proof of Proposition 3.1.11 We must show that $\sqrt{\mathfrak{a}} = \bigcap \{\mathfrak{p} \in \operatorname{Spec} A \colon \mathfrak{a} \subseteq \mathfrak{p}\}$. The inclusion $\subseteq$ is clear. For the reverse inclusion, let $f \in A$ with $f \notin \sqrt{\mathfrak{a}}$. By Zorn's Lemma, there exists and ideal $\mathfrak{p}$ that contains $\mathfrak{a}$ and that is maximal for the property $\mathfrak{p} \cap \{1, f, f^2, \dots\} = \varnothing$. Then $\mathfrak{p}$ is a prime ideal by Exercise 3.1.10, and $f \notin \mathfrak{p}$. $\square$

For $\mathfrak{a} = 0$, Proposition 3.1.11 says that $\operatorname{Nil} A$ is the intersection of all the prime ideals of A. In particular, A and A_{red} have "the same" Zariski spectrum (more precisely: the map $A \to A_{\mathrm{red}}$ induces a homeomorphism $\operatorname{Spec} A_{\mathrm{red}} \to \operatorname{Spec} A$).

Definition 3.1.13 A topological space X is called *irreducible* if it is not the union of two proper closed subsets. A point $x \in X$ is called a *generic point* of X if $X = \overline{\{x\}}$.

If X contains a generic point, then X is irreducible. One is often interested in the converse. Note that in case X is a T_0 space (for example a subspace of some $\operatorname{Spec} A$), then X has at most one generic point.

Proposition 3.1.14 *Every nonempty closed irreducible subspace of $\operatorname{Spec} A$ contains a (unique) generic point. Hence, the nonempty closed irreducible subsets of $\operatorname{Spec} A$ are precisely the sets $\mathcal{V}(\mathfrak{p})$ where $\mathfrak{p} \in \operatorname{Spec} A$.*

Proof Let $\varnothing \neq Y \subseteq \operatorname{Spec} A$ be closed and irreducible, and let $\mathfrak{a} := \mathfrak{I}(Y)$. By Proposition 3.1.7 we must show that $\mathfrak{a}$ is a prime ideal. Since $Y \neq \varnothing$, we have $\mathfrak{a} \neq A$. If $f, g \in A$ are such that $fg \in \mathfrak{a}$, then $Y \subseteq \mathcal{V}(f) \cup \mathcal{V}(g)$. Since Y is irreducible, we obtain without loss of generality that $Y \subseteq \mathcal{V}(f)$, and in particular that $f \in \mathfrak{a}$. $\square$

Recall that a topological space is *quasi-compact* if each of its open coverings has a finite subcovering, and *compact* if it is quasi-compact and Hausdorff.

Proposition 3.1.15 *Let* $f \in A$. *Then* $D(f)$ *is quasi-compact. In particular,* $\operatorname{Spec} A = D(1)$ *is quasi-compact.*

Proof Let T be a subset of A. Then

$$D(f) \subseteq \bigcup_{g \in T} D(g) \Leftrightarrow \mathcal{V}(f) \supseteq \mathcal{V}(T)$$

$$\Leftrightarrow f \in \sqrt{\sum_{g \in T} Ag} \qquad \text{(Corollary 3.1.12)}$$

$$\Leftrightarrow f \in \sqrt{\sum_{g \in T'} Ag} \qquad \text{for some finite } T' \subseteq T$$

$$\Leftrightarrow D(f) \subseteq \bigcup_{g \in T'} D(g) \qquad \text{for some finite } T' \subseteq T. \qquad \square$$

Remark 3.1.16 We use $\overset{\circ}{\mathcal{K}}(\operatorname{Spec} A)$ to denote the set of all finite unions $D(f_1) \cup \cdots \cup D(f_r)$, where $r \geq 1$ and $f_i \in A$. By Proposition 3.1.15, the elements of $\overset{\circ}{\mathcal{K}}(\operatorname{Spec} A)$ are precisely the open quasi-compact subsets of $\operatorname{Spec} A$. It follows in particular that $\overset{\circ}{\mathcal{K}}(\operatorname{Spec} A)$ is completely determined by the topological space $\operatorname{Spec} A$. This is generally not the case for the open basis $\{D(f) : f \in A\}$ of $\operatorname{Spec} A$, cf. Sect. 3.4.

Let us look at the functorial behaviour of the Zariski spectrum. Let $\varphi : A \to B$ be a ring homomorphism. If $q \in \operatorname{Spec} B$, then $\varphi^{-1}(q)$ is a prime ideal of A. Hence φ induces a map of spectra in the opposite direction,

$$\operatorname{Spec} B \to \operatorname{Spec} A, \quad q \mapsto \varphi^{-1}(q),$$

which we denote by $\operatorname{Spec} \varphi$ or φ^*. This map is continuous since $(\operatorname{Spec} \varphi)^{-1}(D_A(f)) = D_B(\varphi(f))$ for all $f \in A$. We conclude that Spec is a functor from the category of commutative rings to the category of topological spaces.

The map φ also induces embeddings of residue fields: if $q \in \operatorname{Spec} B$, then the embedding $A/\varphi^{-1}(q) \hookrightarrow B/q$ induces a field embedding $\varphi_q : \kappa_A(\varphi^*q) \hookrightarrow \kappa_B(q)$. In other words, the diagram

$$
\begin{array}{ccc}
A & \xrightarrow{\;\varphi\;} & B \\
{\scriptstyle \rho_{A,\varphi^*q}}\downarrow & & \downarrow{\scriptstyle \rho_{B,q}} \\
\kappa_A(\varphi^*q) & \xrightarrow{\;\varphi_q\;} & \kappa_B(q)
\end{array}
$$

is commutative.

Thus, interpreting the rings A and B as rings of functions on their spectra (as explained in Remark 3.1.3), the map $\varphi\colon A \to B$ just pulls back functions on Spec A via $\varphi^*\colon$ Spec $B \to$ Spec A and via the embeddings $\varphi_{\mathfrak{q}}$ of the residue fields.

Two special cases are worthy of investigation in more detail. For the first case, let S be a multiplicative subset of A and $i_S\colon A \to S^{-1}A$ the canonical homomorphism $a \mapsto a/1$.

Proposition 3.1.17 *The map*

$$\mathrm{Spec}(i_S) = i_S^*\colon \ \mathrm{Spec}\ S^{-1}A \to \mathrm{Spec}\ A$$

is a homeomorphism from Spec $S^{-1}A$ *to the subspace* $D_A(S)$ *of* Spec A. *The residue homomorphisms* $(i_S)_{\mathfrak{q}}$ *with* $\mathfrak{q} \in$ Spec $S^{-1}A$ *are all isomorphisms.*

Proof The map

$$D_A(S) \to \mathrm{Spec}\ S^{-1}A, \ \mathfrak{p} \mapsto S^{-1}\mathfrak{p} := (S^{-1}A) \cdot i_S(\mathfrak{p}) = \{\tfrac{a}{s}\colon a \in \mathfrak{p},\ s \in S\}$$

is inverse to i_S^*. If $\frac{a}{s} \in S^{-1}A$, then

$$i_S^*\big(D_{S^{-1}A}(\tfrac{a}{s})\big) = D_A(a) \cap D_A(S),$$

which shows that the map i_S^* is open, and thus a homeomorphism. For the second statement, let $\mathfrak{q} \in$ Spec $S^{-1}A$ and $\mathfrak{p} := i_S^*(\mathfrak{q}) = i_S^{-1}(\mathfrak{q})$ (i.e., $\mathfrak{q} = S^{-1}\mathfrak{p}$). It is immediately clear that the embedding $A/\mathfrak{p} \hookrightarrow (S^{-1}A)/(S^{-1}\mathfrak{p})$ induces an isomorphism of the quotient fields. □

By Proposition 3.1.17 we can identify Spec $S^{-1}A$ with the subspace $D_A(S) = \{\mathfrak{p} \in \mathrm{Spec}\ A\colon \mathfrak{p} \cap S = \varnothing\}$ of Spec A. In particular we can interpret every basic open set $D_A(f)$ as the Zariski spectrum of a ring, namely the ring $f^{-\infty}A := S^{-1}A$, where $S := \{1, f, f^2, \dots\}$.

For the second special case, let $\mathfrak{a} \subseteq A$ be an ideal and $\varphi\colon A \to A/\mathfrak{a}$ the residue homomorphism.

Proposition 3.1.18 *The map*

$$\varphi^*\colon \ \mathrm{Spec}\ A/\mathfrak{a} \to \mathrm{Spec}\ A$$

is a homeomorphism from Spec $A/\mathfrak{a}$ *to the closed subspace* $\mathcal{V}_A(\mathfrak{a})$ *of* Spec A. *The residue field embeddings* $\varphi_{\mathfrak{q}}$ *with* $\mathfrak{q} \in$ Spec $A/\mathfrak{a}$ *are all isomorphisms.*

Proof Analogous to the proof of Proposition 3.1.17, where this time $\varphi^*\big(D_{A/\mathfrak{a}}(f + \mathfrak{a})\big) = D_A(f) \cap \mathcal{V}_A(\mathfrak{a})$ for all $f \in A$. □

The observation we made above for $S^{-1}A$ goes through, mutatis mutandis, for the ring $A/\mathfrak{a}$. The closed subspaces of Spec A are thus also spectra of rings.

In the final part of this section we will explain what we mean by affine varieties over a field. Our point of view is in essence Weil's in the sence that all varieties are reduced. We will show that by considering the associated function algebras we obtain a coordinate-free description.

Let k be a field with algebraic closure $\overline{k}$. If $K \supseteq k$ is an overfield, T a subset of $k[t_1, \ldots, t_n]$ and V a subset of K^n, we write

$$Z_K(T) := \{x \in K^n : f(x) = 0 \text{ for all } f \in T\}$$

and

$$I_k(V) := \{f \in k[t_1, \ldots, t_n] : f(x) = 0 \text{ for all } x \in V\}.$$

Definition 3.1.19 An *affine k-variety* is a subset V of $\overline{k}^n$ of the form $V = Z_{\overline{k}}(\mathfrak{a})$, where $\mathfrak{a}$ is an ideal of $k[t_1, \ldots, t_n]$. If $k \subseteq K \subseteq \overline{k}$ is an intermediate field, then the elements of $V(K) := V \cap K^n$ are called the *K-rational points* of V. If $V \subseteq \overline{k}^m$ and $W \subseteq \overline{k}^n$ are affine k-varieties, then a map $\varphi : V \to W$ is called a *k-morphism* from V to W if there exist polynomials $f_1, \ldots, f_n \in k[t_1, \ldots, t_m]$ such that $\varphi(x) = (f_1(x), \ldots, f_n(x))$ for all $x \in V$.

Remark 3.1.20 It is not sufficient to just consider the k-rational points $V(k)$ of V as these usually do not contain enough information about V. For example, $V(k)$ can be empty even though V is not the empty variety (concrete example: the $\mathbb{R}$-variety in the plane defined by $1 + t_1^2 + t_2^2 = 0$).

Remark 3.1.21 If $V \subseteq \overline{k}^n$ is a k-variety, then $I_k(V)$ is clearly the largest ideal $\mathfrak{a}$ of $k[t_1, \ldots, t_n]$ such that $V = Z_{\overline{k}}(\mathfrak{a})$.

Remark 3.1.22 The definition of affine k-variety depends on a choice of coordinates, which is unsatisfactory for many reasons. A coordinate-free definition is possible though, as we will explain next.

Let $V \subseteq \overline{k}^n$ be an affine k-variety. We call

$$k[V] := k[t_1, \ldots, t_n]/I_k(V)$$

the *affine (k-)algebra of* V. It is clear that $k[V]$ is an affine (i.e., a finitely generated) reduced k-algebra, which can be interpreted as the algebra of k-morphisms $V \to \overline{k}$. If $\varphi : V \to W$ is a k-morphism of affine k-varieties, then φ induces a homomorphism $\varphi^* : k[W] \to k[V]$ of k-algebras by "pulling back".

The point is now that up to isomorphism, V and $k[V]$ are the "same". More precisely, V is the same as $k[V]$ plus a choice of coordinates. Let us elaborate this. Consider any overfield K of k, and denote the set of k-algebra homomorphisms from $k[V]$ to K by $\text{Hom}_k(k[V], K)$. Writing $\overline{t}_i := t_i + I_k(V) \in k[V]$, there is a canonical bijection

$$\text{Hom}_k\big(k[V], K\big) \xrightarrow{\sim} Z_K\big(I_k(V)\big), \ x \mapsto \big(x(\overline{t_1}), \dots, x(\overline{t_n})\big).$$

In particular, if $K = \overline{k}$, there is a bijection $\text{Hom}_k\big(k[V], \overline{k}\big) \to V$. This motivates the following definition:

Definition 3.1.23 Given an affine k-algebra A and an overfield K of k, we write $V_A(K) := \text{Hom}_k(A, K)$. If $K = \overline{k}$, then $V_A := V_A(\overline{k}) = \text{Hom}_k(A, \overline{k})$ is called the (abstract) *variety of* A. Every k-algebra homomorphism $\varphi\colon A \to B$ induces a map $\varphi_K^*\colon V_B(K) \to V_A(K)$, namely $y \mapsto y \circ \varphi$, and so in particular a map $\varphi^*\colon V_B \to V_A$ of varieties.

In analogy to the Zariski spectrum we interpret the elements $f \in A$ as functions (with values in $\overline{k}$) on the variety V_A by setting $f(x) := x(f)$ for all $x \in V_A$.

If we choose a system of generators $u_1, \dots, u_n$ of A, then we can identify V_A with an affine k-variety in the sense of Definition 3.1.19: the evaluation map

$$V_A \to \overline{k}^n, \ x \mapsto \big(u_1(x), \dots, u_n(x)\big)$$

is a bijection from V_A to the k-variety $Z_{\overline{k}}(\mathfrak{a}) \subseteq \overline{k}^n$, where $\mathfrak{a}$ is the kernel of the homomorphism $k[t_1, \dots, t_n] \to A$, $t_i \mapsto u_i$. The same construction with a different system of generators results in a k-isomorphic k-variety. In other words, by considering V_A, we have associated a $(k\text{-})$isomorphism class of affine k-varieties to the k-algebra A, and choosing a representative of this isomorphism class corresponds to choosing a system of generators of A.

It is clear that if $\varphi\colon A \to B$ is a homomorphism of affine k-algebras, then $\varphi^*\colon V_B \to V_A$ is a k-morphism of k-varieties (for every choice of coordinates).

The assignments $V \mapsto k[V]$ and $A \mapsto V_A$ are inverse to each other, at least if we consider only *reduced* k-algebras for the second assignment. Indeed, let $V \subseteq \overline{k}^n$ be an affine k-variety and let $A := k[V] = k[t_1, \dots, t_n]/I_k(V)$. If we choose $\overline{t_i} = t_i + I_k(V)$ for $i = 1, \dots, n$ as a system of generators of $A = k[V]$, then the evaluation map from $V_A = \text{Hom}_k(A, \overline{k})$ to $\overline{k}^n$ gives a bijection from V_A to $Z_{\overline{k}}\big(I_k(V)\big) = V$. In this way the k-variety V_A can be canonically identified with V.

To establish conversely that $k[V_A]$ is isomorphic to A if A is reduced, we need the following fundamental result from algebraic geometry, which we state without proof:

Theorem 3.1.24 (Hilbert's Nullstellensatz) *Let A be an affine k-algebra. Then* $\text{Nil}\,A$ *is the intersection of all maximal ideals* $\mathfrak{m}$ *of A, and* $[A/\mathfrak{m}\colon k]$ *is finite for each such* $\mathfrak{m}$.

To give an idea of the proof, the essential step consists of showing that $A/\mathfrak{m}$ is a finite extension field of k for every maximal ideal $\mathfrak{m}$ of the finitely generated k-algebra A (this result is often also referred to as Hilbert's Nullstellensatz). It then follows easily via localization that the closed points of Spec A are dense in Spec A and the theorem then follows by the Weak Nullstellensatz (Proposition 3.1.11). For a complete proof of Hilbert's Nullstellensatz we refer to [68] or [13, Ch. V, §3.3].

Assume now that A is an affine reduced k-algebra. Then Hilbert's Nullstellensatz says that for every $0 \neq f \in A$ there exists $x \in V_A$ such that $f(x) \neq 0$, which in turn signifies that the canonical homomorphism $A \to k[V_A] = A / \bigcap_{x \in V_A} \mathrm{Ker}(x)$ is an isomorphism.

Summary We have seen that an affine k-variety $V \subseteq \overline{k}^n$ is "the same" as an affine reduced k-algebra A together with a system of generators of A. Henceforth, when talking about an affine k-variety we will thus usually mean an affine reduced k-algebra A together with the set $V_A = \mathrm{Hom}_k(A, \overline{k})$, which offers the advantage of a coordinate-free approach.

Remark 3.1.25 Grothendieck already recognized that restricting to reduced affine algebras is artificial. In order to understand non-reduced k-algebras geometrically, we need to move away from the point of view that an algebraic function $V \to \overline{k}$ (on a k-variety $V \subseteq \overline{k}^n$) is already determined by its values at the points of V. The right tool in this situation, that would also allow us to consider non-affine varieties, is sheaf theory. However, for the objectives we have in mind the "naive" interpretation presented in this section suffices.

A final comment we want to make is that the k-rational points of an affine k-variety can be interpreted as a subset of the Zariski spectrum in a canonical way. Namely, if A is an affine k-algebra, then

$$V_A(k) \to \mathrm{Spec}\, A, \ x \mapsto \mathrm{Ker}\, x$$

is a bijection from $V_A(k) = \mathrm{Hom}_k(A, k)$ to the set of those maximal ideals $\mathfrak{m}$ of A for which the map $k \to A \to A/\mathfrak{m}$ is an isomorphism (i.e., $A = \mathfrak{m} + k \cdot 1$). The topology induced by Spec A on $V_A(k)$ is also called the Zariski topology on $V_A(k)$.

However, if $k \neq \overline{k}$, then the variety $V_A = V_A(\overline{k})$ associated to A cannot in general be squeezed into Spec A: if $\mathfrak{m} \in \mathrm{Spec}\, A$ is a closed point there exist in general several k-embeddings $A/\mathfrak{m} \hookrightarrow \overline{k}$ and accordingly several points in V_A associated to $\mathfrak{m}$.

On the other hand, if $k = \overline{k}$ is algebraically closed, then $x \mapsto \mathrm{Ker}\, x$ is a bijection from V_A to the space of closed points in Spec A. Therefore, the topology induced on V_A by $V_A \hookrightarrow \mathrm{Spec}\, A$ is also called the Zariski topology on the variety V_A. Of course, historically speaking the situation was reversed since O. Zariski studied the topology on k-varieties $V \subseteq \overline{k}^n$ that was named after him long before the spectrum was conceived in the 1950s.

3.2 Reality for Commutative Rings

Let A be a ring and $\Sigma A^2 := \{a_1^2 + \cdots + a_n^2 : n \geq 1, a_i \in A\}$ the semiring generated by all squares of A.

When attempting to extend the notion of reality from fields to arbitrary rings, two distinct concepts emerge from the two equivalent field characterizations:[1]

Definition 3.2.1

(a) A is called *real* if $-1 \notin \Sigma A^2$. If A is not real (i.e., if there exist elements $a_1, \ldots, a_n \in A$ such that $1 + a_1^2 + \cdots + a_n^2 = 0$), then A is called *nonreal*.

(b) A is called *real reduced* if for all $a_1, \ldots, a_n \in A$, $a_1^2 + \cdots + a_n^2 = 0 \Rightarrow a_1 = \cdots = a_n = 0$.

(c) An ideal $\mathfrak{a} \subseteq A$ is called *real* (resp. *real reduced*) if the ring $A/\mathfrak{a}$ is real (resp. real reduced).

Remarks 3.2.2

(1) The trivial ring is real reduced, but not real. Every nontrivial real reduced ring is also real. The ring $\mathbb{R}[x, y]/(x^2 + y^2)$ is real (cf. (5) below), but not real reduced.

(2) If A is real reduced, then A is reduced, and every subring of A is also real reduced.

(3) An integral domain is real reduced if and only if its field of fractions is real. A field K is real (reduced) if and only if it has an ordering (cf. Sect. 1.1).

(4) A valuation ring A is real reduced if and only if it is real. (Proof: Real reduced implies real is clear from (1). Let A be real and let $a_i \in A$, $a_i \neq 0$ such that $a_1^2 + \cdots + a_n^2 = 0$. We may assume that $v(a_1) \leq \cdots \leq v(a_n) < \infty$. Letting $b_i := a_i/a_1 \in A$ for $i = 1, \ldots, n$, we then have $1 + b_2^2 + \cdots + b_n^2 = 0$, contradiction.) For example, if A is a residually real valuation ring, then A is real (reduced). This follows from Corollary 2.7.2 and (5) below. A real (reduced) valuation ring is in general not residually real though.

(5) If $\varphi \colon A \to B$ is a homomorphism and B is real, then A is also real.

(6) If $\varphi \colon A \to B$ is a homomorphism and $\mathfrak{b} \subseteq B$ is a real (resp. real reduced) ideal, then $\varphi^{-1}(\mathfrak{b})$ is also real (resp. real reduced). If φ is surjective, then $\mathfrak{b}$ is real (resp. real reduced) if and only if $\varphi^{-1}(\mathfrak{b})$ is real (resp. real reduced).

(7) Let $A = A_1 \times \cdots \times A_r$ be a direct product of rings with $A_i \neq 0$. Then A is real reduced if and only if the factors $A_1, \ldots, A_r$ are all real reduced, and A is real if and only if at least one of the factors $A_1, \ldots, A_r$ is real.

---

[1] In *Einführung in die reelle Algebra* we followed a suggestion of T.Y. Lam [71], and used "semireal" and "real" instead of "real" and "real reduced", respectively. In the meantime the terminology we use here has become standard.

(8) If $S \subseteq A$ is a multiplicative subset and $0 \notin S$, then the following implications hold:

$$A \text{ real reduced} \Rightarrow S^{-1}A \text{ real reduced} \Rightarrow S^{-1}A \text{ real} \Rightarrow A \text{ real},$$

but their converses do not hold in general. (Proof of the first implication: If $\sum_i (a_i/s_i)^2 = 0$ in $S^{-1}A$, then $\sum_i (a_i t_i)^2 = 0$ in A for certain $t_i \in S$. If A is real reduced, it follows that $a_i t_i = 0$ and so $a_i/s_i = 0$ for all i.)

Notation 3.2.3 We denote the set of all *real reduced* prime ideals of A by $(\text{Spec } A)_{\text{re}}$, i.e.,

$$(\text{Spec } A)_{\text{re}} := \{\mathfrak{p} \in \text{Spec } A : \kappa(\mathfrak{p}) \text{ is real}\}.$$

Various characterizations of real reduced rings are given in:

Proposition 3.2.4 *Let A be a ring. The following statements are equivalent:*

 (i) *A is real reduced;*
 (ii) *A is reduced, and all minimal prime ideals of A are real reduced;*
(iii) *A is reduced, and $(\text{Spec } A)_{\text{re}}$ is dense in $\text{Spec } A$;*
(iv) *the intersection of all $\mathfrak{p} \in (\text{Spec } A)_{\text{re}}$ is $\{0\}$.*

Proof (i) $\Rightarrow$ (ii): Let $\mathfrak{p}$ be a prime ideal of A. Then $A_{\mathfrak{p}}$ is real reduced by Remark 3.2.2(8), hence reduced by Remark 3.2.2(2). If $\mathfrak{p}$ is minimal, then $\mathfrak{p}A_{\mathfrak{p}}$ is the only prime ideal of $A_{\mathfrak{p}}$. Hence, $\mathfrak{p}A_{\mathfrak{p}} = 0$ and thus $A_{\mathfrak{p}} = \kappa(\mathfrak{p})$, i.e., $\mathfrak{p}$ is real reduced.

(ii) $\Rightarrow$ (iii): This is clear since the set of minimal prime ideals of an arbitrary ring A is dense in $\text{Spec } A$.

(iii) $\Rightarrow$ (iv): Since $(\text{Spec } A)_{\text{re}}$ is dense in $\text{Spec } A$, Proposition 3.1.11 yields

$$\bigcap\{\mathfrak{p} : \mathfrak{p} \in (\text{Spec } A)_{\text{re}}\} = \bigcap\{\mathfrak{p} : \mathfrak{p} \in \text{Spec } A\} = \text{Nil } A$$

and $\text{Nil } A = (0)$ since A is reduced.

(iv) $\Rightarrow$ (i): Let $a_1, \ldots, a_r \in A \setminus \{0\}$. By assumption (iv) there exists $\mathfrak{p} \in (\text{Spec } A)_{\text{re}}$ such that $a_1 \notin \mathfrak{p}$. Since $\mathfrak{p}$ is real reduced, we have $a_1^2 + \cdots + a_r^2 \notin \mathfrak{p}$, so in particular $\neq 0$. □

Corollary 3.2.5 *A ring is real reduced if and only if it is a subring of a direct product of real fields (with infinitely many factors in general).* □

For the characterization of real rings, we introduce:

Definition 3.2.6 We call an ideal $\mathfrak{a} \in A$ *maximally real reduced* if $\mathfrak{a}$ is real reduced, $\mathfrak{a} \neq A$, and $\mathfrak{a}$ is maximal with respect to these two properties (i.e., if $\mathfrak{a} \subseteq \mathfrak{a}'$ and $\mathfrak{a}'$ is real reduced, then $\mathfrak{a}' = \mathfrak{a}$ or $\mathfrak{a}' = A$). A *maximally real* ideal is defined similarly.

Proposition 3.2.7 *An ideal of A is maximally real reduced if and only if it is maximally real. The maximally real (reduced) ideals of A are precisely those (prime) ideals $\mathfrak{p}$ of A that are maximal for the property $\mathfrak{p} \cap (1 + \Sigma A^2) = \varnothing$.*

Proof Let $S = 1 + \Sigma A^2$. An ideal $\mathfrak{a}$ of A is real if and only if $\mathfrak{a} \cap S = \varnothing$. Without loss of generality we may assume that $0 \notin S$, i.e., that A is real, since otherwise there is nothing to prove. Every proper ideal $\mathfrak{b}$ of $S^{-1}A$ is real (in $S^{-1}A$) since $1 + \left(\frac{a_1}{s_1}\right)^2 + \cdots + \left(\frac{a_r}{s_r}\right)^2 \in \mathfrak{b}$ (with $a_i \in A$, $s_i \in S$) implies $b := s^2 + a_1'^2 + \cdots + a_r'^2 \in \mathfrak{b}$ for certain $s \in S$ and $a_i' \in A$ (common denominator), a contradiction since $b \in S$. If $\mathfrak{a}$ is a maximally real ideal of A, then $\mathfrak{a}$ is a prime ideal by Exercise 3.1.10. Since $S^{-1}\mathfrak{a}$ is a maximal ideal of $S^{-1}A$, the field $S^{-1}A/S^{-1}\mathfrak{a}$ is real by the observation above. The embedding $A/\mathfrak{a} \hookrightarrow S^{-1}A/S^{-1}\mathfrak{a}$ then implies that $\mathfrak{a}$ is real reduced. Conversely, if $\mathfrak{a}$ is a maximally real reduced ideal, then $\mathfrak{a}$ is also maximally real by the direction already established. $\square$

Corollary 3.2.8 *An ideal $\mathfrak{a}$ of A is real if and only if there exists an ideal $\mathfrak{p} \in (\operatorname{Spec} A)_{\mathrm{re}}$ such that $\mathfrak{a} \subseteq \mathfrak{p}$. In particular, a ring A is real if and only if it contains a real reduced prime ideal (i.e., if and only if $\operatorname{Sper} A \neq \varnothing$, cf. Sect. 3.3).*

Proof If $\mathfrak{a}$ is real, then $\mathfrak{a}$ is contained in a maximally real ideal $\mathfrak{p}$, and $\mathfrak{p} \in (\operatorname{Spec} A)_{\mathrm{re}}$ by Proposition 3.2.7. The converse is clear. $\square$

A (surprising) reformulation is given by:

Corollary 3.2.9 *If -1 is a sum of squares in every residue field of A, then -1 is already a sum of squares in A itself.* $\square$

Section 3.1 featured the weak and a strong Nullstellensatz from commutative algebra. Real algebra has its own versions of these theorems. The weak real Nullstellensatz describes the intersection of all *real reduced* prime ideals in any ring:

Proposition 3.2.10 (Weak Real Nullstellensatz) *For any ideal $\mathfrak{a} \subseteq A$ and any $f \in A$, the following statements are equivalent:*

(i) *$f \in \mathfrak{p}$ for all real reduced prime ideals $\mathfrak{p} \supseteq \mathfrak{a}$;*
(ii) *there exist $N \in \mathbb{N}$ and $a_1, \ldots, a_r \in A$ such that $f^{2N} + a_1^2 + \cdots + a_r^2 \in \mathfrak{a}$.*

Proof It suffices to prove the theorem for the ring $A/\mathfrak{a}$. So we may assume that $\mathfrak{a} = (0)$.

(i) $\Rightarrow$ (ii): By assumption we have $D_A(f) \cap (\operatorname{Spec} A)_{\mathrm{re}} = \varnothing$, thus $(\operatorname{Spec} f^{-\infty}A)_{\mathrm{re}} = \varnothing$. By Proposition 3.2.7 an equation of the form

$$1 + \left(\frac{a_1}{f^n}\right)^2 + \cdots + \left(\frac{a_r}{f^n}\right)^2 = 0$$

is satisfied in $f^{-\infty}A$. Hence, $f^{2m}(f^{2n} + a_1^2 + \cdots + a_r^2) = 0$ in A for some $m \geq 0$, which proves (ii).

(ii) $\Rightarrow$ (i): Let $\mathfrak{p} \in (\operatorname{Spec} A)_{\mathrm{re}}$. Then by (ii) we have $f^{2N} \in \mathfrak{p}$, and so $f \in \mathfrak{p}$. $\square$

Definition 3.2.11 If $\mathfrak{a} \subseteq A$ is an ideal, its *real radical* $\sqrt[re]{\mathfrak{a}}$ is defined as the intersection of all real reduced prime ideals $\mathfrak{p} \supseteq \mathfrak{a}$.

Remarks 3.2.12

(1) A ring is real reduced if and only if $\sqrt[re]{(0)} = (0)$, cf. Proposition 3.2.4.
(2) By Proposition 3.2.10 we have

$$\sqrt[re]{\mathfrak{a}} = \{f \in A : \exists\, N \in \mathbb{N}, \exists\, a_1, \ldots, a_r \in A \text{ such that } f^{2N} + a_1^2 + \cdots + a_r^2 \in \mathfrak{a}\}.$$

When giving a direct proof that the right hand side is an ideal, showing closure under addition requires some effort: assume $f^{2M} + a_1^2 + \cdots + a_r^2 \in \mathfrak{a}$ and $g^{2N} + b_1^2 + \cdots + b_s^2 \in \mathfrak{a}$. We may take $M = N$. Then $(f+g)^{4N} + (f-g)^{4N}$ is a sum of elements of the form $f^{2m} g^{2n}$ with $m + n = 2N$. Therefore, $m \geq N$ or $n \geq N$ in every such term. It follows that $-f^{2m} g^{2n}$ is a sum of squares modulo $\mathfrak{a}$, which yields $(f+g)^{4N} + c_1^2 + \cdots + c_t^2 \in \mathfrak{a}$.

Let R be a real closed field, $C = R(\sqrt{-1})$, A an affine R-algebra, and $V_A = \mathrm{Hom}_R(A, C)$ the variety associated to A. As before, we interpret $V_A(R) = \mathrm{Hom}_R(A, R)$ as a subspace of $\mathrm{Spec}\, A$ (cf. Sect. 3.1), as a matter of fact even of $(\mathrm{Spec}\, A)_{\mathrm{re}}$ and present new formulations of the theorems of Artin and Lang that were established in Chap. 2:

Theorem 3.2.13 (Stellensatz) *An affine R-algebra A is real if and only if its associated variety V_A has real points, i.e., if and only if $V_A(R) \neq \varnothing$.*

Proof The sufficient direction is clear. For the necessary direction, assume that A is real. By Corollary 3.2.8 there exists $\mathfrak{p} \in (\mathrm{Spec}\, A)_{\mathrm{re}}$, and $K := \kappa(\mathfrak{p})$ is a real function field over R. Let $f_1, \ldots, f_r$ be generators of A, and $\overline{f_1}, \ldots, \overline{f_r}$ their images in K. By Theorem 2.11.8 there exists a place $\lambda: K \to R \cup \infty$ over R such that $\lambda(\overline{f_i}) \neq \infty$ for $i = 1, \ldots, r$. The composition $\lambda \circ \rho_{\mathfrak{p}}$ is an element of $V_A(R)$. $\qquad\square$

In fact, the following stronger statement is true:

Corollary 3.2.14 *Let A be an affine R-algebra. Then $V_A(R)$ is Zariski dense in $(\mathrm{Spec}\, A)_{\mathrm{re}}$.*

Proof Let $f \in A$ such that $D_A(f) \cap (\mathrm{Spec}\, A)_{\mathrm{re}} \neq \varnothing$. Then $f^{-\infty} A$ is a real affine R-algebra, and so $\mathrm{Hom}_R(f^{-\infty} A, R) \neq \varnothing$ by Theorem 3.2.13. Every element of this set yields an element of $D_A(f) \cap V_A(R)$ (via composition with $A \to f^{-\infty} A$). $\qquad\square$

As a combination of the weak real Nullstellensatz and the Artin–Lang Stellensatz we obtain the following (strong) real Nullstellensatz:

Theorem 3.2.15 (Real Nullstellensatz, D.W. Dubois, J.-J. Risler, 1969/70) *Let A be an affine R-algebra and $f \in A$. Then f vanishes on $V_A(R)$ if and only if there exist $N \in \mathbb{N}$ and $a_1, \ldots, a_r \in A$ such that $f^{2N} + a_1^2 + \cdots + a_r^2 = 0$.*

Proof By definition f vanishes on $V_A(R)$ if and only if $f \in \mathfrak{I}(V_A(R))$. By Corollary 3.2.14 we have $\mathfrak{I}(V_A(R)) = \mathfrak{I}((\operatorname{Spec} A)_{\text{re}})$, and the statement follows from the description of the second set given in Proposition 3.2.10. $\square$

3.3   Definition of the Real Spectrum

Let A be a ring.

Definition 3.3.1 (M. Coste, M.-F. Roy, 1979)

(a) The *real spectrum* Sper A *of* A is the set of all pairs $\alpha = (\mathfrak{p}, T)$ with $\mathfrak{p} \in \operatorname{Spec} A$ and T an ordering of the field $\kappa(\mathfrak{p}) = \operatorname{Quot} A/\mathfrak{p}$. (Other notations in use are $\operatorname{Spec}_r A$ and $\operatorname{Spec}_R A$. In Sect. 3.6 we will motivate our notation Sper A.) The prime ideal $\mathfrak{p}$ is called the *support* of α, denoted $\mathfrak{p} = \operatorname{supp} \alpha$.

(b) For $\alpha = (\mathfrak{p}, T) \in \operatorname{Sper} A$, $k(\alpha)$ denotes the real closure of $\kappa(\mathfrak{p})$ with respect to T. There is a canonical homomorphism

$$r_\alpha \colon A \xrightarrow{\rho_\mathfrak{p}} \kappa(\mathfrak{p}) \hookrightarrow k(\alpha).$$

Given $f \in A$, we write $f(\alpha)$ instead of $r_\alpha(f)$. Relations such as "$f(\alpha) \geq 0$" are always with respect to the unique ordering of $k(\alpha)$.

(c) Given any subset T of A, we define

$$\mathring{H}_A(T) := \{\alpha \in \operatorname{Sper} A \colon f(\alpha) > 0 \text{ for all } f \in T\},$$

$$\overline{H}_A(T) := \{\alpha \in \operatorname{Sper} A \colon f(\alpha) \geq 0 \text{ for all } f \in T\},$$

$$Z_A(T) := \{\alpha \in \operatorname{Sper} A \colon f(\alpha) = 0 \text{ for all } f \in T\}.$$

Usually we drop the index A. If $T = \{f_1, \ldots, f_r\}$ is finite, we just write $\mathring{H}_A(f_1, \ldots, f_r)$ instead of $\mathring{H}_A(\{f_1, \ldots, f_r\})$, etc.

(d) The *Harrison topology* on Sper A is the topology with $\mathfrak{H}_A := \{\mathring{H}_A(f) \colon f \in A\}$ as subbasis of open sets. We call $\mathfrak{H}_A$ the *Harrison subbasis* of the topological space Sper A.

Remarks 3.3.2

(1) A difference with the Zariski spectrum is that the Harrison subbasis is usually not an open basis. An example of an open basis for the Harrison topology is given by

$$\{\mathring{H}(f_1, \ldots, f_r) \colon r \geq 1, f_i \in A\}.$$

Note that $Z(f) = \overline{H}(-f^2)$ and $Z(f_1, \ldots, f_r) = Z(f_1^2 + \cdots + f_r^2) = \overline{H}(-f_1^2 - \cdots - f_r^2)$. The sets $\overline{H}(T)$ and $Z(T)$ are closed for any $T \subseteq A$.

(2) In analogy with the Zariski spectrum (cf. Sect. 3.1), writing $f(\alpha)$ instead of $r_\alpha(f)$ signifies that we interpret the elements $f \in A$ as "functions" on Sper A. The value fields $k(\alpha)$ vary again from point to point. This interpretation is "faithful" (in the sense that $r = (r_\alpha): A \to \prod_{\alpha \in \mathrm{Sper}\, A} k(\alpha)$ is injective) if and only if A is real reduced (cf. Proposition 3.2.4).

(3) Every homomorphism $\varphi: A \to K$ into an ordered field (K, P) defines an element α_φ in Sper A, namely $\alpha_\varphi := (\mathrm{Ker}\,\varphi, P')$, where P' is the ordering of $\kappa(\mathrm{Ker}\,\varphi)$, induced by P. Therefore Sper A can also be defined as the set of all homomorphisms of A into ordered (or: real closed) fields, modulo an appropriate equivalence relation. We will not pursue this approach (cf. [11, §7.1]).

(4) If $A = k$ is a field, then Sper k is the space of orderings of k. Since $\overline{H}_k(f) = \overset{\circ}{H}_k(f)$ for all $f \in k^*$, all the sets of the Harrison subbasis are open and closed. Sper k is in particular a totally disconnected Hausdorff space. (Since Sper A is always quasi-compact—as we will show in Sect. 3.4—Sper k is a Boolean space, i.e., compact and totally disconnected.) This topology was discovered by Harrison (in the case of fields) around 1970. This motivates our use of the term "Harrison topology" also in the general case.

(5) A ring A is real if and only if Sper $A \neq \varnothing$. This follows from Corollary 3.2.8.

Proposition 3.3.3 *The map* supp: Sper $A \to$ Spec A *is continuous. Its image is the subspace* (Spec $A)_{\mathrm{re}}$ *of real reduced prime ideals.*

Proof We have $\mathrm{supp}^{-1}(D_A(f)) = \overset{\circ}{H}_A(f^2)$. The second statement is clear. $\square$

An ordering T of $\kappa(\mathfrak{p})$ is determined by $T \cap (A/\mathfrak{p})$, i.e., by $\rho_\mathfrak{p}^{-1}(T)$. Consequently the elements of Sper A can also be interpreted as certain subsets of A. Given $\alpha = (\mathfrak{p}, T) \in$ Sper A, let $P_\alpha := \rho_\mathfrak{p}^{-1}(T) = \{f \in A: f(\alpha) \geq 0\}$. Then $P = P_\alpha$ satisfies the following properties:

(RO1) $P + P \subseteq P$, $PP \subseteq P$;

(RO2) $P \cup (-P) = A$;

(RO3) supp $P := P \cap (-P)$ is a prime ideal of A.

Since (RO1) and (RO2) already imply that supp P is an ideal of A, we may replace (RO3) by

(RO3') $-1 \notin P$, and: $a \notin P$, $b \notin P \Rightarrow -ab \notin P$ $(a, b \in A)$.

It is easy to see that every subset $P \subseteq A$ that satisfies properties (RO1) to (RO3) defines an element $\alpha_P \in$ Sper A, namely $\alpha_P = (\mathfrak{p}, \overline{P})$, where $\mathfrak{p} := \mathrm{supp}\, P$ and $\overline{P}$ is the ordering on $\kappa(\mathfrak{p})$ induced by P (for $a, b \in A$, $b \notin \mathfrak{p}$ we have: $\rho_\mathfrak{p}(a)/\rho_\mathfrak{p}(b) \in \overline{P} \Leftrightarrow ab \in P$). This motivates:

Definition 3.3.4 An *ordering* of A is a subset P of A that satisfies properties (RO1) to (RO3) above. The terms *cone* and *prime cone* are also used.

The above argument proves:

Proposition 3.3.5 Sper A *can be interpreted as the set of all orderings of* A. *The topology is generated by the open subbasis* $\{\mathring{H}_A(f): f \in A\}$, *where* $\mathring{H}_A(f)$ *denotes the set of all orderings* P *of* A *for which* $-f \notin P$. $\square$

The points of Sper A can thus be viewed in two possible ways. If $P \subseteq A$ is an ordering, we denote the real closure of $\kappa(\text{supp } P)$ with respect to $\overline{P}$ again by $k(P)$. We also write $f(P)$ for the image of $f \in A$ in $k(P)$, and so on.

Definition 3.3.6 Let X be a topological space and $x, y \in X$. If $y \in \overline{\{x\}}$, we call y a *specialization* of x and x a *generalization* of y, denoted $x \succ y$. If X is a T_0 space, then $x \leq y :\Leftrightarrow x \succ y$ defines a (partial) order relation on X.

If $X = \text{Spec } A$, then $\mathfrak{p}, \mathfrak{q} \in X$ satisfy $\mathfrak{p} \succ \mathfrak{q} \Leftrightarrow \mathfrak{p} \subseteq \mathfrak{q}$. In other words, the specialization relation is given by inclusion. A similar result is true for the real spectrum:

Proposition 3.3.7 *Let* $P, Q \in \text{Sper } A$ *be orderings of* A. *Then* $P \succ Q$ (*in* Sper A) *if and only if* $P \subseteq Q$.

Proof

$$P \succ Q \Leftrightarrow \text{for any } f \in A \text{ such that } Q \in \mathring{H}(f) \text{ we have } P \in \mathring{H}(f)$$

$$\Leftrightarrow \text{for any } g \in A \text{ such that } g \in P \text{ we have } g \in Q$$

$$\Leftrightarrow P \subseteq Q.$$ $\square$

Corollary 3.3.8 Sper A *is a* T_0 *space.* $\square$

Corollary 3.3.9 *If* $\alpha \in \text{Sper } A$ *and* supp (α) *is a maximal ideal of* A, *then* α *is a closed point of* Sper A. $\square$

We will present some examples shortly. In particular we note that Example 3.3.14 below shows that the converse of Corollary 3.3.9 is false. Let us first discuss the functorial behaviour of Sper, which is similar to the functorial behaviour of Spec seen in Sect. 3.1.

If $\varphi\colon A \to B$ is a ring homomorphism and $Q \subseteq B$ is an ordering of B, then $\varphi^{-1}(Q)$ is an ordering of A. Hence φ induces a map

$$\text{Sper } \varphi = \varphi^*\colon \text{Sper } B \to \text{Sper } A, \quad Q \mapsto \varphi^{-1}(Q),$$

which is continuous since $(\text{Sper } \varphi)^{-1}\big(\mathring{H}_A(f)\big) = \mathring{H}_B\big(\varphi(f)\big)$. Thus Sper, like Spec, is a functor from commutative rings to topological spaces. (Moreover, supp is a morphism Sper $\to$ Spec of functors since $\text{supp}_A \circ(\text{Sper } \varphi) = (\text{Spec } \varphi) \circ \text{supp}_B$.) The map φ also yields embeddings of real closed fields as follows: let $\beta \in \text{Sper } B$ and $\alpha = \varphi^*(\beta) \in \text{Sper } A$, then φ induces an order preserving (with respect to α, β)

embedding $\varphi_{\text{supp}\beta}\colon \kappa(\text{supp}\,\alpha)\hookrightarrow\kappa(\text{supp}\,\beta)$, and hence $\varphi_\beta\colon k(\alpha)\hookrightarrow k(\beta)$ (cf. Sect. 1.11). Letting $\mathfrak{p}:=\text{supp}\,\alpha$ and $\mathfrak{q}:=\text{supp}\,\beta$ we thus obtain the commutative diagram

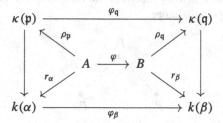

Continuing the analogy with the Zariski spectrum we can think of φ as pulling back the "functions" $g\in B$ that are defined on Sper B via Sper φ (and via the embeddings φ_β).

Proposition 3.3.10 *Consider a multiplicative subset $S\subseteq A$ and its associated canonical homomorphism $i_S\colon A\to S^{-1}A$. Then $\mathrm{Sper}(i_S)$ is a homeomorphism from* Sper $S^{-1}A$ *to the subspace* $\{\alpha\in\mathrm{Sper}\,A\colon S\cap\text{supp}\,\alpha=\varnothing\}=\bigcap_{s\in S}\mathring{H}_A(s^2)$ *of* Sper A. *If $\beta\in$* Sper $S^{-1}A$ *and $\alpha=(\mathrm{Sper}\,i_S)(\beta)$, then $(i_S)_\beta\colon k(\alpha)\to k(\beta)$ is an isomorphism.*

Proposition 3.3.11 *Consider an ideal $\mathfrak{a}\subseteq A$ and its associated residue map $\pi\colon A\to A/\mathfrak{a}$. Then* Sper π *is a homeomorphism from* Sper $A/\mathfrak{a}$ *to the closed subspace* $Z_A(\mathfrak{a})=\{\alpha\in\mathrm{Sper}\,A\colon\mathfrak{a}\subseteq\text{supp}\,\alpha\}$ *of* Sper A. *If $\beta\in$* Sper $A/\mathfrak{a}$ *and $\alpha=(\mathrm{Sper}\,\pi)(\beta)$, then $\pi_\beta\colon k(\alpha)\to k(\beta)$ is an isomorphism.*

The proof of Proposition 3.3.11 is similar to:

Proof of Proposition 3.3.10 Let $X:=\bigcap_{s\in S}\mathring{H}_A(s^2)$. The inverse $X\to$ Sper $S^{-1}A$ of Sper(i_S) is $P\mapsto S^{-2}P:=\{f/s^2\colon f\in P,s\in S\}$ $(P\in X)$. The homeomorphy follows from

$$(\mathrm{Sper}\,i_S)\,\mathring{H}_{S^{-1}A}\Big(\frac{f_1}{s_1},\dots,\frac{f_r}{s_r}\Big)=X\cap\mathring{H}_A(s_1f_1,\dots,s_rf_r).$$

Finally, $(i_S)_\beta$ is an isomorphism since $(i_S)_{\text{supp}\,\beta}$ is an isomorphism (cf. Sect. 3.1) and order preserving. $\qquad\square$

Corollary 3.3.12 *For any $\mathfrak{p}\in$ Spec A there exists a natural homeomorphism*

$$\mathrm{Sper}\,\kappa(\mathfrak{p})\to\{\alpha\in\mathrm{Sper}\,A\colon\text{supp}\,\alpha=\mathfrak{p}\},$$

induced by $\rho_\mathfrak{p}\colon A\to\kappa(\mathfrak{p})$. $\qquad\square$

Example 3.3.13 Let R be a real closed field and $A=R[\![t]\!]$ the ring of formal power series over R. The only prime ideals of A are (0) and $\mathfrak{m}_A=(t)$, and $\kappa(A)=$

$A/\mathfrak{m}_A \cong R$. By Sect. 2.9, Sper A has precisely three elements, namely the orderings P_0, P_-, P_+ of A, where supp $P_0 = \mathfrak{m}_A$ and supp $P_- =$ supp $P_+ = (0)$. From $P_- \cup P_+ \subseteq P_0$ it follows that P_0 is closed and has P_- and P_+ as generalizations. The situation can be illustrated by means of the following symbolic diagram, where the arrows represent specialization:

Example 3.3.14 Assume again that R is a real closed field and let $A = R[t]$. The real reduced prime ideals of $R[t]$ are (0) and all $(t-c)$, where $c \in R$. The orderings of $R(t)$ were determined in Sect. 2.9. The following list therefore gives a complete description of all points of Sper $R[t]$:

1. for every $c \in R$, a closed point P_c and also two generalizations $P_{c,-}$ and $P_{c,+}$ of P_c (simply written as c, c_-, c_+);
2. the closed points $P_{-\infty}$, $P_{+\infty}$ with support (0) (simply written as $-\infty$, $+\infty$);
3. (only if $R \neq \mathbb{R}$) for every free Dedekind cut ξ of R, a closed point P_ξ with support (0) (simply written as ξ).

As in the previous example, we illustrate the situation with a symbolic diagram:

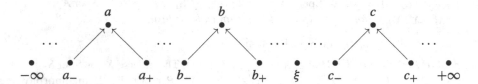

The diagram shows that Sper $R[t]$ possesses a natural linear order relation "$\leq$" that extends the ordering of R.

An open basis of the topology of Sper $R[t]$ is given by all open intervals

$$]a, b[:= \{\alpha \in \text{Sper } R[t] \colon a < \alpha < b\}$$

and all $[-\infty, b[$, $]a, \infty]$ where $a, b \in R$, defined similarly. Note that $]a, b[$ contains the points a_+ and b_-, but not the points a_-, a, b, b_+, and similarly for the other intervals.

In this concrete example it is not difficult to see that Sper $R[t]$ is quasi-compact and that the subspace topology induced on $R \subset$ Sper $R[t]$ is precisely the strong topology. In fact, this observation is true in general, as we will show now.

Let R be a real closed field, $C = R(\sqrt{-1})$ and A an affine R-algebra with associated variety $V := V_A = \text{Hom}_R(A, C)$ and set of real points $V(R) := V_A(R)$.

In Sect. 3.1 we interpreted $V(R)$ as a subset of Spec A via the identification of $x \in V(R)$ with $\mathfrak{m}_x := \mathrm{Ker}(x)$. Since $\kappa(\mathfrak{m}_x) = R$ possesses a unique ordering we can view $V(R)$ as a subset of Sper A as well: there is a canonical injection

$$V(R) \hookrightarrow \mathrm{Sper}\, A, \quad x \mapsto P_x := \{f \in A : f(x) \geq 0\}$$

(whose image consists of all $P \in \mathrm{Sper}\, A$ for which $R \to A \xrightarrow{r_P} k(P)$ is an isomorphism). Furthermore, $V(R)$ consists of closed points of Sper A by Corollary 3.3.9.

Let $u_1, \dots, u_n$ be generators of A and let $j: V(R) \hookrightarrow R^n$ be the associated embedding, i.e., $j(x) = \big(u_1(x), \dots, u_n(x)\big)$. The strong topology on R^n (cf. Sect. 2.6) induces a topology on $V(R)$ via j, which we also call the *strong topology* (and which is independent of the choice of j, as we will see shortly). The following fact is an indicator of how useful the real spectrum is when studying the geometry of real varieties:

Proposition 3.3.15 *The relative topology of $V_A(R)$ in* Sper A *coincides with the strong topology on $V_A(R)$, which is in particular independent of the choice of embedding $V_A(R) \hookrightarrow R^n$.*

Proof A basis of the strong topology on $V(R)$ is given by the system of sets $\{x \in V(R): f_1(x) > 0, \dots, f_r(x) > 0\}$ (where $r \geq 1$ and $f_i \in A$), which can be rewritten as $V(R) \cap \mathring{H}_A(f_1, \dots, f_r)$ upon identification of $V(R)$ with its image in Sper A. $\square$

Remark 3.3.16 The real points of V, equipped with the strong topology, therefore form a subspace of Sper A. In other words, we do not loose any information about $V(R)$ after the transition to Sper A. Furthermore, Sper A has a great advantage compared to $V(R)$ as $V(R)$ has bad topological properties when $R \neq \mathbb{R}$ (totally disconnected, almost never (locally) compact), whereas Sper A is quasi-compact and only has finitely many connected components (cf. [11, §7.5]), which facilitates the definition of a non-trivial concept of connectedness even when $R \neq \mathbb{R}$. (A different approach that does not use the real spectrum can be found in [31].) Thus, the addition of "ideal" points results in a much more geometric space. The example of the affine line $A = R[t]$ above (Example 3.3.14) already illustrates nicely how the additional points in Sper A "close the gaps" (if $R \neq \mathbb{R}$) and make the space quasi-compact.

An immediate consequence of the Artin–Lang theorem is that $V(R)$ is dense as a subspace of Sper A!

Theorem 3.3.17 $V_A(R)$ *is a dense subspace of* Sper A.

Proof Let $f_1, \dots, f_r \in A$ such that $\mathring{H}(f_1, \dots, f_r) \neq \varnothing$ and let $\alpha = (\mathfrak{p}, T) \in \mathring{H}(f_1, \dots, f_r)$. Then $\kappa(\mathfrak{p})$ is a function field over R and T is an ordering of $\kappa(\mathfrak{p})$ for which $\overline{f_1}, \dots, \overline{f_r} > 0$, where $\overline{f_i} := \rho_\mathfrak{p}(f_i) \in \kappa(\mathfrak{p})$. By Theorem 2.11.9 there exists an R-homomorphism $\varphi\colon A/\mathfrak{p} \to R$ such that $\varphi(\overline{f_i}) > 0$ for $i = 1, \ldots, r$. The composition $x\colon A \to A/\mathfrak{p} \xrightarrow{\varphi} R$ is a point of $V(R) \cap \mathring{H}(f_1, \ldots, f_r)$. $\square$

We will present a substantial generalization of this fact in Theorem 3.5.3.

3.4   Constructible Subsets and Spectral Spaces

Let A be a ring. In this section we will introduce a second topology on $\mathrm{Sper}\, A$, the *constructible topology*. We will always think of the Harrison topology as the "proper" topology of $\mathrm{Sper}\, A$, while the constructible topology will rather play an auxiliary role. In situations where we do not specify which topology is used for $\mathrm{Sper}\, A$, it will always be the Harrison topology.

Definition 3.4.1

(a) A partially ordered set $(L, \leq)$ is called a *lattice* (with 0 and 1) if every finite subset $X \subseteq L$ has a greatest lower bound $\inf X$ and a least upper bound $\sup X$ in L. In particular, $0 := \sup \varnothing = \inf L$ and $1 := \inf \varnothing = \sup L$ exist. For $x, y \in L$ we use the notation

$$x \wedge y := \inf\{x, y\}, \quad x \vee y := \sup\{x, y\}.$$

(b) A *Boolean lattice* is a lattice $(L, \leq)$ that satisfies the following properties:

(BL1) Distributivity: $x \wedge (y \vee z) = (x \wedge y) \vee (x \wedge z)$ for all $x, y, z \in L$;
(BL2) Complement: for every $x \in L$, there exists $x' \in L$ such that $x \wedge x' = 0$ and $x \vee x' = 1$.

Remarks 3.4.2 (cf. [43])

(1) In a Boolean lattice L the complement x' of x is uniquely determined and the dual identity of (BL1),

$$x \vee (y \wedge z) = (x \vee y) \wedge (x \vee z),$$

is satisfied for all $x, y, z \in L$.
(2) Boolean lattices can also be defined as those $\mathbb{Z}/2\mathbb{Z}$-algebras B that satisfy $x^2 = x$ for all $x \in B$, the *Boolean algebras*. Endowing a Boolean lattice with the operations

$$x + y := (x \wedge y') \vee (x' \wedge y) \quad \text{and} \quad xy := x \wedge y$$

turns it into a Boolean algebra, and conversely every Boolean algebra becomes a Boolean lattice via

$$x \wedge y := xy, \quad x \vee y := x + y + xy, \quad x' := 1 + x.$$

Definition 3.4.3

(a) $\mathcal{K}(\operatorname{Sper} A)$ denotes the Boolean sublattice of the power set $2^{\operatorname{Sper} A}$ generated by the Harrison subbasis $\overset{\circ}{\mathcal{H}}_A$ of $\operatorname{Sper} A$. The elements of $\mathcal{K}(\operatorname{Sper} A)$ are called the *constructible subsets* of $\operatorname{Sper} A$. We let

$$\overset{\circ}{\mathcal{K}}(\operatorname{Sper} A) := \{Y \in \mathcal{K}(\operatorname{Sper} A) \colon Y \text{ is open}\},$$

$$\overline{\mathcal{K}}(\operatorname{Sper} A) := \{Y \in \mathcal{K}(\operatorname{Sper} A) \colon Y \text{ is closed}\}.$$

(b) The *constructible topology* on $\operatorname{Sper} A$ is the topology generated by $\mathcal{K}(\operatorname{Sper} A)$ as basis of open sets.

Remarks 3.4.4

(1) $\mathcal{K}(\operatorname{Sper} A)$ is the smallest subset $\mathcal{K}$ of $2^{\operatorname{Sper} A}$ such that $\overset{\circ}{\mathcal{H}}_A \subseteq \mathcal{K}$ and which is stable under taking finite intersections and complements. The elements of $\mathcal{K}(\operatorname{Sper} A)$ are precisely the finite unions of sets of the form

$$\overset{\circ}{H}_A(f_1, \ldots, f_r) \cap \overline{H}_A(g_1, \ldots, g_s) \quad (f_i, g_j \in A),$$

or those of the form

$$\overset{\circ}{H}_A(f_1, \ldots, f_r) \cap Z_A(g) \quad (f_i, g \in A),$$

or those of the form

$$\{\alpha \in \operatorname{Sper} A \colon \operatorname{sgn} f_1(\alpha) = \varepsilon_1, \ldots, \operatorname{sgn} f_r(\alpha) = \varepsilon_r\} \quad (f_i \in A, \varepsilon_i \in \{-1, 0, 1\}).$$

(2) The constructible topology is finer than the Harrison topology. If $A = k$ is a field, then both topologies coincide since the Harrison subbasis is stable under taking complements: $(\operatorname{Sper} k) \setminus \overline{H}_k(f) = \overset{\circ}{H}_k(-f)$ for all $f \in K^*$.

The constructible topology is still coarse enough to possess pleasing properties, as we will see next. We note that the proof of Theorem 3.4.5 belongs to the realm of model theory rather than real algebra.

Theorem 3.4.5 *$\operatorname{Sper} A$, endowed with the constructible topology, is a totally disconnected compact (Hausdorff) space. The constructible subsets of $\operatorname{Sper} A$ are precisely the clopen sets of the topology.*

Proof Let us prove the second statement first. Consider the constructible topology on Sper A. Then all $Y \in \mathcal{K}(\text{Sper } A)$ are clopen by definition. Conversely, let $Y \subseteq$ Sper A be clopen. Since Sper A is compact (see below), Y is the union of *finitely* many constructible sets, and thus constructible itself.

To prove compactness, consider $Z = \prod_{f \in A}\{0, 1\}$, endowed with the product topology. Then Z is totally disconnected and also compact by Tykhonov's theorem. We identify an element $(\varepsilon_f)_{f \in A}$ with the subset $\{f \in A: \varepsilon_f = 1\}$ of A, in other words we identify Z with the power set of A. It follows that $j(\alpha) := \{f \in A: f(\alpha) \geq 0\} = P_\alpha$ defines an injection $j\colon \text{Sper } A \hookrightarrow Z$, and the topology induced by Z (via j) on Sper A is precisely the constructible topology, cf. Remark 3.4.4(1). It remains to show that $j(\text{Sper } A)$ is closed in Z, or equivalently that $Z \setminus j(\text{Sper } A)$ is open. To this end, let $S \subseteq A$ be a subset such that $S \notin j(\text{Sper } A)$, i.e., such that S is *not* and ordering of A. Then one of the following axioms (cf. Sect. 3.3) is violated by S:

(1) $S + S \subseteq S$,
(2) $SS \subseteq S$,
(3) $S \cup (-S) = A$,
(4) $-1 \notin S$,
(5) $a \notin S, b \notin S \Rightarrow -ab \notin S \ (a, b \in A)$.

But then this particular axiom is also violated in a neighbourhood of S in Z! For example, if S violates axiom (5), then there exist $a, b \in A \setminus S$ such that $-ab \in S$. Then $U := \{T \subseteq A: a \notin T, b \notin T, -ab \in T\}$ is a neighbourhood of S in Z, and each $T \in U$ violates axiom (5) for the same reason that S violates axiom (5).   □

Since the Harrison topology is coarser than the constructible topology, we obtain:

Corollary 3.4.6 *The Harrison topology on* Sper A *is quasi-compact.*   □

Let $\varphi\colon A \to B$ be a ring homomorphism. Then Sper $\varphi\colon$ Sper $B \to$ Sper A is continuous in the constructible topology since $(\text{Sper } \varphi)^{-1}\big(\mathring{H}_A(f)\big) = \mathring{H}_B\big(\varphi(f)\big)$ for all $f \in A$ and taking inverse images commutes with the Boolean operations. Nevertheless, the image of a constructible set is in general not constructible anymore (consider for example the map $R[t] \hookrightarrow R(t)$ and the set Sper $R(t)$). Consequently it is often necessary to consider more general subspaces.

Definition 3.4.7 A subset Y of Sper A is called *proconstructible* if Y is the intersection of (arbitrarily many) constructible subsets of Sper A or, equivalently, if Y is closed in the constructible topology.

Remarks 3.4.8

(1) The equivalence of the conditions in the definition follows from the fact that the complement of a subset Y which is closed in the constructible topology can be written as the union of constructible subsets of Sper A.

(2) Some examples of proconstructible subspaces of Sper A are given by Sper $S^{-1}A$ and Sper $A/\mathfrak{a}$. If S, resp. $\mathfrak{a}$, is finitely generated as semigroup,

resp. ideal, then it is constructible. A further example is given by the sets $\overset{\circ}{H}_A(T)$ and $\overline{H}_A(T)$ for arbitrary $T \subseteq A$.

Proposition 3.4.9 Sper φ *is continuous in the constructible topology for every* $\varphi: A \to B$. *Preimages* and *images of proconstructible subsets are again procon-structible.*

Proof The first statement was remarked earlier. For the second statement, assume that $Y \subseteq$ Sper B is proconstructible. Then Y is compact in the constructible topology. The same is then true for (Sper φ)(Y) and so (Sper φ)(Y) is closed in Sper A for the constructible topology. $\quad\square$

Corollary 3.4.10 (of Theorem 3.4.5)

(a) *Let Y be a proconstructible subset of* Sper A. *Then every covering of Y by constructible subsets of* Sper A *has a finite subcovering.*
(b) *If a family of proconstructible subsets of* Sper A *has empty intersection, this is already the case for a finite subfamily.* $\quad\square$

For example, an open subset of Sper A is proconstructible if and only if it is quasi-compact, and thus constructible (i.e., in $\mathcal{K}$(Sper A)). These sets are thus precisely the finite unions of sets of the form $\overset{\circ}{H}(f_1, \ldots, f_r)$.

A very important consequence is given by:

Proposition 3.4.11 *Let Y be any proconstructible subset of* Sper A, *then* $\overline{Y} = \bigcup_{y \in Y} \overline{\{y\}}$ *(with respect to the Harrison topology).*

Corollary 3.4.12 *A proconstructible subset of* Sper A *is (Harrison) closed if and only if it is stable under specialization (in* Sper A). *A constructible subset is open if and only if it is stable under generalization.*

Proof of Proposition 3.4.11 Only the inclusion "$\subseteq$" is nontrivial. Let $z \in \overline{Y}$ and let $\{U_i : i \in I\}$ be the family of open constructible neighbourhoods of z in Sper A. Then $Y \cap U_i \neq \varnothing$ for all i and so $Y \cap \bigcap_{i \in I} U_i \neq \varnothing$ by Corollary 3.4.10. For $y \in Y \cap \bigcap_{i \in I} U_i$ we have $z \in \overline{\{y\}}$. $\quad\square$

Whereas the closure of any proconstructible set is trivially proconstructible again, there are examples of constructible sets whose closure is no longer con-structible. However, for real spectra of affine algebras over fields (or more generally excellent rings) this does not happen, cf. [1, Theorem 3.1].

Corollary 3.4.13 *If Y is a closed irreducible subset of* Sper A *and $Y \neq \varnothing$, then Y contains a (unique) generic point.*

Proof Let Z be the intersection of Y with all $U \in \mathcal{K}$(Sper A) that satisfy $Y \cap U \neq \varnothing$. Since Y is irreducible, Y meets the intersection of only finitely many such U. Hence, $Z \neq \varnothing$ by Corollary 3.4.10. For any $z \in Z$ we have $Y = \overline{\{z\}}$. Indeed, otherwise there exists $y \in Y \setminus \overline{\{z\}}$ and then y has a neighbourhood V such that $z \notin V$, a contradiction. $\square$

Remark 3.4.14 If A is a noetherian ring, for instance an affine algebra over a field, then $X = \operatorname{Spec} A$ has only finitely many irreducible components (i.e., A has only finitely many minimal prime ideals). These play an important role in algebraic geometry. In contrast, the spaces $X = \operatorname{Sper} A$ almost always have infinitely many minimal points, also in the "geometric setting", which is why the concept of irreducible components is less useful in this context.

We take a temporary leave of real algebra and embark on an excursion into point-set topology. Specifically we will have a closer look at spectral spaces. These satisfy all the topological properties that we already observed for the Zariski spectrum and the real spectrum. Spectral spaces elucidate the boundary between real algebraic and purely topological arguments. Moreover, in one swoop we obtain the properties of many new spaces that occur naturally in the study of spectra (for example, proconstructible subspaces are spectral). The main examples of spectral spaces are the real spectra and especially the Zariski spectra. In fact, Hochster has shown that these are actually *all* the examples, cf. [51].

Definition 3.4.15

(a) A *spectral space* is a topological space X that satisfies the following properties:

  (1) X is a quasi-compact T_0 space;
  (2) the intersection of any two open quasi-compact subsets of X is again quasi-compact;
  (3) the open quasi-compact sets form a basis of the topology on X;
  (4) every nonempty closed irreducible subspace of X has a (unique) generic point.

(b) Let X and Y be spectral spaces. A *spectral map* is a map $f \colon X \to Y$ such that for all $V \subseteq Y$, if V is open and quasi-compact, then so is $f^{-1}(V)$. (In particular, f is continuous.)

Definition 3.4.16 Let X be a spectral space.

(a) We denote the Boolean lattice generated by the open quasi-compact subsets of X (in 2^X) by $\mathcal{K}(X)$. Its elements are called the *constructible subsets* of X. The open, resp. closed, constructible subsets are denoted $\overset{\circ}{\mathcal{K}}(X)$, resp. $\overline{\mathcal{K}}(X)$.

(b) The topology on X that has $\mathcal{K}(X)$ as open basis is called the *constructible topology*. The space X endowed with this topology is denoted X_{con}. The original topology on X is then often called the *spectral topology* on X to make the distinction.

Examples 3.4.17

(1) The spectral maps are precisely those continuous maps $X \to Y$ for which $X_{\mathrm{con}} \to Y_{\mathrm{con}}$ is also continuous.
(2) Sper A is a spectral space, and the current definition of constructible subsets is the same as the original one. If φ is a ring homomorphism, the map Sper φ is spectral. In a similar fashion, Spec A is a spectral space and Spec φ is a spectral map.
(3) Every *Boolean space* (i.e., a compact and totally disconnected topological space) is spectral. Its constructible sets are precisely the clopen sets, which means that in this case the spectral and constructible topologies coincide. The Boolean spaces are precisely the spectral spaces that are Hausdorff.

Proposition 3.4.18 *Consider a spectral space X. Then X_{con} is compact and totally disconnected. The identity map $X_{\mathrm{con}} \to X$ is spectral, and $(X_{\mathrm{con}})_{\mathrm{con}} = X_{\mathrm{con}}$.*

Proof The only nontrivial statement that needs a proof is the quasi-compactness of X_{con}, i.e., the finite intersection property for $\mathcal{K}(X)$. We start by showing the following

Claim: If $\mathcal{Y} = \{Y_i : i \in I\}$ is a family of either open quasi-compact or closed sets in X such that $\bigcap_{j \in J} Y_j \neq \varnothing$ for all finite $J \subseteq I$, then $\bigcap_{i \in I} Y_i \neq \varnothing$.

Proof of the claim: Without loss of generality we may assume that $\mathcal{Y}$ is maximal with respect to the indicated property (by Zorn's Lemma). Let Z be the intersection of the closed sets in $\mathcal{Y}$. Then $Z \neq \varnothing$ (since X is quasi-compact), hence also $Z \in \mathcal{Y}$. Assume for the sake of contradiction that Z is reducible. Then $Z = Z_1 \cup Z_2$ for some closed $Z_i \subsetneq Z$. Since $Z_i \notin \mathcal{Y}$, there exist open $U_1, U_2 \in \mathcal{Y}$ such that $U_i \cap Z_i = \varnothing$ for $i = 1, 2$. In particular, $(U_1 \cap U_2) \cap Z = \varnothing$. But $U_1 \cap U_2 \in \mathcal{Y}$ since $\mathcal{Y}$ is maximal, a contradiction. Hence Z is irreducible. It is now clear that the generic point of Z is an element of $\bigcap_i Y_i$. This proves the claim.

The remainder of the proof is a modification of an argument provided by Marcus Tressl. Consider the set Ultra $\mathcal{K}(X)$ of all ultrafilters in the lattice $\mathcal{K}(X)$, cf. Definition 3.5.10. For $x \in X$, let

$$F_x := \{U \in \mathcal{K}(X) : x \in U\}$$

denote the principal ultrafilter generated by x. If F is an arbitrary ultrafilter in $\mathcal{K}(X)$, then the intersection $\bigcap_{U \in F} U$ of all sets in F is nonempty by the claim. Let x be in this intersection. Then $F \subseteq F_x$, and so $F = F_x$. It follows that the map

$$X \to \mathrm{Ultra}\,\mathcal{K}(X), \quad x \mapsto F_x \tag{3.1}$$

is surjective, hence also bijective since the injectivity is clear.

If we interpret $\mathcal{K}(X)$ in the usual way as a (Boolean) ring, then a subset P of $\mathcal{K}(X)$ is a prime ideal of $\mathcal{K}(X)$ if and only if $\{X \setminus K : K \in P\}$ is an ultrafilter in

$\mathcal{K}(X)$. Thus (3.1) yields a bijective map

$$X \to \operatorname{Spec} \mathcal{K}(X), \quad x \mapsto P_x := \{K \in \mathcal{K}(X) \colon x \notin K\}. \tag{3.2}$$

Transferring the Zariski topology from Spec $\mathcal{K}(X)$ to the set X via the bijection (3.2) yields the constructible topology on X, which can be seen by direct verification. Hence the map (3.2) is a homeomorphism $X_{\mathrm{con}} \approx \operatorname{Spec} \mathcal{K}(X)$ of topological spaces. Since every Zariski spectrum is quasi-compact (cf. Proposition 3.1.15) we conclude that X_{con} is also quasi-compact. □

Corollary 3.4.19 *The elements of $\overset{\circ}{\mathcal{K}}(X)$ are precisely the open quasi-compact subsets of X.* □

Definition 3.4.20 Let X be a spectral space. A subset Y of X is called *proconstructible* if Y is the intersection of constructible subspaces (or equivalently, if Y is closed in X_{con}).

As before we have:

Corollary 3.4.21 *Let X be a spectral space.*

(a) *Every covering of a proconstructible set by constructible sets has a finite subcovering.*
(b) *If a family of proconstructible sets in X has empty intersection, this is already the case for a finite subfamily.*
(c) *If Y is a proconstructible subset of X, then $\overline{Y} = \bigcup_{y \in Y} \overline{\{y\}}$.* □

Suitably adapted, Corollary 3.4.12 also holds in this context.

Proposition 3.4.22 *Let X be a spectral space and Y a subspace of X.*

(a) *Y is proconstructible in X if and only if Y itself is a spectral space and the inclusion $Y \hookrightarrow X$ is spectral.*
(b) *Assume that Y is proconstructible. Then*

$$\mathcal{K}(Y) = Y \cap \mathcal{K}(X), \quad \overset{\circ}{\mathcal{K}}(Y) = Y \cap \overset{\circ}{\mathcal{K}}(X), \quad \overline{\mathcal{K}}(Y) = Y \cap \overline{\mathcal{K}}(X),$$

where $Y \cap \mathcal{K}(X)$ denotes $\{Y \cap K \colon K \in \mathcal{K}(X)\}$, etc. The constructible topology on Y is the subspace topology of the constructible topology on X.

Proof The statements are easy to prove with the help of Proposition 3.4.18. The details are left to the reader. □

We finish this section with another useful concept:

Definition 3.4.23 Let X be a spectral space. The set X, endowed with the topology generated by $\overline{\mathcal{K}}(X)$ as basis of open sets is denoted X^* and called the *inverse spectral space* of X.

The terminology is motivated by:

Proposition 3.4.24 *Let X be a spectral space. Then X^* is also a spectral space, and for all $x, y \in X$ we have: $x \succ y$ in $X \Leftrightarrow y \succ x$ in X^*. Furthermore, $\mathcal{K}(X^*) = \mathcal{K}(X)$, $\mathring{\mathcal{K}}(X^*) = \overline{\mathcal{K}}(X)$ and $\overline{\mathcal{K}}(X^*) = \mathring{\mathcal{K}}(X)$, hence also $(X^*)_{con} = X_{con}$ and $X^{**} = X$.*

Proof For all $x, y \in X$ we have: $x \succ y$ in $X \Leftrightarrow y \in Y$ for all $Y \in \overline{\mathcal{K}}(X)$ such that $x \in Y \Leftrightarrow y \succ x$ in X^*. Hence X^* is a T_0 space. Furthermore, $\overline{\mathcal{K}}(X)$ consists precisely of the sets that are open and quasi-compact in X^* since every $Y \in \overline{\mathcal{K}}(X)$ is quasi-compact in X^* by the compactness of X_{con}. This establishes the validity of the spectral space axioms (1) to (3) for X^*. For axiom (4): Let $\varnothing \neq Y \subseteq X$ be closed and irreducible in X^*, and let $\mathcal{Z} := \{Z \in \overline{\mathcal{K}}(X): Y \cap Z \neq \varnothing\}$. For every finite collection of $Z_i \in \mathcal{Z}$ we have $Y \cap Z_1 \cap \cdots \cap Z_r \neq \varnothing$ (irreducibility). Moreover, Y is an intersection of sets from $\mathring{\mathcal{K}}(X)$, and is thus proconstructible in X. By Corollary 3.4.21(b) there is then an element $y \in Y \cap \bigcap_{Z \in \mathcal{Z}} Z$, and y is the generic point for the subspace Y of X^*. We conclude that X^* is a spectral space, and the remaining statements follow immediately. $\qquad\square$

3.5 The Geometric Setting: Semialgebraic Sets and Filter Theorems

The "geometric setting" refers to the case of an affine variety (or rather an affine algebra) over a real closed field. In this context there exists quite a concrete geometric interpretation of Sper A and its constructible subspaces via semialgebraic subsets of $V_A(R)$.

We assume throughout this section that R is a real closed field, $C = R(\sqrt{-1})$, A is an affine R-algebra, $V = V_A = \mathrm{Hom}_R(A, C)$ is its associated variety and $V(R) = V_A(R) = \mathrm{Hom}_R(A, R)$ is the set of real points of the variety. We view $V(R)$, endowed with the strong topology, as a subspace of Sper A, as explained in Sect. 3.3. Given any subset Y of Sper A we may then consider the subset $Y \cap V(R)$ of $V(R)$.

Definition 3.5.1

(1) We let

$$\mathfrak{S}(V(R)) := V(R) \cap \mathcal{K}(\mathrm{Sper}\, A) = \{Y \cap V(R): Y \in \mathcal{K}(\mathrm{Sper}\, A)\}$$

and call the sets $M \in \mathfrak{S}(V(R))$ the *semialgebraic subsets* of $V(R)$. More generally, for $M \in \mathfrak{S}(V(R))$ we denote the set of semialgebraic subsets of M by

$$\mathfrak{S}(M) := M \cap \mathcal{K}(\mathrm{Sper}\, A) = M \cap \mathfrak{S}(V(R)) = \{N \in \mathfrak{S}(V(R)): N \subseteq M\}.$$

Clearly, $\mathfrak{S}(M)$ is a Boolean sublattice of 2^M.

(2) For any semialgebraic subset M of $V(R)$ we let

$$\mathring{\mathfrak{S}}(M) := \big\{N \in \mathfrak{S}(M) : N \text{ is open in } M\big\},$$

$$\overline{\mathfrak{S}}(M) := \big\{N \in \mathfrak{S}(M) : N \text{ is closed in } M\big\},$$

with respect to the strong topology.

Remark 3.5.2 The semialgebraic subsets of $V(R)$ are thus precisely those subsets that can be described by finitely many (equalities and) inequalities. In other words, they are the finite unions of sets of the form

$$\{x \in V(R) : f(x) = 0, \ g_1(x) > 0, \ldots, g_r(x) > 0\},$$

where $r \geq 0$ and $f, g_1, \ldots, g_r \in A$ (cf. Sect. 3.4). In the special case $A = R[t]$, where $V(R) = R$, they are thus precisely the unions of finitely many (possibly degenerate) intervals.

A crucial observation is now that in the transition from a constructible subset Y of Sper A to the semialgebraic subset $Y \cap V(R)$ of $V(R)$ no information is lost. In other words, Y can be recovered! This is again a consequence of the Artin–Lang Theorem:

Theorem 3.5.3 *Let Y be a constructible subset of* Sper A. *Then $Y \cap V(R)$ is dense in Y with respect to the constructible topology (and a fortiori with respect to the Harrison topology).*

Proof It suffices to show that if $Y \neq \varnothing$, then also $Y \cap V(R) \neq \varnothing$. Assume without loss of generality that $Y = Z_A(f) \cap \mathring{H}_A(g_1, \ldots, g_r)$ with $f, g_i \in A$. Let $B := A/Af$ and $W(R) := V_B(R) = \mathrm{Hom}_R(B, R)$ (the real points of the variety associated to B). We must show that $W(R) \cap \mathring{H}_B(\overline{g_1}, \ldots, \overline{g_r}) \neq \varnothing$, where $\overline{g_i}$ denotes the image of g_i in B. By assumption we have $\mathring{H}_B(\overline{g_1}, \ldots, \overline{g_r}) \neq \varnothing$. The statement thus follows from the density of $W(R)$ in Sper B, cf. Theorem 3.3.17. □

Corollary 3.5.4 *Let $x \in$ Sper A. Then $\{x\}$ is constructible if and only if $x \in V(R)$.* □

Corollary 3.5.5 *The map $Y \mapsto Y \cap V(R)$ defines an isomorphism of Boolean lattices $\mathcal{K}(\mathrm{Sper}\, A) \to \mathfrak{S}\big(V(R)\big)$. More generally, if Z is a constructible subset of* Sper A *and $M := Z \cap V(R)$, then the map $Y \mapsto Y \cap V(R)$ defines a lattice isomorphism from $\mathcal{K}(Z)$ to $\mathfrak{S}(M)$.* □

Definition 3.5.6 We use the tilde $\sim$ to denote the map $\mathfrak{S}\big(V(R)\big) \to \mathcal{K}(\mathrm{Sper}\, A)$ which is the inverse of $Y \mapsto Y \cap V(R)$. Thus, for a semialgebraic subset M of $V(R)$, $\widetilde{M}$ denotes the (uniquely determined) constructible subset of Sper A that satisfies

$M = \widetilde{M} \cap V(R)$. In fact, $\widetilde{M}$ is the closure of M in Sper A with respect to the constructible topology.

In this context the following fact is very important but beyond the scope of this book:

Theorem 3.5.7 *Let M and N be semialgebraic subsets of $V(R)$ such that $N \subseteq M$. If N is open in M, then $\widetilde{N}$ is also open in $\widetilde{M}$ (the reverse is trivially true). Hence the following equalities also hold:*

$$\overset{\circ}{\mathfrak{S}}(M) = M \cap \overset{\circ}{\mathcal{K}}(\text{Sper } A) \quad and \quad \overline{\mathfrak{S}}(M) = M \cap \overline{\mathcal{K}}(\text{Sper } A).$$

An equivalent formulation is given by:

Theorem 3.5.8 (Finiteness Theorem) *Let $U, M \subseteq V(R)$ be semialgebraic with $U \subseteq M$ and U open in M. Then U is the union of finitely many sets of the form*

$$\{x \in M : f_1(x) > 0, \ldots, f_r(x) > 0\},$$

where $r \in \mathbb{N}$ and $f_1, \ldots, f_r \in A$.

Example 3.5.9 Let $A = R[t]$, hence $V(R) = R$. We refer to Sect. 3.3 for an explicit description of Sper $R[t]$ and use the natural total order relation $\leq$ on Sper $R[t]$. The images of the intervals $M \subseteq R$ under the map $\sim \colon \mathfrak{S}(V(R)) \to \mathcal{K}(\text{Sper } R[t])$ are as follows:

$$M = [a, b] \quad \rightsquigarrow \quad \widetilde{M} = [a, b],$$
$$M = [a, b[\quad \rightsquigarrow \quad \widetilde{M} = [a, b[= [a, b_-],$$
$$M =]a, b[\quad \rightsquigarrow \quad \widetilde{M} =]a, b[= [a_+, b_-],$$
$$M =]{-}\infty, a[\quad \rightsquigarrow \quad \widetilde{M} = [-\infty, a_-] = [-\infty, a[,$$
$$\vdots$$

where the intervals $\widetilde{M}$ are described in terms of the order relation $\leq$ on Sper $R[t]$. We should pay attention to the fact that there are other, *non*-constructible, subsets $Y \neq \widetilde{M}$ of Sper $R[t]$ such that $M = V(R) \cap Y$ (for example, $Y := [a_-, b_+]$ is such a set for $M = [a, b]$, and is actually also closed). Thus, we only obtain a bijection when restricting ourselves to constructible subspaces of Sper A.

For the remainder of this section we require some additional concepts. Let X be a set and L a sublattice of 2^X (i.e., $\{\emptyset, X\} \subseteq L$ and L is closed under finite intersections and unions).

Definition 3.5.10 Let L be a sublattice of 2^X.

(a) A *filter* of L is a nonempty subset F of L with $\varnothing \notin F$ that satisfies the following properties:

    (1) $A, B \in F \Rightarrow A \cap B \in F$;
    (2) $A \in F, B \in L, A \subseteq B \Rightarrow B \in F$.

(b) A filter F is called a *prime filter* of L if for all $A, B \in L$,

$$A \cup B \in F \Rightarrow A \in F \text{ or } B \in F.$$

(c) A maximal filter of L is called an *ultrafilter*.
(d) The sets of filters, prime filters and ultrafilters of L are denoted Filt(L), Prime(L) and Ultra(L), respectively.

Remarks 3.5.11

(1) If E is a subset of L such that $\bigcap_{A \in E'} A \neq \varnothing$ for all finite $E' \subseteq E$, then E is contained in a filter of L. The smallest such filter is

$$\{B \in L : \text{there exists a finite subset } E' \subseteq E \text{ such that } \bigcap_{A \in E'} A \subseteq B\}.$$

Thus, a filter F is ultra if and only if for all $B \in L \setminus F$ there exists an $A \in F$ such that $A \cap B = \varnothing$. Every ultrafilter is a prime filter.

(2) If L is a *Boolean* sublattice of 2^X (i.e., if $A \in L$, then also $X \setminus A \in L$), then prime filters and ultrafilters in L coincide. These are precisely the filters F such that $A \in F$ or $X \setminus A \in F$, for all $A \in L$.

The following terminology is chosen ad hoc:

Definition 3.5.12 A sublattice L of 2^X is called *well-behaved* if for all subsets E of L and all $B \in L$ such that $\bigcap_{A \in E} A \subseteq B$, there exists a finite subset E' of E such that $\bigcap_{A \in E'} A \subseteq B$. We write

$$\text{pro-}L := \left\{ Y \subseteq X : \text{there exists } E \subseteq L \text{ such that } Y = \bigcap_{A \in E} A \right\}$$

for the lattice of *pro-L sets* in X.

Example 3.5.13 Let X be a spectral space. For example, $X = \text{Sper } A$. Then $\mathcal{K}(X)$, $\overset{\circ}{\mathcal{K}}(X)$ and $\overline{\mathcal{K}}(X)$ are well-behaved sublattices of 2^X. Furthermore, $\mathcal{K}(X)$ is Boolean. We will shortly see that the filters of these lattices offer a different way of describing X.

Lemma 3.5.14 *Let X be a set and L a well-behaved sublattice of 2^X. Then the map*

$$\text{Filt}(L) \rightarrow (\text{pro-}L) \setminus \{\varnothing\}, \quad F \mapsto \bigcap_{A \in F} A$$

is an inclusion reversing bijection. The inverse map is given by

$$Y \mapsto F_Y := \{B \in L : Y \subseteq B\}.$$

Proof It is clear that F_Y is a filter of L for $\varnothing \neq Y \subseteq X$. If $F \in \mathrm{Filt}(L)$ and $Y = \bigcap_{A \in F} A$, then we must show that $F_Y = F$. The inclusion "$\supseteq$" is trivial. The reverse inclusion follows from the fact that L is well-behaved. □

An immediate consequence is:

Corollary 3.5.15 *Let L be a well-behaved sublattice of 2^X. Under the above bijection* $\mathrm{Filt}(L) \to (\mathrm{pro}\text{-}L) \setminus \{\varnothing\}$,

(a) *the prime filters of L correspond to the pro-L sets $\varnothing \neq Y \subseteq X$ that satisfy*

$$A, B \in L, \ Y \subseteq A \cup B \Rightarrow Y \subseteq A \ or \ Y \subseteq B;$$

(b) *the ultrafilters of L correspond to the minimal nonempty pro-L sets.* □

Definition 3.5.16 Let X be a spectral space. We let

$$X^{\max} := \{x \in X : y \in \overline{\{x\}} \Rightarrow y = x\} \quad \text{and} \quad X^{\min} := \{x \in X : x \in \overline{\{y\}} \Rightarrow y = x\}.$$

These are thus the sets of maximal, resp. minimal points with respect to the specialization relation. We observe that $X^{\max}$ is precisely the subset of closed points of X, and that $X^{\min} = (X^*)^{\max}$ and $X^{\max} = (X^*)^{\min}$, where X^* is the inverse spectral space of X, cf. Sect. 3.4.

Now we are ready to describe X in terms of filters of lattices of constructible subsets.

Proposition 3.5.17 *Let X be a spectral space.*

(a) *There exists a canonical inclusion reversing bijection*

$$\mathrm{Filt}\,\overline{\mathcal{K}}(X) \to \{Y \subseteq X : Y \text{ is closed and } Y \neq \varnothing\},$$

namely $F \mapsto \bigcap_{A \in F} A$. Its inverse is given by $Y \mapsto F_Y = \{A \in \overline{\mathcal{K}}(X) : Y \subseteq A\}$.

(b) *There exists a canonical bijection*

$$X \to \mathrm{Prime}\,\overline{\mathcal{K}}(X), \ x \mapsto F_x = \{A \in \overline{\mathcal{K}}(X) : x \in A\}.$$

(c) *The map from (b) induces a bijection*

$$X^{\max} \to \mathrm{Ultra}\,\overline{\mathcal{K}}(X), \ x \mapsto F_x.$$

Proof (a) is a consequence of Lemma 3.5.14 since pro-$\overline{\mathcal{K}}(X) = \{Y \subseteq X : Y$ is closed$\}$. It follows from Corollary 3.5.15 that Prime $\overline{\mathcal{K}}(X)$ corresponds to the closed irreducible subsets $Y \neq \varnothing$ of X under the bijection (a). Upon identification of Y with its generic point we obtain (b) and (c).   □

Interpreting $\overset{\circ}{\mathcal{K}}(X)$ as $\overline{\mathcal{K}}(X^*)$ and $\mathcal{K}(X)$ as $\overline{\mathcal{K}}(X_{\text{con}})$ yields results such as:

Proposition 3.5.18 *Let X be a spectral space. There exists a bijection*

$$X \to \text{Prime} \overset{\circ}{\mathcal{K}}(X), \quad x \mapsto F_x = \{A \in \overset{\circ}{\mathcal{K}}(X) : x \in A\},$$

that induces a bijection $X^{\min} \to \text{Ultra} \overset{\circ}{\mathcal{K}}(X)$.   □

Proposition 3.5.19 *Let X be a spectral space. There exists a bijection*

$$\text{Filt} \, \mathcal{K}(X) \to \{Y \subseteq X : Y \text{ is proconstructible and } Y \neq \varnothing\},$$

namely $F \mapsto \bigcap_{A \in F} A$, with inverse $Y \mapsto F_Y = \{A \in \mathcal{K}(X) : Y \subseteq A\}$, which in turn induces a bijection

$$X \to \text{Prime} \, \mathcal{K}(X) = \text{Ultra} \, \mathcal{K}(X), \quad x \mapsto F_x = \{A \in \mathcal{K}(X) : x \in A\}. \quad \square$$

This last theorem in particular is often used in the geometric setting:

Corollary 3.5.20 *Let R be a real closed field, A an affine R-algebra and $V(R)$ the set of real points of the variety $V = V_A$ associated to A.*

(a) (Ultrafilter Theorem, L. Bröcker, 1981) *There exists a bijection*

$$\text{Sper} \, A \to \text{Ultra} \, \mathfrak{S}\big(V(R)\big), \quad x \mapsto F_x = \{M : x \in \widetilde{M}\}$$

with inverse $F \mapsto \bigcap_{M \in F} \widetilde{M}$.
(b) *There exists a bijection from $\text{Filt} \, \mathfrak{S}\big(V(R)\big)$ to the set of nonempty proconstructible subsets of $\text{Sper} \, A$, namely $F \mapsto \bigcap_{M \in F} \widetilde{M}$.*   □

Let A be an affine algebra over a real closed field R with associated variety $V = V_A$. We illustrate the correspondence between the real spectrum and ultrafilters in several examples below. For $x \in \text{Sper} \, A$, we consider the corresponding ultrafilter

$$F_x = \{M \in \mathfrak{S}\big(V(R)\big) : x \in \widetilde{M}\}.$$

Given a semialgebraic subset $M \subseteq V(R)$ we thus have

$$\widetilde{M} = \{F : M \in F\}$$

upon identification of $\text{Sper} \, A$ with $\text{Ultra} \, \mathfrak{S}\big(V(R)\big)$.

Example 3.5.21 Let $A = R[t]$, hence $V(R) = R$ (cf. Example 3.3.14). The correspondence Sper $A \to$ Ultra $\mathfrak{S}(R)$ is as follows:

for $x \in$ Sper A,	F_x consists of all M such that	
$x = c \in R$	$c \in M$	
$x = c_- \ (c \in R)$	$]c - \varepsilon, c[\subseteq M$ for some $\varepsilon > 0$	
$x = c_+ \ (c \in R)$	$]c, c + \varepsilon[\subseteq M$ for some $\varepsilon > 0$	
$x = -\infty$	$]-\infty, a[\subseteq M$ for some $a \in R$	
$x = +\infty$	$]a, \infty[\subseteq M$ for some $a \in R$	
$x = \xi$ (free cut)	$\exists a, b$ such that $a < \xi < b$ and $]a, b[\subseteq M$	

Example 3.5.22 If we view the elements of Sper A as ultrafilters of $\mathfrak{S}(V(R))$, the associated orderings of A are easy to describe. For $x \in$ Sper A and $f \in A$ we have

$$f(x) \geq 0 \Leftrightarrow \text{there exists } M \in F_x \text{ such that } f|_M \geq 0,$$
$$f(x) > 0 \Leftrightarrow \text{there exists } M \in F_x \text{ such that } f|_M > 0,$$
$$\vdots$$

The support $\mathfrak{p} = \mathrm{supp}(x)$ of x is described as follows: the subvariety $W := V_{A/\mathfrak{p}}$ of V associated to $\mathfrak{p}$ is the smallest closed subvariety of V such that $W(R) \in F_x$; in other words,

$$W(R) = \bigcap_{W'} W'(R),$$

where the intersection is over all closed subvarieties W' of V such that $W'(R) \in F_x$.

Example 3.5.23 Let $x, y \in$ Sper A. Then

$$x \succ y \ (\text{i.e., } y \in \overline{\{x\}}) \Leftrightarrow F_x \cap \overline{\mathfrak{S}}(V(R)) \subseteq F_y \Leftrightarrow F_y \cap \mathring{\mathfrak{S}}(V(R)) \subseteq F_x.$$

If $M \subseteq V(R)$ is semialgebraic and $x \in \tilde{M}$, then $x \in \tilde{M}^{\max}$ if and only if for every $N \in \mathfrak{S}(M)$ with $N \in F_x$ there exists an $N' \in F_x$ which is closed in M and such that $N' \subseteq N$.

Example 3.5.24 Let $A = R[x, y]$. Consider the homomorphisms

$$\varphi_0 \colon A \to R, \qquad\qquad f \mapsto f(0, 0)$$
$$\varphi_1 \colon A \to R(x), \qquad\quad f \mapsto f(x, 0)$$
$$\varphi_2 \colon A \hookrightarrow R(x, y), \qquad f \mapsto f.$$

Assume that the ordering of $R(x, y)$ is such that x compared to R and y compared to $R(x)$ are infinitely small and positive, and assume that $R(x)$ is endowed with the induced ordering. Let $\alpha_i \in \operatorname{Sper} A$ be the point defined by φ_i (and the indicated orderings) for $i = 0, 1, 2$ (cf. Remark 3.3.2(3)). We have

$$\operatorname{supp}(\alpha_2) = (0) \subseteq \operatorname{supp}(\alpha_1) = (y) \subseteq \operatorname{supp}(\alpha_0) = (x, y),$$

and it is not difficult to see that $\alpha_2 \succ \alpha_1 \succ \alpha_0$ in $\operatorname{Sper} A$. The associated ultrafilters are described as follows. For semialgebraic $M \subseteq R^2$ we have

$$\alpha_0 \in \tilde{M} \iff \alpha_0 = (0, 0) \in M;$$

$$\alpha_1 \in \tilde{M} \iff \exists \varepsilon > 0 \text{ such that }]0, \varepsilon[\times \{0\} \subseteq M;$$

$$\alpha_2 \in \tilde{M} \iff \exists \varepsilon > 0, \ \exists g \in R[t] \text{ with } g(t) > 0 \text{ for } t \in \,]0, \varepsilon[\,,$$

$$\text{such that } \{(a, b) \in R^2 \colon 0 < a < \varepsilon, \, 0 < b < g(a)\} \subseteq M.$$

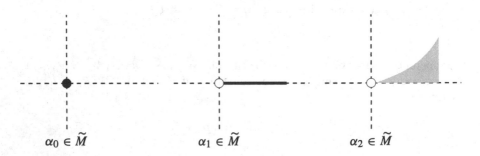

$$\alpha_0 \in \tilde{M} \qquad\qquad \alpha_1 \in \tilde{M} \qquad\qquad \alpha_2 \in \tilde{M}$$

(The verification only requires some effort in the case of α_2. Since $\alpha_2 \in (\operatorname{Sper} A)^{\min}$, we have: $\alpha_2 \in \tilde{M} \iff$ there exist $f_1, \ldots, f_n \in A$ such that $\alpha_2 \in \mathring{H}_A(f_1, \ldots, f_n)$ and $\bigcap_i \{x \in R^2 \colon f_i(x) > 0\} \subseteq M$. Then one needs to consider that for every $f \in A$ with $f(\alpha_2) > 0$, the set $\{x \in R^2 \colon f(x) > 0\}$ contains a set of the form described above.)

We can picture α_1 as the bud of the positive branch of the x-axis at the origin, and α_2 as the tip of the upper shoreline of the half-plane to the right of the x-axis. Thus we see that $\alpha_0 = (0, 0)$ has a chain of generalizations of (maximal possible) length 2. It is a straightforward exercise (cf. Example 3.7.7) to show that every point $x \in R^n$ has a chain of generalizations of length n in $\tilde{R}^n$. (In more generality the analogous statement holds for smooth R-varieties.)

Example 3.5.25 Assume that A is an integral domain and let $K = \text{Quot}\,A$. The orderings of K can be visualized in $V(R)$ as follows.

We call a semialgebraic subset $M \subseteq V(R)$ *thin* if M is not Zariski dense in $\text{Spec}\,A$. (If we define the dimension of a semialgebraic subset as the dimension of its Zariski closure in $\text{Spec}\,A$, then M is thin if and only if $\dim M < \dim A = \text{tr.deg.}(K/R) = \dim V$.)

Let $x \in \text{Sper}\,A$. Then $\text{supp}(x) = 0$ (i.e., $x \in \text{Sper}\,K \subseteq \text{Sper}\,A$) if and only if F_x contains no thin sets. Expressed in a different way: there is an equivalence relation on $\mathfrak{S}(V(R))$ defined by

$$M \sim N :\Leftrightarrow M \bigtriangleup N := (M \setminus N) \cup (N \setminus M) \text{ is thin,}$$

and the quotient set $\mathfrak{S}(V(R))/ \sim$ carries a lattice structure induced by $\mathfrak{S}(V(R))$. The ultrafilters of $\mathfrak{S}(V(R))/ \sim$ then correspond precisely to those ultrafilters of $\mathfrak{S}(V(R))$ that do not contain any thin sets. This proves:

Proposition 3.5.26 (G.W. Brumfiel) *There exists a canonical bijection*

$$\text{Sper}\,K \to \text{Ultra}\big(\mathfrak{S}(V(R))/ \sim\big). \qquad \square$$

Remark 3.5.27 We have shown that for every spectral space X there exists a natural bijection $X \to \text{Prime}\,\overset{\circ}{\mathcal{K}}(X)$. In this way it is possible to "recover" the set X from the lattice $\overset{\circ}{\mathcal{K}}(X)$. In fact this relationship goes much further. Indeed, since we can determine the topology of X from $\overset{\circ}{\mathcal{K}}(X)$, (the topological space) X and (the distributive lattice) $\overset{\circ}{\mathcal{K}}(X)$ are essentially the same. We give a short sketch (without proof) of this duality.

If L is a distributive lattice (with 0 and 1), then the set $\text{Prime}\,L$ carries a natural topology in which the sets $U_x := \{F \in \text{Prime}\,L : x \in F\}$ ($x \in L$) form an open basis. This topological space is called the Stone space $\text{St}(L)$ of L. It is a spectral space and L is (via $x \mapsto U_x$) canonically isomorphic to $\overset{\circ}{\mathcal{K}}(\text{St}(L))$. Conversely, if X is a spectral space, then X is canonically homeomorphic to $\text{St}\big(\overset{\circ}{\mathcal{K}}(X)\big)$ (we did describe the bijection). The following holds (see for instance [43, p. 103] or [56, p. 65 ff.]):

Stone Duality *The assignments $X \mapsto \overset{\circ}{\mathcal{K}}(X)$ and $L \mapsto \text{St}(L)$ determine a duality (i.e., anti-equivalence) between the category of spectral spaces (and spectral maps) and the category of distributive lattices.*

3.6 The Space of Closed Points

This section deals with properties of real spectra that are quite false for general spectral spaces. As usual, A denotes an arbitrary commutative unitary ring.

Proposition 3.6.1 *If X is a quasi-compact T_0 space and $X^{\max}$ is its subspace of closed points, then $X^{\max}$ is quasi-compact, and for all $x \in X$ we have $\overline{\{x\}} \cap X^{\max} \neq \varnothing$.*

Proof It suffices to show that $\overline{\{x\}} \cap X^{\max} \neq \varnothing$ for $x \in X$, since this implies that X is the only neighbourhood of $X^{\max}$ in X. Thus, let $x \in X$ and let $Y := \overline{\{x\}}$. Then Y is ordered via $y \leq y' \Leftrightarrow y \succ y'$. If M is a totally ordered subset of Y, then $\bigcap_{y \in M} \overline{\{y\}} \neq \varnothing$ since Y is quasi-compact. By Zorn's Lemma, Y contains a maximal element. $\qquad\square$

Remarks 3.6.2

(1) If X is spectral, the above statement already follows from Proposition 3.5.17. After passing to the inverse spectral space X^*, we see that $X = \bigcup_{x \in X^{\min}} \overline{\{x\}}$ (but $X^{\min}$ is in general not quasi-compact!).
(2) By Corollary 3.4.21(c) the subspace $X^{\max}$ of a spectral space X is proconstructible if and only if it is closed. Usually this is not the case however. For example, if R is a real closed field and A an affine R-algebra with associated variety V, then we know from Sect. 3.5 that $V(R)$ is already dense in Sper A, hence a fortiori in $(\text{Sper } A)^{\max}$.
(3) Let R be a real closed field and $X = \text{Sper } R[t]$. Then $X^{\max}$ consists of $R \cup \{-\infty, +\infty\}$ together with all free Dedekind cuts of R. In the special case that $R = \mathbb{R}$, the space $X^{\max} = \mathbb{R} \cup \{\pm\infty\}$ is the natural compactification (with two end points) of $\mathbb{R}$.

We have reached the point mentioned at the start of the chapter, where the real spectra of rings exhibit special features that general spectral spaces do not. Despite its simplicity, the consequences of the next lemma are far reaching!

Lemma 3.6.3 *Let P and Q be orderings of A. The following statements are equivalent:*

(i) *$P \not\subseteq Q$ and $Q \not\subseteq P$ (i.e., P and Q are incomparable with respect to specialization);*
(ii) *there exists $f \in A$ such that $P \in \mathring{H}_A(f)$ and $Q \in \mathring{H}_A(-f)$;*
(iii) *there exist open $U, V \subseteq \text{Sper } A$ such that $U \cap V = \varnothing$ and $P \in U, Q \in V$.*

Proof (i) $\Rightarrow$ (ii): Choose $a \in P \setminus Q$ and $b \in Q \setminus P$. Then $f := a - b$ yields the desired property. Indeed, arguing by contradiction, if $-f \in P$, then also $b = a - f \in P$, and if $f \in Q$, then also $a = b + f \in Q$. Either conclusion is false.

The implications (ii) $\Rightarrow$ (iii) $\Rightarrow$ (i) are trivial. $\qquad\square$

Theorem 3.6.4 *Let X be a proconstructible subspace of Sper A. Then $X^{\max}$ is Hausdorff, hence compact.*

Proof This follows immediately from Lemma 3.6.3 and Proposition 3.6.1. $\qquad\square$

Remark 3.6.5 Theorem 3.6.4 is in general completely false for the Zariski spectrum. For example, if A is an affine $\mathbb{C}$-algebra with associated variety $V = V(\mathbb{C})$, then $(\operatorname{Spec} A)^{\max} = V(\mathbb{C})$ carries the (induced) Zariski topology, which is profoundly non-Hausdorff whenever $\dim A > 0$.

Remark 3.6.6 Let R be a real closed field and A an affine R-algebra with associated variety $V = V_A$. It follows from the density of $V(R)$ in $(\operatorname{Sper} A)^{\max}$ that $(\operatorname{Sper} A)^{\max}$ is a natural compactification of $V(R)$.

If $x_1, \ldots, x_n$ is a system of generators of A and $\alpha \in (\operatorname{Sper} A)^{\max}$, then $k(\alpha)$ is archimedean over R if and only if there exists $c \in R$ such that $x_1(\alpha)^2 + \cdots + x_n(\alpha)^2 < c$, as we will see later. (The condition is clearly equivalent to $A/\operatorname{supp}(\alpha)$ being archimedean over R; we will see in Sect. 3.7 that $k(\alpha)$ is archimedean over $A/\operatorname{supp}(\alpha)$ for all $\alpha \in (\operatorname{Sper} A)^{\max}$.)

We call those $\alpha \in (\operatorname{Sper} A)^{\max}$ for which $k(\alpha)/R$ is archimedean the *finite points* and all the remaining $\alpha \in (\operatorname{Sper} A)^{\max}$ the *infinitely distant points* of $(\operatorname{Sper} A)^{\max}$. The set of finite points thus forms an *open* dense subspace of $(\operatorname{Sper} A)^{\max}$ (equal to $(\operatorname{Sper} A)^{\max} \cap \bigcup_{c \in R} \mathring{H}_A(c - x_1^2 - \cdots - x_n^2)$). This is interesting in the particular case $R = \mathbb{R}$, since the set of finite points is already equal to $V(\mathbb{R})$ in this case (cf. Sect. 2.1). Thus we see that $V(\mathbb{R})$ is contained as an open dense subspace in its compactification $(\operatorname{Sper} A)^{\max}$. This fact has been used by G.W. Brumfiel, who gave an interpretation of the Thurston compactification of Teichmüller spaces by means of the real spectrum, cf. [20].

A further and equally fundamental consequence is

Theorem 3.6.7 *Let $x \in \operatorname{Sper} A$. Then $\overline{\{x\}}$ forms a chain with respect to specialization. In other words, for all $y, z \in \operatorname{Sper} A$ we have*

$$x \succ y \text{ and } x \succ z \Rightarrow y \succ z \text{ or } z \succ y.$$

Corollary 3.6.8 *Let $X \subseteq \operatorname{Sper} A$ be proconstructible and let $x \in X$. Then $\overline{\{x\}} \cap X^{\max}$ contains just one element.* □

Proof of Theorem 3.6.7 For any neighbourhoods U of y and V of z we have $x \in U \cap V$. Conclude with Lemma 3.6.3. □

Remark 3.6.9 For every $x \in \operatorname{Sper} A$ we can visualize $\overline{\{x\}}$ as a "spear", with the closed specialization of x as its tip. The notation Sper for real spectrum alludes to this picture and also to the French term "spectre réel".

Remark 3.6.10 A spectral space X is said to be *completely normal* if it has the property of Theorem 3.6.7, i.e., if for any $x, y, z \in X$ with $y, z \in \overline{\{x\}}$ one has $y \in \overline{\{z\}}$ or $z \in \overline{\{y\}}$. Real spectra of rings are completely normal spectral spaces (Theorem 3.6.7). As mentioned before Definition 3.4.15, Hochster [51] showed that every spectral space is homeomorphic to the Zariski spectrum of some (commutative) ring. In view of this result it is natural to ask whether every completely normal spectral space is homeomorphic to the real spectrum of some ring. However, this turns out not to be true since Delzell and Madden [34] constructed a counter-example (of combinatorial dimension one). See also Marshall [80, Sect. 8.8]. As a comprehensive work on all aspects of spectral spaces we recommend [35]. (Note that in op. cit. the term "spectral root system" is used instead of "completely normal space".)

Remark 3.6.11 As mentioned in Sect. 3.5 we will show in Sect. 3.7 that every point $x_0 \in R^n$ is the tip of a spear

$$x_n \succ x_{n-1} \succ \cdots \succ x_1 \succ x_0 \qquad (x_i \neq x_{i-1}, \ i = 1, \ldots, n)$$

of length n in $\widetilde{R^n} = \operatorname{Sper} R[t_1, \ldots, t_n]$. Conversely, any point $\alpha \in \operatorname{Sper} R[t_1, \ldots, t_n]$ with $\operatorname{supp}(\alpha) = (0)$ has at most n distinct specializations. This follows easily from Theorem 3.6.7 and some results from commutative algebra. (One considers the support ideals of the specializations; they form a chain.) The spear $\overline{\{\alpha\}}$ can indeed be *shorter*, for two reasons in fact. First of all, the closed specialization of α need not be in R^n (e.g., $\alpha = \pm\infty$ in $\widetilde{R}$). Secondly, the spear can have "gaps". We illustrate this for $\mathbb{R}^2$ in the following example.

Example 3.6.12 Let

$$P := \{f \in \mathbb{R}[x, y] : \exists \, \varepsilon > 0 \text{ such that } f(t, e^t - 1) \geq 0 \text{ for } 0 < t < \varepsilon\}.$$

Then P is an ordering of $\mathbb{R}[x, y]$ with $\operatorname{supp}(P) = P \cap (-P) = (0)$, and it is clear that $(0, 0) \in \mathbb{R}^2$ is the closed specialization of P in $\operatorname{Sper} \mathbb{R}[x, y]$. We should now convince ourselves that $\mathbb{R}(x, y)$ contains *just one* nontrivial convex subring (with respect to the ordering induced by P), namely the convex hull $o(\mathbb{R}(x, y)/\mathbb{R})$ of $\mathbb{R}$. This implies (as we will show in Sect. 3.7) that P actually has *just one* proper specialization. Heuristically this is pretty clear: were P to have a one-dimensional specialization, its support could only be the curve $y + 1 = e^x$, which is not an algebraic curve.

Proposition 3.6.13 ("Normality" of $\operatorname{Sper} A$**)** *If Y and Z are closed disjoint subsets of $\operatorname{Sper} A$, there exist open quasi-compact neighbourhoods U of Y and V of Z such that $U \cap V = \varnothing$.*

Proof If $X \subseteq \operatorname{Sper} A$ is closed and $y \in \operatorname{Sper} A$ is such that $\overline{\{y\}} \cap X = \varnothing$, then every $x \in X$ has an open quasi-compact neighbourhood U_x such that $\overline{\{y\}} \cap U_x = \varnothing$. Since X is quasi-compact, X is covered by finitely many U_x, and so there exists an open quasi-compact $U \subseteq \operatorname{Sper} A$ such that $X \subseteq U$ and $\overline{\{y\}} \cap U = \emptyset$. Assume now that $\{U_i : i \in I\}$, resp. $\{V_j : j \in J\}$, is the system of open quasi-compact neighbourhoods of Y, resp. Z. Then by the above, $\bigcap_i U_i = \{x \in \operatorname{Sper} A : \overline{\{x\}} \cap Y \neq \emptyset\}$ and similarly for V_j. It follows that

$$\bigcap_{i \in I} U_i \cap \bigcap_{j \in J} V_j = \emptyset.$$

Indeed, if the intersection were to contain an element x with closed specialization $\overline{x}$, then we would have $\overline{x} \in Y \cap Z$, a contradiction. By Corollary 3.4.10 there exist finite $I' \subseteq I$ and $J' \subseteq J$ such that for $U := \bigcap_{i \in I'} U_i$ and $V := \bigcap_{j \in J'} V_j$ we already have $U \cap V = \emptyset$. $\qquad\square$

Remark 3.6.14 The proof shows that a spectral space X satisfies the normality property from Proposition 3.6.13 only if $|\overline{\{x\}} \cap X^{\max}| = 1$ for all $x \in X$. The converse is also true, cf. [23]. In the same way Propositions 3.6.16 and 3.6.17 remain valid for these so-called *normal spectral spaces*. Note that the property from Theorem 3.6.7 (which is always satisfied for $\operatorname{Sper} A$) is even more specific.

Definition 3.6.15 Let $X \subseteq \operatorname{Sper} A$ be proconstructible. The *canonical retraction* from X to $X^{\max}$ is the map $\rho = \rho_X \colon X \to X^{\max}$ that is uniquely determined by $\rho(x) \in \overline{\{x\}} \cap X^{\max}$ for all $x \in X$. Note that $\rho^2 = \rho$.

Proposition 3.6.16 *Let $X \subseteq \operatorname{Sper} A$ be proconstructible. The canonical retraction $\rho \colon X \to X^{\max}$ is a continuous and closed map.*

Proof Let $Y \subseteq X$ be closed. Then $\rho(Y) = Y \cap X^{\max}$ is closed in $X^{\max}$. To prove the continuity, let $Y' \subseteq X^{\max}$ be closed in $X^{\max}$, say $Y' = Y \cap X^{\max}$ for a closed $Y \subseteq X$. Then

$$\rho^{-1}(Y') = \{x \in X : \overline{\{x\}} \cap Y \neq \emptyset\} = \bigcap_{i \in I} U_i$$

is the intersection of all quasi-compact open neighbourhoods of Y (see the proof of Proposition 3.6.13). Thus $\rho^{-1}(Y')$ is in particular proconstructible and hence closed since $\rho^{-1}(Y')$ is stable under specialization. $\qquad\square$

Proposition 3.6.17 *Let $X \subseteq \operatorname{Sper} A$ be proconstructible. Then $Y \mapsto \rho(Y)$ is a bijection from the set of connected components Y of X to the set of connected components Y' of $X^{\max}$. The inverse map is given by $Y' \mapsto \rho^{-1}(Y')$.*

Proof It suffices to show that every closed $Y \subseteq X$ for which $\rho(Y)$ is connected is itself connected. Assume that $Y = Y_1 \cup Y_2$ with Y_1, Y_2 closed and $Y_1 \cap Y_2 = \emptyset$. Then also $\rho(Y_1) \cap \rho(Y_2) = \varnothing$ and $\rho(Y_1), \rho(Y_2)$ are closed in $X^{\max}$ by Proposition 3.6.16, from which the assertion follows.   □

We recognize now that the spaces $X^{\max}$ (with $X \subseteq \operatorname{Sper} A$ proconstructible) are topologically pleasing, not only because they are compact, but also because they retain some information about the space X. For example, one can use ρ to reduce the cohomology of X to the cohomology of $X^{\max}$; both have isomorphic cohomology, cf. [23]. In the geometric setting it seems important to treat $\widetilde{M}^{\max}$ as the natural compactification of $M \subseteq V(R)$. In spite of everything however, $\operatorname{Sper} A$ should be considered as the central object, and $(\operatorname{Sper} A)^{\max}$ rather as a useful auxiliary space.

3.7   Specializations and Convex Ideals

As usual, we assume that A is an arbitrary (commutative, unitary) ring. Let $P \subseteq A$ be an ordering of A and $\mathfrak{p} := \operatorname{supp}(P)$. We denote the natural homomorphism $A \to A/\mathfrak{p}$ by π and write $\overline{P}$ for the ordering of $\kappa(\mathfrak{p})$ induced by P. When studying the ideals of $A/\mathfrak{p}$ that are convex with respect to $\overline{P}$, the following concept presents itself:

Definition 3.7.1 An ideal $\mathfrak{a}$ of A is called *P-convex* if for all $f, g \in P$,

$$f + g \in \mathfrak{a} \Rightarrow f, g \in \mathfrak{a}.$$

Lemma 3.7.2 *The map* $\mathfrak{a} \mapsto \pi(\mathfrak{a})$ *defines a bijection from the set of P-convex ideals of A to the set of ideals of $A/\mathfrak{p}$ that are convex with respect to $\overline{P}$. The inverse is given by* $\overline{\mathfrak{a}} \mapsto \pi^{-1}(\overline{\mathfrak{a}})$.

Proof We endow $A/\mathfrak{p}$ with the total order relation induced by $\overline{P}$. Since we have $f \in P$ and $-f \in P$ for every $f \in \mathfrak{p}$, it follows that $\mathfrak{p}$ is contained in every P-convex ideal of A. This shows that the two described maps are inverse to each other. If $\mathfrak{a} \subseteq A$ is a P-convex ideal, then $\pi(\mathfrak{a})$ is convex in $A/\mathfrak{p}$. Indeed, if $a \in \mathfrak{a}$ and $f \in A$ such that $0 \le \pi(f) \le \pi(a)$, then $f \in P$ and $a - f \in P$, hence $f \in \mathfrak{a}$ by assumption. Conversely, if $\overline{\mathfrak{a}} \subseteq A/\mathfrak{p}$ is a convex ideal and if $f, g \in P$ are such that $f + g \in \mathfrak{a} := \pi^{-1}(\overline{\mathfrak{a}})$, then $0 \le \pi(f) \le \pi(f + g) \in \overline{\mathfrak{a}}$, and thus also $\pi(f) \in \overline{\mathfrak{a}}$.   □

Corollary 3.7.3

(a) *The P-convex ideals form a chain, and $\mathfrak{p} = \operatorname{supp}(P)$ is the smallest ideal in it.*
(b) *For every P-convex ideal $\mathfrak{a}$ of A we have $P + \mathfrak{a} = P \cup \mathfrak{a}$.*

Proof (a) See Remark 2.1.4(4).
(b) Let $p \in P$ and $a \in \mathfrak{a}$. If $p + a \notin P$, then $-(p+a) \in P$, and $-a = p - (p+a) \in \mathfrak{a}$ implies $p + a \in \mathfrak{a}$.   □

Theorem 3.7.4 *Let $P \in \operatorname{Sper} A$ be an ordering of A and $\mathfrak{p} = \operatorname{supp}(P)$. Then the following sets are in canonical bijection:*

(1) $\overline{\{P\}}$ *(the set of specializations Q of P in $\operatorname{Sper} A$);*
(2) *the set of P-convex prime ideals $\mathfrak{q}$ of A;*
(3) *the set of prime ideals $\overline{\mathfrak{q}}$ of $A/\mathfrak{p}$ that are convex with respect to $\overline{P}$.*

The bijection (1) $\to$ (2) is given by $Q \mapsto \operatorname{supp}(Q)$, its inverse by $\mathfrak{q} \mapsto P + \mathfrak{q} = P \cup \mathfrak{q}$. The bijection (2) $\to$ (3) is given by $\mathfrak{q} \mapsto \pi(\mathfrak{q})$, its inverse by $\overline{\mathfrak{q}} \mapsto \pi^{-1}(\overline{\mathfrak{q}})$.

Proof The bijection between (2) and (3) was established in Lemma 3.7.2. Let $Q \supseteq P$ be an ordering of A and $\mathfrak{q} := Q \cap (-Q)$. Let $f, g \in P$ such that $f + g \in \mathfrak{q}$. Then $f \in \mathfrak{q}$ since $-f = g - (f + g) \in Q$. Hence $\mathfrak{q}$ is P-convex and $Q = P \cup \mathfrak{q}$ (to see $\subseteq$: if $f \in Q \setminus \mathfrak{q}$, then $-f \notin Q$ and so $-f \notin P$ which yields $f \in P$). Conversely, let $\mathfrak{q}$ be a P-convex prime ideal. Then $Q := P + \mathfrak{q} = P \cup \mathfrak{q}$ is an ordering of A with support $\mathfrak{q}$ since $Q \cap (-Q) = (P \cap -P) \cup (P \cap \mathfrak{q}) \cup (\mathfrak{q} \cap -P) \cup \mathfrak{q} = \mathfrak{q}$ is a prime ideal. $\qquad\square$

Once more we see that $\overline{\{P\}}$ forms a chain.

Example 3.7.5 Let $\lambda \colon K \to L \cup \infty$ be a surjective place with valuation ring $B = \mathfrak{o}_\lambda$. Upon identifying $\operatorname{Sper} K$, resp. $\operatorname{Sper} L$, with those points of $\operatorname{Sper} B$ whose support is (0), resp. $\mathfrak{m}_B$, the following holds for all $x \in \operatorname{Sper} K$: λ is compatible with x if and only if x has a specialization y with support $\mathfrak{m}_B$; when this happens, y is the ordering of $L \cong B/\mathfrak{m}_B$ induced by x in the sense of Sect. 2.2. As a consequence we can formulate the Baer–Krull Theorem (cf. Sect. 2.7) as follows:

Let Γ be the value group of λ and let $y \in \operatorname{Sper} B$ with $\operatorname{supp}(y) = \mathfrak{m}_B$. Then $\widehat{\Gamma/2\Gamma}$ acts transitively and free on the (nonempty) set of generalizations of y with support (0).

Thus, if $\Gamma \cong \mathbb{Z}^n_{\text{lex}}$ and if $(0) = \mathfrak{p}_0 \subset \cdots \subset \mathfrak{p}_n = \mathfrak{m}_B$ are the prime ideals of B, then every $y \in \operatorname{Sper} B$ with $\operatorname{supp}(y) = \mathfrak{m}_B$ has precisely 2^{n-i} generalizations with support $\mathfrak{p}_i$ for $i = 0, \ldots, n$.

Definition 3.7.6 Let B be a valuation ring and A a subring of B. The prime ideal $A \cap \mathfrak{m}_B$ of A is called the *centre of B on A*. Similarly, given a place $\lambda \colon K \to L \cup \infty$ and a subring A of K on which λ is finite, we call the kernel of the homomorphism $\lambda|_A \colon A \to L$ the *centre of λ on A*.

Example 3.7.7 Let $\lambda \colon K \to L \cup \infty$ be a place and A a subring of K on which λ is finite and has centre $\mathfrak{p}$. The homomorphisms $A \subseteq \mathfrak{o}_\lambda \xrightarrow{\overline{\lambda}} L$ induce maps $\operatorname{Sper} L \to \operatorname{Sper} \mathfrak{o}_\lambda \to \operatorname{Sper} A$ whose composition we denote by λ^*. Then it follows from Example 3.7.5 that for every $z \in \operatorname{Sper} L$ the element $\lambda^*(z) \in \operatorname{Sper} A$ (with support $\mathfrak{p}$) has a generalization with support (0) in $\operatorname{Sper} A$.

In more generality, let $\lambda = \lambda_n \circ \cdots \circ \lambda_1 \colon K \to L \cup \infty$ be a composition of places which is finite on $A \subseteq K$ and has centre $\mathfrak{p}$, and let $\mathfrak{p}_i$ be the centre of $\lambda_i \circ \cdots \circ \lambda_1$ on A for $i = 0, \ldots, n$. Then $(0) = \mathfrak{p}_0 \subseteq \mathfrak{p}_1 \subseteq \cdots \subseteq \mathfrak{p}_n = \mathfrak{p}$, and for every $z \in$

Sper L, $\lambda^*(z)$ has a chain of generalizations with the $\mathfrak{p}_i$ as supports, i.e., there exist $y_0, \ldots, y_n \in$ Sper A with $\text{supp}(y_i) = \mathfrak{p}_i$ such that $y_0 \succ y_1 \succ \cdots \succ y_n = \lambda^*(z)$.

This shows by way of example that for a real closed field R every point $a \in R^n$ has chains of generalizations of length n in Sper $R[t_1, \ldots, t_n]$: namely, in Sect. 2.10 we constructed a place $\lambda = \lambda_n \circ \cdots \circ \lambda_1 \colon R(t_1, \ldots, t_n) \to R \cup \infty$ over R for which $\lambda_i \circ \cdots \circ \lambda_1$ has centre $(t_n - a_n, \ldots, t_{n-i+1} - a_{n-i+1})$ on $R[t_1, \ldots, t_n]$ for $i = 0, \ldots, n$.

Definition 3.7.8 A valuation ring B is called *real closed* if Quot B is a real closed field and B is residually real. (The second condition is equivalent to: B is convex in Quot B, cf. Sect. 2.5.)

Proposition 3.7.9 *If B is a real closed valuation ring, then* supp: Sper $B \to$ Spec B *is a homeomorphism, and $\kappa(\mathfrak{p})$ is real closed for all $\mathfrak{p} \in$ Spec B (thus, $k(\alpha) = \kappa(\text{supp}\,\alpha)$ for all $\alpha \in$ Sper B).*

Proof Let $\mathfrak{p} \in$ Spec B. Then B, hence also $B_\mathfrak{p}$, is convex in $R := $ Quot B and it follows that $\kappa(\mathfrak{p}) = B_\mathfrak{p}/\mathfrak{p}B_\mathfrak{p} = \kappa(B_\mathfrak{p})$ is real closed by Theorem 2.5.1. The bijectivity of supp follows. The map supp is also open: Let $f \in B$. If $f < 0$ (in R), then $\overset{\circ}{H}_B(f) = \varnothing$. Otherwise there exists $g \in B$ such that $f = g^2$ and it follows that $\overset{\circ}{H}_B(f) = \{\alpha \in$ Sper $B \colon f(\alpha) \neq 0\}$, hence supp $\overset{\circ}{H}_B(f) = D_B(f)$.   □

Now consider a real closed field R and a subring $A \subseteq R$. We denote the convex hull of a subset $M \subseteq R$ in R by M^c. In particular, let $B := A^c$ denote the convex hull of A. Let $P_0 := \{a \in A \colon a \geq 0\} \in$ Sper A. The P_0-convex ideals of A are simply the convex ideals of A (with respect to the total order relation induced by R).

Let $\mathfrak{a} \subseteq A$ be an ideal. Then $\mathfrak{a}^c$ is an ideal of B and $\mathfrak{a}$ is P_0-convex if and only if $\mathfrak{a} = A \cap \mathfrak{a}^c$. Thus $\mathfrak{b} \mapsto A \cap \mathfrak{b}$ defines a surjection

$$\{\text{ideals of B}\} \longrightarrow \{P_0\text{-convex ideals of } A\}.$$

If $\mathfrak{p} \subseteq A$ is a P_0-convex *prime* ideal, then $\mathfrak{p}^c$ is not necessarily prime. Nevertheless the following holds:

Lemma 3.7.10 *The map*

$$\text{Spec } B \to \{P_0\text{-}convex\ prime\ ideals\ of A\}, \quad \mathfrak{q} \mapsto A \cap \mathfrak{q}$$

is surjective.

Proof If $\mathfrak{p}$ is a P_0-convex prime ideal of A, then $\sqrt{\mathfrak{p}^c}$ is a prime ideal of B by Proposition 2.4.17(e), and we have $A \cap \sqrt{\mathfrak{p}^c} = \sqrt{\mathfrak{p}} = \mathfrak{p}$.   □

Corollary 3.7.11 *If C is a convex overring of A in R, then $A \cap \mathfrak{m}_C$ is a P_0-convex prime ideal of A and the image of the map $\operatorname{Spec} C \to \operatorname{Spec} A$ consists precisely of those P_0-convex prime ideals $\mathfrak{p}$ of A that satisfy $\mathfrak{p} \subseteq A \cap \mathfrak{m}_C$.*

Proof This is clear since $C = B_{\mathfrak{m}_C}$, cf. Sect. 2.2.   □

Returning to an arbitrary ring A, Theorem 3.7.4 and Proposition 3.7.9 allow us to reformulate the above facts:

Proposition 3.7.12 *Let $\varphi\colon A \to R$ be a homomorphism into a real closed field, B the convex hull of $\varphi(A)$ in R and $\varphi_0\colon A \to B$ the induced homomorphism. Then the image of $\operatorname{Sper} \varphi_0$ consists precisely of the specializations of α_φ in $\operatorname{Sper} A$, where α_φ denotes the element in $\operatorname{Sper} A$ defined by φ (cf. Remark 3.3.2(3)).*

Proof Let β_0 be the generic point of $\operatorname{Sper} B$. Then $\alpha_\varphi = (\operatorname{Sper} \varphi_0)(\beta_0)$ and so $\operatorname{Im}(\operatorname{Sper} \varphi_0) \subseteq \overline{\{\alpha_\varphi\}}$. Let $P = \varphi^{-1}[0, \infty[_R$ be the ordering of A corresponding to α_φ. Given any P-convex prime ideal $\mathfrak{q}$ of A there exists $\beta \in \operatorname{Sper} B$ such that $\operatorname{supp}((\operatorname{Sper} \varphi_0)(\beta)) = \mathfrak{q}$ by Lemma 3.7.10 and Proposition 3.7.9. The statement then follows by Theorem 3.7.4.   □

Proposition 3.7.13 *Let $\varphi\colon A \to B$ be a homomorphism into a real closed valuation ring B, let β_0 be the generic point and β_1 the closed point of $\operatorname{Sper} B$, and let $\alpha_i = (\operatorname{Sper} \varphi)(\beta_i)$ for $i = 0, 1$. Then the image of $\operatorname{Sper} \varphi$ is precisely the "interval" $\{\alpha \in \operatorname{Sper} A\colon \alpha_0 \succ \alpha \succ \alpha_1\}$.*

Proof Similar to the proof of Proposition 3.7.12, using Corollary 3.7.11.   □

Let A and B be arbitrary commutative rings. If $\varphi\colon A \to B$ is an *integral* homomorphism (i.e., B is integral over $\varphi(A)$), it is well-known that $\operatorname{Spec} \varphi$ is a closed map. (This is a reformulation of the so-called "Going Up" Theorem of Cohen–Seidenberg which follows immediately from Theorem 2.3.5, cf. [13, Ch. V, §2, no. 1] or [68, p. 49].) As an application of the previous results we show that the same holds for $\operatorname{Sper} \varphi$:

Proposition 3.7.14 *If $\varphi\colon A \to B$ is an integral ring homomorphism, then $\operatorname{Sper} \varphi$ is a closed map.*

Proof Let $\varphi^* := \operatorname{Sper} \varphi$. If $Y \subseteq \operatorname{Sper} B$ is closed, then $\varphi^*(Y)$ is proconstructible and it suffices to show stability under specialization. Thus, let $\beta \in \operatorname{Sper} B$ and $\alpha = \varphi^*(\beta)$. We must show that $\varphi^*\overline{\{\beta\}} = \overline{\{\alpha\}}$. Let $r_\beta\colon B \to k(\beta)$, $b \mapsto b(\beta)$ denote the evaluation map associated to β and C the convex hull of $r_\beta \varphi(A)$ in $k(\beta)$. Since B is integral over A, hence also $r_\beta(B)$ integral over $r_\beta \varphi(A)$, we have $r_\beta(B) \subseteq C$. We define $\psi\colon A \to C$ and $r\colon B \to C$ via the commutative diagram

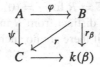

Then $r^*(\mathrm{Sper}\, C) = \overline{\{\beta\}}$ and $\psi^*(\mathrm{Sper}\, C) = \overline{\{\alpha\}}$ by Proposition 3.7.12, from which $\varphi^*\overline{\{\beta\}} = \overline{\{\alpha\}}$ follows.   □

We want to make the statements of Lemma 3.7.10 and Proposition 3.7.12 a bit more precise. Lemma 3.7.10 can also be formulated as follows: for every P_0-convex prime ideal $\mathfrak{p}$ of A there exists a convex overring C of A with centre $\mathfrak{p}$ on A. More accurately:

Proposition 3.7.15 *Let R be a real closed field, $A \subseteq R$ a subring and $P = A \cap [0, \infty[_R$. If $\mathfrak{p} \subseteq A$ is a P-convex prime ideal, then $C := (A_{\mathfrak{p}})^c$ is the smallest and $C_{\mathfrak{c}}$, where $\mathfrak{c} := \sqrt{\mathfrak{p}^c}$, the largest convex overring of A in R with centre $\mathfrak{p}$ on A.*

Proof Let $B \subseteq R$ be a convex overring of A with centre $\mathfrak{p}$ on A. Then $A_{\mathfrak{p}} \subseteq B$, thus also $C = (A_{\mathfrak{p}})^c \subseteq B$, and $B = C_{\mathfrak{q}}$ for some prime ideal $\mathfrak{q}$ of C. Since $\mathfrak{p} \subseteq \mathfrak{m}_B = \mathfrak{q}$ it must also be the case that $\mathfrak{p}^c \subseteq \mathfrak{q}$ and $\mathfrak{c} = \sqrt{\mathfrak{p}^c} \subseteq \mathfrak{q}$, i.e., $B \subseteq C_{\mathfrak{c}}$ (note that $\mathfrak{c}$ is a prime ideal of C, cf. Lemma 3.7.10).   □

Corollary 3.7.16 *Let A be an arbitrary commutative ring. Given $\alpha, \beta \in \mathrm{Sper}\, A$ with $\alpha \succ \beta$, there exists a homomorphism $\varphi \colon A \to B$ into a real closed valuation ring B such that the image of $\mathrm{Sper}\, \varphi$ is precisely the "interval"*

$$\{\gamma \in \mathrm{Sper}\, A : \alpha \succ \gamma \succ \beta\}.$$

Proof Let $\mathfrak{q} := \mathrm{supp}\, \beta$, $A' := r_\alpha(A)$ and $\mathfrak{q}' := r_\alpha(\mathfrak{q})$. It suffices to take for B the convex hull of $A'_{\mathfrak{q}'}$ in $k(\alpha)$ (and for φ the map induced by $r_\alpha \colon A \to k(\alpha)$).   □

Corollary 3.7.17 ("Convexity" of $\mathrm{Sper}\, \varphi$) *Let $\varphi \colon A' \to A$ be a homomorphism of commutative rings and let $\alpha, \beta \in \mathrm{Sper}\, A$ such that $\alpha \succ \beta$. Then the image of the interval $\{\gamma \in \mathrm{Sper}\, A : \alpha \succ \gamma \succ \beta\}$ in $\mathrm{Sper}\, A'$ is again an interval, i.e., given $\gamma' \in \mathrm{Sper}\, A'$ such that*

$$(\mathrm{Sper}\, \varphi)(\alpha) \succ \gamma' \succ (\mathrm{Sper}\, \varphi)(\beta),$$

there exists a preimage γ of γ' such that $\alpha \succ \gamma \succ \beta$.

Proof This follows immediately from Corollary 3.7.16 and Proposition 3.7.13.   □

Corollary 3.7.18 *Let $\alpha \in \mathrm{Sper}\, A$. Then α is closed in $\mathrm{Sper}\, A$ if and only if $k(\alpha)$ is archimedean over $r_\alpha(A)$.*

Proof We may assume without loss of generality that $\mathrm{supp}\, \alpha = (0)$, i.e., that $R := k(\alpha)$ is the real closure of $K := \mathrm{Quot}\, A$. Since R is archimedean over K (cf. Proposition 1.7.1), we have $C = R$ for $\mathfrak{p} = (0)$ in Proposition 3.7.15, i.e., every convex overring B of A which is distinct from R induces a proper specialization of α. Hence, $\overline{\{\alpha\}} = \{\alpha\}$ if and only if there are no such overrings.   □

Since convex valuation rings of a real closed field R and real places $R \to S \cup \infty$ are essentially the same thing, the previous statements can also be formulated in the language of places. Let us give an example.

Let R and S be real closed fields, $\varphi: A \to R$ a homomorphism and $\lambda: R \to S \cup \infty$ a place which is finite on $\varphi(A)$. Then the element $\alpha_{\lambda \circ \varphi}$ is a specialization of α_φ in Sper A, and for every specialization of α_φ there is such a place λ. More precisely:

Definition 3.7.19 A homomorphism $\varphi: A \to R$ into a real closed field is called *tight* if R is archimedean over Quot $\varphi(A)$.

Proposition 3.7.20 *Let $\varphi: A \to R$ be a homomorphism into a real closed field. For every specialization β of α_φ there exists a surjective place $\lambda = \lambda_\beta: R \to S \cup \infty$ to a real closed field S which is finite on $\varphi(A)$ and such that $\lambda \circ \varphi$ is tight and $\alpha_{\lambda \circ \varphi} = \beta$. If γ is another specialization of α_φ, then: $\beta \succ \gamma \Leftrightarrow \lambda_\gamma$ factors through λ_β:*

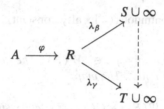

Proof This is just a reformulation of the results already obtained.   □

3.8   The Real Spectrum and the Reduced Witt Ring of a Field

In the previous section we made a detailed study of the specialization chains (the "spears") in the real spectrum of a ring. Visually speaking these are transversal to the fibres of the support map supp: Sper $A \to$ Spec A since every spear has at most one point in common with such a fibre. The fibres are precisely the real spectra of the residue fields of A (cf. Corollary 3.3.12). Thus, in order to gain a better understanding of the spectral space Sper A, a deeper study of the real spectrum of fields suggests itself. In this section we will initiate this study.

Throughout, F denotes a field, without loss of generality assumed to be a real. For the sake of brevity we write X_F instead of Sper F. The spectral space X_F is Hausdorff, thus compact and totally disconnected, its (Harrison) topology coincides with the constructible topology, and the constructible subsets of X_F are precisely the clopen subsets (cf. Remark 3.3.2(4) and Sect. 3.4). If $a_1, \ldots, a_n$ are elements of F^*, we simply write $H_F(a_1, \ldots, a_n)$ or $H(a_1, \ldots, a_n)$ instead of $\mathring{H}_F(a_1, \ldots, a_n) = \overline{H}_F(a_1, \ldots, a_n)$.

Thus $X_F = \mathrm{Sper}\, F$ is not particularly interesting as a topological space. However, a stronger structure is provided by the field elements, or rather by the sign distributions on X_F that they determine. In order to utilize this structure we return to the theory of quadratic forms, which has shown itself to be an invaluable tool for real algebra and geometry. One should then also consider quadratic forms over more general (commutative) rings. In this section we will just cover a number of elementary observations about the interplay of quadratic forms and the real spectrum of a field.

We remind ourselves that every ordering P of F defines a ring homomorphism $\mathrm{sign}_P \colon W(F) \to \mathbb{Z}$, the signature at P, (cf. Sect. 1.2).

Definition 3.8.1 For $\varphi \in W(F)$ the map

$$\mathrm{sign}\,\varphi \colon X_F \to \mathbb{Z}, \quad P \mapsto \mathrm{sign}_P(\varphi)$$

is called the *total signature* of φ.

We denote the ring of continuous (= locally constant) functions from X_F to $\mathbb{Z}$ by $C(X_F, \mathbb{Z})$.

Proposition 3.8.2 *Consider any $\varphi \in W(F)$. The total signature of φ is a continuous function from X_F to $\mathbb{Z}$. The map*

$$\mathrm{sign} \colon W(F) \to C(X_F, \mathbb{Z}), \quad \varphi \mapsto \mathrm{sign}\,\varphi$$

(the total signature of F) is a ring homomorphism with kernel equal to the nilradical of $W(F)$.

Proof Since all sign_P are homomorphisms, the map $\varphi \mapsto \mathrm{sign}\,\varphi$ is a homomorphism from $W(F)$ to the ring of all maps $X_F \to \mathbb{Z}$. Given $\varphi = \langle a \rangle$ with $a \in F^*$, the continuity of $\mathrm{sign}\,\varphi$ is clear. It follows that $\mathrm{sign}\,\varphi$ is continuous for all $\varphi \in W(F)$. The kernel of sign is the intersection of the kernels of all the sign_P, and these are precisely the minimal prime ideals of $W(F)$ (cf. Sect. 1.4). The statement follows.
$\square$

Definition 3.8.3 We call

$$\overline{W}(F) := W(F)_{\mathrm{red}} = W(F)/\mathrm{Nil}\,W(F)$$

the *reduced Witt ring* of F and denote the image of the fundamental ideal $I(F)$ in $\overline{W}(F)$ by $\overline{I}(F) = I(F)/\mathrm{Nil}\,W(F)$.

Given $\varphi \in W(F)$, we denote its image in $\overline{W}(F)$ by $\overline{\varphi}$. By Proposition 3.8.2 we may identify $\overline{W}(F)$ canonically with a subring of $C(X_F, \mathbb{Z})$, and usually this is what we will do. Under this identification we have $\overline{\varphi} = \mathrm{sign}\,\varphi$.

We denote the characteristic function of a subset $Y \subseteq X_F$ by χ_Y. Given a 2-dimensional form $\varphi = \langle 1, a \rangle$ with $a \in F^*$, we then have $\overline{\varphi} = 2 \cdot \chi_{H(a)}$. Since $I(F)$

is generated as an additive group by the forms $\langle 1, a \rangle$ (note: $\langle 1, a \rangle \perp \langle -1, b \rangle \sim \langle a, b \rangle$), it follows that in $C(X_F, \mathbb{Z})$, $\overline{I}(F)$ is generated as an additive group by the functions $2 \cdot \chi_{H(a)}$ with $a \in F^*$. We observe that $\overline{W}(F) = \mathbb{Z} + \overline{I}(F)$ and so $\overline{W}(F) \subseteq \mathbb{Z} + 2 \cdot C(X_F, \mathbb{Z})$.

Definition 3.8.4 An *n-fold Pfister form* over F is a quadratic form that can be written as $\langle 1, a_1 \rangle \otimes \cdots \otimes \langle 1, a_n \rangle$ with $a_i \in F^*$.

Let $\varphi = \langle 1, a_1 \rangle \otimes \cdots \otimes \langle 1, a_n \rangle$. Then clearly, $\overline{\varphi} = 2^n \cdot \chi_{H(a_1, \ldots, a_n)}$. Since $I(F)$ is additively generated by the 1-fold Pfister forms, the n-th power $I^n(F) := I(F)^n$ is additively generated by the n-fold Pfister forms. Hence, $\overline{I}^n(F) := \overline{I}(F)^n$ is generated by the functions $2^n \cdot \chi_{H(a_1, \ldots, a_n)}$ with $a_1, \ldots, a_n \in F^*$.

Lemma 3.8.5 *Let Y be a constructible subset of X_F. Then there exists a finite sequence $a_1, \ldots, a_n$ of elements in F^* such that Y is the (disjoint) union of sets of the form $H(\varepsilon_1 a_1, \ldots, \varepsilon_n a_n)$ with $\varepsilon_i \in \{\pm 1\}$.*

Proof There exist finite subsets $S_1, \ldots, S_r$ of F^* such that $Y = H(S_1) \cup \cdots \cup H(S_r)$. Let $\{a_1, \ldots, a_n\} := S_1 \cup \cdots \cup S_r$. Since the elements of S_i have constant sign on each set $H(\varepsilon_1 a_1, \ldots, \varepsilon_n a_n)$, every $H(S_i)$ (and thus also Y) is a union of such sets. $\square$

Proposition 3.8.6 *The abelian group $C(X_F, \mathbb{Z})/\overline{W}(F)$, i.e., the cokernel of the total signature of F, is a 2-primary torsion group.*

Proof Every element of $C(X_F, \mathbb{Z})$ can be written in the form $m_1 \chi_{Y_1} + \cdots + m_r \chi_{Y_r}$ with $r \geq 1$, $m_i \in \mathbb{Z}$ and $Y_i \subseteq X_F$ constructible. It follows from Lemma 3.8.5 that $C(X_F, \mathbb{Z})$ is generated as an additive group by the $f = \chi_{H(a_1, \ldots, a_n)}$ with $n \geq 1$ and $a_i \in F^*$. For such an f we have $2^n f \in \overline{W}(F)$. The statement follows. $\square$

Definition 3.8.7 The smallest integer $s \geq 0$ such that $2^s \cdot C(X_F, \mathbb{Z}) \subseteq \overline{W}(F)$ (or $s = \infty$ if there is no such integer) is called the *stability index* of F and is denoted $\mathrm{st}(F)$.

In real algebra and geometry the stability index has turned out to be a significant invariant of real fields. In many important cases $\mathrm{st}(F)$ is finite and often lends itself to computation.

Since $\overline{W}(F) \subseteq \mathbb{Z} + 2 \cdot C(X_F, \mathbb{Z})$, the condition $\mathrm{st}(F) = 0$ is clearly equivalent to the existence of a (unique) ordering of F. We continue with a characterization of those fields for which $\mathrm{st}(F) \leq 1$.

Lemma 3.8.8 *If $Y \subseteq X_F$ is constructible and if $2 \cdot \chi_Y \in \overline{W}(F)$, then there exists an $a \in F^*$ such that $Y = H(a)$.*

Proof Let $\varphi = \langle a_1, \ldots, a_{2n} \rangle$ be a form such that $\overline{\varphi} = 2 \cdot \chi_Y$ (note that φ then necessarily has even dimension). Then $a := a_1 \cdots a_{2n}$ has sign $(-1)^{n-1}$ on Y and sign $(-1)^n$ on $X_F \setminus Y$. Hence $Y = H(a)$ or $Y = H(-a)$. $\square$

Definition 3.8.9 A real field F is called a *SAP field* if, given any two disjoint closed subsets Y and Y' of X_F, there exists an $a \in F^*$ such that $a|_Y > 0$ and $a|_{Y'} < 0$. (SAP stands for "strong approximation property", which is unrelated to the number theoretic concept of strong approximation.)

Proposition 3.8.10 *Let F be a real field. The following statements are equivalent:*

(i) $\mathrm{st}(F) \leq 1$;
(ii) *every constructible subset of X_F is of the form $H(a)$ for some $a \in F^*$;*
(iii) $\overline{W}(F) = \mathbb{Z} + 2 \cdot C(X, \mathbb{Z})$;
(iv) *F is a SAP field.*

Proof (i) $\Rightarrow$ (ii): If $Y \subseteq X_F$ is constructible, then $2 \cdot \chi_Y \in \overline{W}(F)$ by (i) and so $Y = H(a)$ for some $a \in F^*$ by Lemma 3.8.8.

(ii) $\Rightarrow$ (iii): For every constructible $Y \subseteq X_F$ we have $2 \cdot \chi_Y \in \overline{W}(F)$. It follows that $2 \cdot C(X_F, \mathbb{Z}) \subseteq \overline{W}(F)$.

(iii) $\Rightarrow$ (i): This is trivial.

(ii) $\Leftrightarrow$ (iv): Let $Y, Y' \subseteq X_F$ be closed and disjoint. Then there exists a constructible neighbourhood Z of Y such that $Z \cap Y' = \varnothing$. Every $a \in F^*$ such that $Z = H(a)$ separates Y and Y'. Conversely, (ii) follows for a constructible $Y \subseteq X_F$ upon applying the SAP property to the sets Y and $Y' := X_F \setminus Y$. □

To convey a better feeling for the stability index we state the following two theorems without proof:

Theorem 3.8.11 (Special case of [15, Satz 4.8]) *Let R be a real closed field and K a real n-dimensional function field over R. Then $\mathrm{st}(K) = n$.* □

Theorem 3.8.12 (Special case of [9, Satz 15]) *Let F be a real field and $n \in \mathbb{N}$ such that $n \geq \mathrm{st}(F)$. For every $(n + 1)$-fold Pfister form φ over F there exists an $(n$-fold$)$ Pfister form ψ over F such that $\overline{\varphi} = 2\overline{\psi}$.* □

It is not difficult to deduce new characterizations of $\mathrm{st}(F)$ from Theorem 3.8.12. Since we do not prove the theorem, nor have any use for its corollary below in the remainder of this book, we leave the proof of the corollary as an exercise to the reader.

Corollary 3.8.13 *Let F be a real field and $s \in \mathbb{N}$. The following statements are equivalent:*

(i) $\mathrm{st}(F) \leq s$;
(ii) *every set $H(a_1, \ldots, a_N)$ with $a_i \in F^*$ and $N \geq 1$ can be written in the form $H(b_1, \ldots, b_s)$ with $b_j \in F^*$;*
(iii) *for every $(s + 1)$-fold Pfister form φ over F there exists an $(s$-fold$)$ Pfister form ψ over F such that $\overline{\varphi} = 2\overline{\psi}$;*
(iv) $\overline{I}^{s+1}(F) = 2\overline{I}^s(F)$. □

The term stability index for $\text{st}(F)$ comes from characterization (iv): $\text{st}(F)$ is the smallest index from which point onwards the sequence $(\bar{I}^n(F) : n = 1, 2, \ldots)$ is stable in the sense of (iv). (This is a K-theoretic notion of stability, cf. [83].)

In the final part of this section we turn our attention to finitely generated field extensions. If $F \subseteq K$ is a field extension we write $r_{K/F}$ for the restriction map from $X_K := \text{Sper } K$ to $X_F = \text{Sper } F$. We consider finite extensions first:

Proposition 3.8.14 *Let $K \supseteq F$ be a field extension of degree $n < \infty$. Then:*

(a) *$r := r_{K/F}: X_K \to X_F$ is an open (and closed) map, i.e., the images of constructible sets are constructible.*

(b) *For each $x \in X_F$, the fibre $r^{-1}(x)$ is finite and its cardinality t satisfies $t \le n$ and $t \equiv n \bmod 2$.*

(c) *Let $U_t := \{x \in X_F : \#r^{-1}(x) = t\}$ for $t \ge 0$. Then every U_t is constructible and r is topologically trivial over every U_t (i.e., there exists a decomposition of $r^{-1}(U_t)$ into t disjoint open subsets $W_1, \ldots, W_t$ such that $r|_{W_i}$ is a homeomorphism from W_i to U_t for every $i = 1, \ldots, t$).*

(d) *The map $f: X_F \to \mathbb{Z}, x \mapsto \#r^{-1}(x)$ is an element of $\overline{W}(F)$.*

Proof Consider the trace form $\varphi := \text{tr}_*(\langle 1 \rangle_K)$ associated to the extension K/F. Then $f = \overline{\varphi}$ by Corollary 1.12.10, which proves (d) as well as the constructibility of the U_t. Furthermore, (b) follows immediately since $\dim \varphi = n$.

Let us prove (a) next: r is closed since it is a continuous map between compact spaces. In addition, $r(X_K) = U_1 \cup \cdots \cup U_n$ is constructible in X_F. Let $a_1, \ldots, a_m \in K^*$. Then $H_K(a_1, \ldots, a_m) = r_{L/K}(X_L)$ for $L := K(\sqrt{a_1}, \ldots, \sqrt{a_m})$ (cf. Proposition 1.3.3). Consequently, $r_{K/F}(H_K(a_1, \ldots, a_m)) = r_{K/F}(r_{L/K}(X_L)) = r_{L/F}(X_L)$ is also constructible in X_F. Since these $H_K(a_1, \ldots, a_m)$ form a basis of the topology of X_K, (a) follows.

It remains to prove the second statement of (c). Let $t \ge 0$ and fix an $x \in U_t$. If $r^{-1}(x) = \{y_1, \ldots, y_t\}$, there exist pairwise disjoint constructible neighbourhoods W_i' of the y_i in $r^{-1}(U_t)$ for $i = 1, \ldots, t$. Then $V(x) := r(W_1') \cap \cdots \cap r(W_t')$ is a constructible neighbourhood of x in U_t and the restriction of r to every $W_i(x) := W_i' \cap r^{-1}(V(x))$ is a homeomorphism to $V(x)$. Then we cover U_t with finitely many $V(x_i)$ $(i = 1, \ldots, N)$ and make this covering disjoint by shrinking the $V(x_i)$ if necessary. For every $1 \le i \le N$ we have $r^{-1}(V(x_i)) = \bigcup_{j=1}^{t} W_j(x_i)$ (disjoint union), and $r|_{W_j}(x_i)$ is a homeomorphism from $W_j(x_i)$ to $V(x_i)$. Finally, letting $W_j := \bigcup_{i=1}^{N} W_j(x_i)$ $(j = 1, \ldots, t)$, the sets $W_1, \ldots, W_t$ yield the desired result. $\qquad \square$

With help from the sign change criterion (Theorem 2.12.8), we can generalize statement (a) from Proposition 3.8.14 to finitely generated field extensions:

Proposition 3.8.15 (R. Elman, T.Y. Lam, A. Wadsworth [38]) *Let $K \supseteq F$ be a finitely generated field extension. Then $r_{K/F}: X_K \to X_F$ is an open (and closed) map.*

Proof The argument used in the proof of Proposition 3.8.14(a) shows that it suffices to prove that $r_{K/F}(X_K)$ is open in X_F. Let $\alpha_1, \ldots, \alpha_d$ be a transcendence basis of K over F, and let $\beta \in K$ be such that $K = F(\alpha_1, \ldots, \alpha_d, \beta)$ (Primitive Element Theorem). If $p \in F[t_1, \ldots, t_d, t_{d+1}] = F[t]$ is an irreducible polynomial such that $p(\alpha_1, \ldots, \alpha_d, \beta) = 0$, then the field of fractions of $F[t]/(p)$ is isomorphic over F to K (cf. Proposition 2.12.9).

Consider $x \in r_{K/F}(X_K)$ and let $k(x)$ be the associated real closure of K. By the sign change criterion (Theorem 2.12.8) there exist $a, b \in k(x)^{d+1}$ such that $p(a) < 0 < p(b)$. Let $L = F(a, b)$ be the finite extension of F that is generated over F by the coordinates of a and b in $k(x)$. Let x' be the restriction of the ordering of $k(x)$ to L. Then $x' \in H_L(-p(a), p(b))$. By Proposition 3.8.14, $Y := r_{L/F}\big(H_L(-p(a), p(b))\big)$ is a constructible neighbourhood of x in X_F. The sign change criterion then yields $Y \subseteq r_{K/F}(X_K)$. Indeed, if $y' \in H_L(-p(a), p(b))$ and $y = r_{L/F}(y') \in Y$, then $k(y) = k(y')$ is an overfield of L, and p is indefinite over $k(y)$ by the choice of y'. This implies that y has an extension to K, as required.

$\square$

Remark 3.8.16 Coste and Roy [28] established a far-reaching generalization of Proposition 3.8.14(a) by means of model theoretic arguments (quantor elimination in the theory of real closed fields): if $\varphi \colon A \to B$ is a ring homomorphism of finite presentation (i.e., B is finitely generated as an A-algebra and $\operatorname{Ker}\varphi$ is a finitely generated ideal of A), then $\operatorname{Sper}\varphi \colon \operatorname{Sper} B \to \operatorname{Sper} A$ is an open map with respect to the constructible topology. Using this result one can also obtain another proof of Proposition 3.8.15.

3.9 Preorderings of Rings and Positivstellensätze

Let A be ring. The purpose of this section is to investigate the following question: let $X \subseteq \operatorname{Sper} A$ be an intersection of sets of the form $\{f > 0\}$, $\{f \geq 0\}$ and $\{f = 0\}$ (where $f \in A$), and explicitly represented in this manner. Determine those functions in A that are positive (resp. are nonnegative, resp. vanish) on X.

Definition 3.9.1 A *preordering of* A is a subset $T \subseteq A$ that satisfies the following properties:

(RP1) $T + T \subset T, TT \subset T$ (i.e., T is a subsemiring of A);
(RP2) $a^2 \in T$ for all $a \in A$.

If in addition

(RP3) $-1 \notin T$

holds, then T is called a *proper* preordering of A. Otherwise T is an *improper* preordering of A.

Examples and Remarks 3.9.2

(1) Preorderings of fields in the sense of Definition 1.1.4 are now called proper preorderings. This slight inconsistency should not lead to confusion.

(2) Preimages of (proper) preorderings under homomorphisms are again (proper) preorderings; images of preorderings under surjective homomorphisms are again preorderings.

(3) Every ordering of A is a proper preordering of A. Arbitrary intersections and upward directed unions of preorderings are again preorderings. The smallest preordering of A is ΣA^2. Thus A has a proper preordering if and only if A is real.

(4) If $1/2 \in A$, then $T = A$ is the only improper preordering of A (since $4a = (a+1)^2 - (a-1)^2$).

(5) For a subset $F \subseteq A$ we denote the preordering generated by F (in A) by $P[F]$. Then $P[F]$ consists of all finite sums of elements of the form $a \cdot f_1 \cdots f_n$, where $a \in \Sigma A^2$, $n \geq 0$ and $f_1, \ldots, f_n \in F$ (without loss of generality assumed to be pairwise distinct). In particular we observe that

$$\overline{H}_A(F) = \overline{H}_A(P[F])$$

(indeed, the inclusion $\supseteq$ is trivial and the inclusion $\subseteq$ follows from the description).

(6) For fields k with char $k \neq 2$ (in particular real fields), Σk^2 is the intersection of all orderings of k (cf. Sect. 1.1). For rings this is false in general. For example, $A = \mathbb{R}[t_1, \ldots, t_n]$ is a real reduced ring and if $n \geq 2$, then ΣA^2 is not equal to the intersection of all orderings. Indeed, letting $K = \operatorname{Quot} A$, we can identify $(\operatorname{Sper} A)^{\min}$ with $\operatorname{Sper} K$ (we did not fully prove this, but compare with Example 3.7.7) and the intersection of all orderings of A is thus equal to $A \cap \Sigma K^2$, the positive semidefinite polynomials. But these are not always sums of squares in A when $n \geq 2$ (cf. Sect. 2.12).

As for fields we nevertheless have:

Proposition 3.9.3 *Every proper preordering of A is contained in an ordering.*

We first prove:

Lemma 3.9.4 *Let $T \subseteq A$ be a proper preordering and $a \in A$. Then:*

(a) $(aT) \cap (1+T) = \varnothing$ *or* $(-aT) \cap (1+T) = \varnothing$;
(b) *if* $(aT) \cap (1+T) = \varnothing$, *then* $T - aT$ *is also a proper preordering of* A.

Proof (a) Assume that $as = 1 + s'$ and $-at = 1 + t'$ for certain $s, s', t, t' \in T$. Then

$$-a^2 st = 1 + s' + t' + s't',$$

hence $-1 = a^2 st + s' + t' + s't' \in T$, a contradiction.

(b) The statements $-1 \in T - aT$ and $aT \cap (1+T) \neq \varnothing$ are equivalent. $\square$

Proof of Proposition 3.9.3 By Zorn's Lemma it suffices to show that maximal proper preorderings are orderings. Thus, let T be a maximal proper preordering. Then $T \cup (-T) = A$ by Lemma 3.9.4. Hence $\mathfrak{p} := T \cap (-T)$ is an ideal of A. Assume for the sake of contradiction that $\mathfrak{p}$ is not a prime ideal. Then there exist $a, b \in A \setminus \mathfrak{p}$ such that $ab \in \mathfrak{p}$. We may take $a, b \in T$, hence $a, b \notin -T$. By Lemma 3.9.4(b) we then have $(aT) \cap (1 + T) \neq \varnothing$ and $(bT) \cap (1 + T) \neq \varnothing$, say

$$as = 1 + s' \quad \text{and} \quad bt = 1 + t'$$

for certain $s, s', t, t' \in T$. Multiplication gives

$$abst = 1 + s' + t' + s't',$$

hence $-1 = (-ab)st + s' + t' + s't' \in T$, a contradiction. □

The following theorems describe those functions that are positive (semi)definite on sets of the form $\overline{H}_A(F)$:

Proposition 3.9.5 *Let F be a subset of A and $T := P[F]$ the preordering it generates. For any $a \in A$, the following statements are equivalent:*

(i) $a > 0$ *on* $\overline{H}_A(F)$ *(i.e., $a(\alpha) > 0$ for all $\alpha \in \overline{H}_A(F)$);*
(ii) *there exist $t, t' \in T$ such that $at = 1 + t'$;*
(iii) *there exist $t, t' \in T$ such that $a(1 + t) = 1 + t'$.*

Proof (i) $\Rightarrow$ (ii): If $-1 \in T$, then $\overline{H}_A(F) = \overline{H}_A(T) = \varnothing$ and we may choose $t = 0$, $t' = -1$. Thus, let $-1 \notin T$. Assume that $(aT) \cap (1 + T) = \varnothing$. By Lemma 3.9.4 and Proposition 3.9.3 there exists an ordering P of A such that $P \supseteq T - aT \supseteq T$. Then $a(P) \leq 0$ and $P \in \overline{H}_A(F)$, a contradiction.

(ii) $\Rightarrow$ (iii): From $at = 1 + t'$ we get $a(1 + t') = a^2 t$ and addition gives $a(1 + t + t') = 1 + (a^2 t + t')$.

(iii) $\Rightarrow$ (i): For any $\alpha \in \overline{H}_A(F)$ we have $t(\alpha) \geq 0$, $t'(\alpha) \geq 0$, hence $a(\alpha) > 0$. □

Corollary 3.9.6 *Let F be a subset of A and $T := P[F]$. For any $a \in A$, the following statements are equivalent:*

(i) *a does not vanish on $\overline{H}_A(F)$;*
(ii) *a divides an element of $1 + T$.*

Proof For (i) $\Rightarrow$ (ii) we may replace a by a^2 and conclude with Proposition 3.9.5, while (ii) $\Rightarrow$ (i) is clear. □

Corollary 3.9.7 *Let R be a real closed field, $A = R[t_1, \ldots, t_n]$ and*

$$B := \{f/g : f, g \in A, \ g(x) \neq 0 \text{ for all } x \in R^n\} \subseteq \text{Quot } A$$

the ring of regular functions on R^n. Then $B = S^{-1}A$, where $S := 1 + \Sigma A^2$. (An analogous result holds for an arbitrary affine R-algebra A with $V_A(R)$ instead of R^n.)

Proof of $\subseteq$ If $g \in A$ is such that $g(x) \neq 0$ on R^n, then g does not vanish on Sper A since $Z_A(g)$ is constructible (cf. Theorem 3.5.3). By Corollary 3.9.6 there exist $s \in S$ and $a \in A$ such that $ag = s$, i.e., such that $\frac{1}{g} = \frac{a}{s}$. $\qquad\square$

Proposition 3.9.8 *Let F be a subset of A and $T := P[F]$. For any $a \in A$, the following statements are equivalent:*

(i) $a \geq 0$ on $\overline{H}_A(F)$;
(ii) *there exist* $t, t' \in T$ *and* $n \geq 0$ *such that* $at = a^{2n} + t'$;
(iii) *there exist* $t, t' \in T$ *and* $n \geq 0$ *such that* $a(a^{2n} + t) = a^{2n} + t'$;
(iv) *there exist* $t, t' \in T$ *and* $n \geq 0$ *such that* $a(a^{2n} + t) = t'$.

Proof (i) $\Rightarrow$ (ii): Let $B := a^{-\infty}A$ and $U := \{t/a^{2n} : n \geq 0, t \in T\}$ the preordering generated by T in B. We identify Sper B with $\mathring{H}_A(a^2) \subseteq$ Sper A (see Sect. 3.3). Since a is positive on $\overline{H}_B(U)$, there exist $m, n \geq 0$ and $t, t' \in T$ such that

$$a \cdot \frac{t}{a^{2m}} = 1 + \frac{t'}{a^{2n}} \quad \text{in } B$$

by Proposition 3.9.5. We may assume without loss of generality that $m = n$, and it follows that

$$at = a^{2n} + t' \quad \text{in } B,$$

i.e.,

$$a \cdot a^{2N}t = a^{2(N+n)} + a^{2N}t' \quad \text{in } A$$

for some $N \geq 0$. This is a relation of the form (ii).

(ii) $\Rightarrow$ (iii): From $at = a^{2n} + t'$ it follows again that $a(a^{2n} + t') = a^2t$, hence $a(a^{2n} + t + t') = a^{2n} + (a^2t + t')$ by addition. This is a relation of the form (iii). The implications (iii) $\Rightarrow$ (iv) and (iv) $\Rightarrow$ (i) are clear. $\qquad\square$

Next we consider generalizations of these theorems that describe those functions that are positive (semi)definite on a class of proconstructible sets that is more general than sets of the form $X = \overline{H}_A(F)$.

Let $F, G, E \subseteq A$ be subsets and consider the proconstructible subset

$$X := \mathring{H}_A(F) \cap \overline{H}_A(G) \cap Z_A(E),$$

of Sper A, i.e.,

$$X = \{\alpha \in \text{Sper } A : f(\alpha) > 0 \; \forall f \in F, \; g(\alpha) \geq 0 \; \forall g \in G, \; e(\alpha) = 0 \; \forall e \in E\}.$$

(Since $Z_A(a) = \overline{H}_A(-a^2) = \overline{H}_A(a) \cap \overline{H}_A(-a)$ the contribution of $Z_A(E)$ is not needed for the description of X, but it often does make the representation easier.) Let S be the semigroup generated by F in A (and note that $1 \in S$), $T := P[F \cup G]$ the preordering generated by F and G, and $\mathfrak{a} := \sum_{e \in E} Ae$ the ideal generated by E.

Theorem 3.9.9 (Positivstellensatz) *For any $a \in A$, the following statements are equivalent:*

(i) $a > 0$ *on* X;
(ii) $\exists s \in S, \exists t, t' \in T$ *such that* $at \equiv s + t' \mod \mathfrak{a}$;
(iii) $\exists s \in S, \exists t, t' \in T$ *such that* $a(s + t) \equiv s + t' \mod \mathfrak{a}$.

Theorem 3.9.10 (Nichtnegativstellensatz) *For any $a \in A$, the following statements are equivalent:*

(i) $a \geq 0$ *on* X;
(ii) $\exists s \in S, \exists t, t' \in T, \exists n \geq 0$ *such that* $at \equiv a^{2n}s + t' \mod \mathfrak{a}$;
(iii) $\exists s \in S, \exists t, t' \in T, \exists n \geq 0$ *such that* $a(a^{2n}s + t) \equiv a^{2n}s + t' \mod \mathfrak{a}$.

Theorem 3.9.11 (Nullstellensatz) *For any $a \in A$, the following statements are equivalent:*

(i) a *vanishes identically on* X;
(ii) $\exists s \in S, \exists t \in T, \exists n \geq 0$ *such that* $a^{2n}s + t \in \mathfrak{a}$.

Theorem 3.9.9 incorporates Proposition 3.9.5, Theorem 3.9.10 incorporates Proposition 3.9.8 and Theorem 3.9.11 incorporates the weak real Nullstellensatz Proposition 3.2.10 as a special case.

Exercise 3.9.12 Show that it suffices to consider the case where $\mathfrak{a} = 0$ for the proofs of these Stellensätze.

Proofs By the exercise, we may assume without loss of generality that $E = \varnothing$. Let $B := S^{-1}A$ and let $U = \{t/s^2 : t \in T, s \in S\}$ be the preordering of B generated by T. Upon canonical identification of Sper B with a subspace of Sper A we have $X \subseteq$ Sper B, more precisely $X = \overline{H}_B(U)$. Thus, for all $a \in A$ we have

$$a > 0 \ (\geq 0, = 0) \text{ on } X \subseteq \text{Sper } A \Leftrightarrow \frac{a}{1} > 0 \ (\geq 0, = 0) \text{ on } X \subseteq \text{Sper } B.$$

Note that all $s \in S$ are positive on X and that all $t \in T$ are nonnegative on X.

Proof of Theorem 3.9.9. For the implication (i) $\Rightarrow$ (ii) we let $a > 0$ on X. By Proposition 3.9.5 (applied to B) there exist $s, s' \in S, t, t' \in T$ such that

$$a \cdot \frac{t}{s^2} = 1 + \frac{t'}{s'^2} \quad \text{in } B.$$

Rewriting yields the existence of an $s'' \in S$ such that

$$a(s's'')^2 t = (ss's'')^2 + (ss'')^2 t' \quad \text{in } A,$$

in other words, an identity of the form (ii). The implication (ii) $\Rightarrow$ (iii) follows using the same trick as in the proof of Proposition 3.9.5 and the implication (iii) $\Rightarrow$ (i) is clear.

Proof of Theorem 3.9.10. For the implication (i) $\Rightarrow$ (ii) we let $a \geq 0$ on X. By Proposition 3.9.8 there exist $s, s' \in S$, $t, t' \in T$ and $n \geq 0$ such that

$$a \cdot \frac{t}{s^2} = a^{2n} + \frac{t'}{s'^2} \quad \text{in } B.$$

Hence there exists $s'' \in S$ such that

$$a(s's'')^2 t = a^{2n}(ss's'')^2 + (ss'')^2 t' \quad \text{in } A,$$

an identity of the form (ii). The implication (ii) $\Rightarrow$ (iii) follows as in the proof of Proposition 3.9.8 and the implication (iii) $\Rightarrow$ (i) is clear.

Proof of Theorem 3.9.11. For the implication (i) $\Rightarrow$ (ii) we let $a = 0$ on X. By Theorem 3.9.10 there exist $s_1, s_2 \in S$, $t_1, t'_1, t_2, t'_2 \in T$ and $m, n \geq 0$ such that

$$at_1 = a^{2m} s_1^2 + t'_1 \quad \text{and} \quad -at_2 = a^{2n} s_2^2 + t'_2.$$

Multiplication then gives

$$-a^2 t_1 t_2 = a^{2(m+n)}(s_1 s_2)^2 + t$$

for some $t \in T$, an identity of the form (ii). The converse implication is clear. $\square$

Remark 3.9.13 Consider a real closed field R and an affine R-algebra A with associated variety V. Letting $F, G, E \subseteq A$ and keeping everything else as it was, we obtain *geometric* Stellensätze as long as F and G are assumed to be *finite*. In this situation X is namely constructible (since A is noetherian, the cardinality of E does not matter) and in (i) of Theorems 3.9.9, 3.9.10 and 3.9.11 we may replace X by $X(R) := X \cap V(R)$.

Example 3.9.14 The solution of Hilbert's 17th Problem can be deduced from Theorem 3.9.10: If $A = R[t_1, \ldots, t_n]$ and $a \in A$ is positive semidefinite on R^n, there exist $f, g \in \Sigma A^2$ and $m \geq 0$ such that

$$a = (a^{2m} + f)/(a^{2m} + g) \quad (\text{and } a^{2m} + g \neq 0).$$

This expression can be rewritten as

$$a = \frac{(a^{2m} + f)(a^{2m} + g)}{(a^{2m} + g)^2},$$

which we recognize as a sum of squares in Quot A. Moreover, it is possible to write this expression in such a way that the denominators of the rational functions that occur in it vanish at most at zeroes of a.

We illustrate the results obtained thus far with two applications of Prestel's Positivstellensatz (Proposition 3.9.5) and one application of the Nullstellensatz (Theorem 3.9.11).

First, let A be an arbitrary commutative ring and P an ordering of A. We give a description of the maximal specialization Q of P in Sper A (cf. Corollary 3.6.8). Let Q^+ denote the complement of $\mathfrak{q} := \operatorname{supp} Q$ in Q. Then A is the disjoint union of Q^+, $-Q^+$ and $\mathfrak{q}$. Thus, Q is in particular determined by Q^+.

Proposition 3.9.15 Q^+ *is the set of all those elements* $a \in P$ *that divide an element of* $1 + P$ *in* A.

Proof Since the spear $\overline{\{P\}} = \overline{H}_A(P)$ consists of generalizations of Q, we have that Q^+ is the set of those $a \in A$ that are positive on $\overline{H}_A(P)$. The statement then follows from Proposition 3.9.5 (applied to $T = P$) since $ap = 1 + p'$ with $a \in A$ and p, $p' \in P$ implies that $a \in P$. $\square$

Next we derive a special case of a theorem of Hörmander concerning the growth of polynomials in several variables (cf. [53], also [112, p. 224]). Refinements of this theorem (loc. cit., [25, 42], ...) are useful in the theory of partial differential equations (with constant coefficients in particular), which is a fertile ground for applications of real algebra in general.

Let R be a real closed field and denote the euclidean norm on R^n by $\|x\|$. In other words, $\|x\|^2 = x_1^2 + \cdots + x_n^2$.

Proposition 3.9.16 *Let* $f \in R[t_1, \ldots, t_n]$ *and let* $r \geq 0$ *in* R *be such that* $f(x) \neq 0$ *for all* $x \in R^n$ *that satisfy* $\|x\| \geq r$. *Then there exists a constant* $C > 0$ *in* R *and a natural number* N *such that*

$$|f(x)| \geq C(1 + \|x\|^2)^{-N}$$

for all $x \in R^n$ *that satisfy* $\|x\| \geq r$.

Proof We let

$$g(t_1, \ldots, t_n) := t_1^2 + \cdots + t_n^2 - r^2$$

and $M := \{x \in R^n : g(x) \geq 0\}$. By assumption, f does not vanish on M. Thus, by Corollary 3.9.6 (applied to $A = R[t_1, \ldots, t_n]$ and $F = \{g\}$) there exists a polynomial $h \in R[t_1, \ldots, t_n]$ such that

$$|f(x)h(x)| \geq 1 \quad \text{for all } x \in M. \tag{3.3}$$

Let N denote the total degree of h. Then h is of the form

$$h(t) = \sum_{|\alpha| \leq N} c_\alpha t^\alpha$$

for certain constants $c_\alpha \in R$. (In this expression, $\alpha = (\alpha_1, \ldots, \alpha_n) \in \mathbb{N}_0^n$ runs through those multi-indices that satisfy $|\alpha| = \alpha_1 + \cdots + \alpha_n \leq N$.) Since $|x_i| \leq 1 + \|x\|^2$, we have

$$|h(x)| \leq \sum_{|\alpha| \leq N} |c_\alpha| \left(1 + \|x\|^2\right)^{|\alpha|} \leq \left(\sum_{|\alpha| \leq N} |c_\alpha|\right) \left(1 + \|x\|^2\right)^N$$

for all $x \in R^n$, and the statement follows from (3.3). $\qquad\square$

Let us return to an arbitrary commutative ring A.

Proposition 3.9.17 *Let X be a closed subset of* Sper A, *and let $f, g \in A$ be such that every $x \in X$ that satisfies $f(x) = 0$ also satisfies $g(x) = 0$. Then there exist an $a \in A$ and an $n \in \mathbb{N}$ such that*

$$g(x)^{2n} \leq \left(1 + a(x)^2\right) f(x)^2$$

for all $x \in X$.

Conversely, if such an inequality is satisfied for all $x \in X$, then of course g vanishes at the zeroes of f in X.

Proof By assumption we have $X \cap Z_A(f) \subseteq Z_A(g)$. Since X is an intersection of closed constructible subsets of Sper A and $Z_A(g)$ is constructible, there exists a constructible closed overset Y of X such that $Y \cap Z_A(f) \subseteq Z_A(g)$ (cf. Corollary 3.4.10). We may replace X by Y. In other words, we may assume without loss of generality that X is constructible. Then there exist (finitely generated) preorderings $T_1, \ldots, T_r$ of A such that

$$X = \overline{H}_A(T_1) \cup \cdots \cup \overline{H}_A(T_r).$$

(To see this, write the complement of X as a finite union of sets of the form $\overset{\circ}{H}_A(f_1, \ldots, f_N)$.) An application of the Nullstellensatz (Theorem 3.9.11) yields equations

$$g^{2n_i} + t_i = a_i f^2$$

where $n_i \in \mathbb{N}$, $t_i \in T_i$ and $a_i \in A$ (for $i = 1, \ldots, r$), and after multiplication with powers of g^2 we obtain $n_1 = \cdots = n_r =: n$. For $x \in \overline{H}_A(T_i)$ we have

$$g(x)^{2n} \leq |a_i(x)| \, f(x)^2 \leq \left(1 + a_i(x)^2\right) f(x)^2.$$

Hence, for every $x \in X$,

$$g(x)^{2n} \leq \left(1 + a_1(x)^2 + \cdots + a_r(x)^2\right) f(x)^2 \leq \left(1 + a(x)^2\right) f(x)^2,$$

where $a := 1 + a_1^2 + \cdots + a_r^2$. $\qquad\qquad\qquad\qquad\qquad\qquad\qquad\qquad\qquad\qquad\qquad\qquad$ □

Historical Remarks Our Theorems 3.9.9–3.9.11 are usually attributed to G. Stengle, but this is only partially correct. Stengle ([111], 1974) worked in a geometric setting. First he proved a "semialgebraic Nullstellensatz" (our Theorem 3.9.11), from which he then deduced the Nichtnegativstellensatz. However, Theorems 3.9.9 and 3.9.11 were essentially already discovered by J.-L. Krivine in 1964 [66]. After introducing preorderings of rings, Krivine worked with maximal preorderings, which in our terminology correspond to closed points in the real spectrum. Krivine's work was largely unknown in the early years of real algebraic and semialgebraic geometry, and was "rediscovered" only much later. For more details see [92, Sect. 4.7]. We also mention the approaches in [17] and [112, §10] that are historically largely independent of those of Stengle and Prestel.

In 1959 Łojasiewicz [76] proved a statement in the geometric setting about semialgebraic functions (these are continuous functions with semialgebraic graph), similar to Proposition 3.9.17. A modern treatment, for arbitrary real closed fields R, of this famous and important result known as the *Łojasiewicz inequality* can be found in [11, §2.6]. The theory of abstract semialgebraic functions (cf. [107] or [108]) leads to the insight that Proposition 3.9.17 is not only related to the Łojasiewicz inequality, but is in fact an abstract version of this result. (Łojasiewicz proved his inequality even for semianalytic functions, over $R = \mathbb{R}$ of course. Also in this case it seems possible to reduce the inequality to Proposition 3.9.17.)

3.10   The Convex Radical Ideals Associated to a Preordering

Let A be a ring. In Sect. 3.7 we studied the ideals of A that are convex with respect to an ordering P. In this section we will more generally consider the notion of convexity with respect to a preordering T. Furthermore, the Stellensätze from Sect. 3.9 will provide a geometric insight in the case of T-convex radical ideals.

Almost all the results concerning T-convex radical ideals proved in this section go back to G. Brumfiel [17, 18], who also conducted fruitful research on more general T-convex ideals ("completely convex ideals") with his student R. Robson, cf. [18, 95].

Let T be a preordering of A. We define the relation $\leq_T$ on A via

$$a \leq_T b :\Leftrightarrow b - a \in T.$$

Then $\leq_T$ is reflexive and transitive, but only antisymmetric (and thus a partial order relation on A) if and only if $T \cap (-T) = 0$. (If $T \cup (-T) = A$, then $\leq_T$ can be interpreted as an ordering of the abelian group $A / T \cap (-T)$.) For every subgroup G of the additive group $(A, +)$ the following properties are equivalent:

(i) From $t, t' \in T$ and $t + t' \in G$ it follows that $t \in G$ (and $t' \in G$);
(ii) from $a, b \in G, c \in A$ and $a \leq_T c \leq_T b$ it follows that $c \in G$;
(iii) from $a \in G, c \in A$ and $0 \leq_T c \leq_T a$ it follows that $c \in G$.

(For (i) $\Rightarrow$ (ii), take $t := c - a$ and $t' := b - c$. For (iii) $\Rightarrow$ (i), take $a := t + t'$ and $c := t$. The multiplicative structure of A does not play a role here. Thus T need not be a preordering either.)

Definition 3.10.1 Let T be a preordering of A. A subgroup G of A (considered as additive group) is called T-*convex* (or *convex with respect to* T) if the equivalent conditions (i)–(iii) are satisfied.

This generalizes Definition 3.7.1, see also Definition 2.1.1. In this section we only consider T-convex *ideals*, while in the next section we focus on T-convex subrings. Since every intersection of T-convex subgroups of A is again T-convex, every subgroup (resp. every ideal, every subring) of A yields a smallest T-convex subgroup (resp. a smallest T-convex ideal, a smallest T-convex subring) of A that contains the given object.

If $T = A$ is the trivial (improper) preordering, then A is the only T-convex subgroup of A, an uninteresting case.

Consider the smallest preordering $T_0 = \Sigma A^2$ of A and an ideal $\mathfrak{a}$ of A. If $\mathfrak{a}$ is real reduced (i.e., if the ring $A/\mathfrak{a}$ is real reduced), then $\mathfrak{a}$ is T_0-convex. The converse is false in general (counterexample: take $A = k[t]/(t^2)$ and $\mathfrak{a} = (0)$ for any real reduced ring k), but is true if $\mathfrak{a} = \sqrt{\mathfrak{a}}$.

Since for preorderings $T \subseteq T'$ every T'-convex subgroup is also T-convex, it follows from Proposition 3.9.3 and Sect. 3.7 that for every proper preordering T of A there exists a T-convex ideal of A, which is proper (i.e., distinct from A).

Let T be any preordering of A. We want to show that if $\mathfrak{p}$ is any ideal of A that is maximal with respect to the property $\mathfrak{p} \cap (1 + T) = \varnothing$, then $\mathfrak{p}$ is T-convex.

Exercise 3.10.2 Let $\mathfrak{a} \neq A$ be a T-convex ideal. Show that $\mathfrak{a} \cap (1 + T) = \varnothing$.

Exercise 3.10.3 Let $S \neq \varnothing$ be a multiplicative subset of A that is contained in $1 + T$. Show that

$$T' := \{a \in A : \text{ there exists } s \in S \text{ such that } as \in T\}$$

is a proper preordering of A that contains T.

Lemma 3.10.4 *Assume that every $a \in A$ satisfies the property "$2a \in T \Rightarrow a \in T$" (e.g., this is the case if 2 is a unit). Then $\mathfrak{a} := T \cap (-T)$ is the smallest T-convex ideal of A.*

Proof $\mathfrak{a}$ is an additive subgroup of A that satisfies "$2c \in \mathfrak{a} \Rightarrow c \in \mathfrak{a}$" and $c^2\mathfrak{a} \subseteq \mathfrak{a}$ for every $c \in A$. All $a \in \mathfrak{a}$ and $b \in A$ satisfy

$$2ba = (1+b)^2 a - b^2 a - a.$$

Since the right hand side is in $\mathfrak{a}$, we also have $ba \in \mathfrak{a}$. Hence $\mathfrak{a}$ is an ideal. The T-convexity of $\mathfrak{a}$ follows immediately (if $t, t' \in T$ such that $t+t' \in \mathfrak{a}$, then $t+t' \in -T$, hence $t = (t + t') - t' \in -T$) and it is clear that $\mathfrak{a}$ is contained in every T-convex ideal of A. $\qquad\square$

Proposition 3.10.5 *Let T be a preordering of A and assume that the ideal $\mathfrak{p}$ of A is maximal with respect to the property $\mathfrak{p} \cap (1 + T) = \varnothing$. Then $\mathfrak{p}$ is a T-convex prime ideal.*

Proof $\mathfrak{p}$ is a prime ideal by Exercise 3.1.10. Consider the proper preordering $U := T + \mathfrak{p}$ of A and the multiplicative subset $S := \{2^n : n \geq 0\} \subseteq 1 + U$. Applying the construction from Exercise 3.10.3 to S and U yields a preordering $U' = \{a \in A : 2^n a \in U$ for some $n \geq 0\}$ that satisfies the property from Lemma 3.10.4. It follows that $\mathfrak{p}' := U' \cap (-U')$ is a U'-convex, hence also T-convex ideal that contains $\mathfrak{p}$. By Exercise 3.10.2 we have $\mathfrak{p}' \cap (1 + T) = \varnothing$, and we conclude that $\mathfrak{p}' = \mathfrak{p}$. $\qquad\square$

We deduce from Proposition 3.10.5 and Exercise 3.10.2 that the proper maximal T-convex ideals of A are precisely those (prime) ideals $\mathfrak{p}$ of A that are maximal for the property $\mathfrak{p} \cap (1+T) = \varnothing$. If $T_0 = \Sigma A^2$ we note that the condition $\mathfrak{a} \cap (1+T_0) = \varnothing$ characterizes the real ideals $\mathfrak{a}$ of A (cf. Sect. 3.2); since the T_0-convex prime ideals are precisely the real reduced ideals we have thus recovered Proposition 3.10.6 in this special case.

Proposition 3.10.6 *Let $\mathfrak{a}$ be a T-convex ideal of A. Then every minimal prime ideal of A that contains $\mathfrak{a}$ is also T-convex.*

Proof Let $\mathfrak{p}$ be a minimal prime ideal of A that contains $\mathfrak{a}$. Then

$$U := \left\{ \tfrac{t}{s^2} : t \in T, s \in A \setminus \mathfrak{p} \right\}$$

is a proper preordering of $A_\mathfrak{p}$. Indeed, if it were the case that $-1 \in U$, then there would exist an $s \in A \setminus \mathfrak{p}$ such that $s^2 \in T \cap (-T) \subseteq \mathfrak{a}$, contradicting $\mathfrak{a} \subseteq \mathfrak{p}$ and $\mathfrak{p}$ prime. The (proper) ideal $\mathfrak{a}A_\mathfrak{p}$ is U-convex in $A_\mathfrak{p}$, hence contained in a U-convex prime ideal of $A_\mathfrak{p}$ by Zorn's Lemma and Proposition 3.10.5. But $\mathfrak{p}A_\mathfrak{p}$ is the only prime ideal of $A_\mathfrak{p}$ that contains $\mathfrak{a}A_\mathfrak{p}$, implying that $\mathfrak{p}A_\mathfrak{p}$ is U-convex. Then it follows easily that $\mathfrak{p}$ is T-convex in A. $\qquad\square$

Corollary 3.10.7 *Let $\mathfrak{a}$ be a proper ideal of A.*

(a) *If $\mathfrak{a}$ is T-convex, then so is $\sqrt{\mathfrak{a}}$.*
(b) *If $\mathfrak{a}$ is a radical ideal, then $\mathfrak{a}$ is T-convex if and only if all minimal prime ideals of A that contain $\mathfrak{a}$ are T-convex.* □

In general, a preordering is not the intersection of orderings, as remarked in Sect. 3.9. Nevertheless we may restrict ourselves to intersections of orderings when studying T-convex radical ideals, as we will show next. By Proposition 3.10.6 it suffices to consider prime ideals.

Lemma 3.10.8 (A. Klapper, cf. [17, p. 63]) *If T and T' are preorderings of A and $\mathfrak{p}$ is a $(T \cap T')$-convex prime ideal, then $\mathfrak{p}$ is T-convex or T'-convex.*

Proof We consider the relation $\leq_T$ from the beginning of this section. If $a, a', b, b' \in A$, and if $0 \leq_T a \leq_T b$ and $0 \leq_T a' \leq_T b'$, then also $0 \leq_T aa' \leq_T bb'$. Assume for the sake of contradiction that $\mathfrak{p}$ is neither T-convex, nor T'-convex. Then there exist $a, b \in \mathfrak{p}$ and $u, v \in A \setminus \mathfrak{p}$ such that

$$0 \leq_T u \leq_T a \quad \text{and} \quad 0 \leq_{T'} v \leq_{T'} b.$$

It follows that $0 \leq_T u^2 \leq_T a^2$ and $0 \leq_{T'} v^2 \leq_{T'} b^2$, hence

$$0 \leq_T u^2 v^2 \leq_T a^2 v^2 \leq_T a^2 v^2 + b^2 u^2 \quad \text{and} \quad 0 \leq_{T'} u^2 v^2 \leq_{T'} u^2 b^2 \leq_{T'} a^2 v^2 + u^2 b^2,$$

and so $0 \leq_{T \cap T'} u^2 v^2 \leq_{T \cap T'} a^2 v^2 + b^2 u^2$. Since $a^2 v^2 + b^2 u^2 \in \mathfrak{p}$ and $u^2 v^2 \notin \mathfrak{p}$, this contradicts the $(T \cap T')$-convexity of $\mathfrak{p}$. □

Definition 3.10.9 For a preordering T of A we denote the intersection of all orderings P of A such that $P \supseteq T$ by $\widehat{T}$ (thus $\widehat{T} = A$ if $T = A$). If $T = \widehat{T}$, we call T *saturated*.

If T is a proper preordering, then so is $\widehat{T}$ by Proposition 3.9.3. For every preordering T of A we have

$$\widehat{T} = \{a \in A : -a^{2n} \in T - aT \text{ for some } n \geq 0\}$$

$$= \left\{a \in A : \text{there exist } t, t' \in T \text{ and } n \geq 0 \text{ such that } a(a^{2n} + t) = t'\right\}$$

by the Nichtnegativstellensatz Proposition 3.9.8.

Proposition 3.10.10 *Let T be a preordering of A and $\mathfrak{a}$ a radical ideal of A. Then $\mathfrak{a}$ is T-convex if and only if $\mathfrak{a}$ is $\widehat{T}$-convex.*

Proof By Corollary 3.10.7(b) we may assume that $\mathfrak{a} = \mathfrak{p}$ is prime. We prove the nontrivial direction. Assume that $\mathfrak{p}$ is T-convex, but not $\widehat{T}$-convex. Then there exist $a, b \in \widehat{T}$ such that $a + b \in \mathfrak{p}$ and $a, b \notin \mathfrak{p}$. There exist $u, u', v, v' \in T$ and $m, n \geq 0$ such that $au = a^{2m} + u'$ and $bv = b^{2n} + v'$. Both elements are in T, but not in $\mathfrak{p}$ since $\mathfrak{p}$ is T-convex. It follows that also $c := a \cdot aubv$ and $d := b \cdot aubv$ are in T, but not in $\mathfrak{p}$. This contradicts the T-convexity of $\mathfrak{p}$ since $c + d = (a + b)aubv \in \mathfrak{p}$. □

Corollary 3.9.6 immediately yields that $1 + \widehat{T}$ consists of divisors of elements of $1 + T$. This result now also follows from Propositions 3.10.5 and 3.10.10. (Let $a \in \widehat{T}$ and assume that $1 + a$ is not a divisor of an element of $1 + T$. Then, by Proposition 3.10.5, there exists a T-convex prime ideal $\mathfrak{p}$ such that $1 + a \in \mathfrak{p}$. This ideal is of course not $\widehat{T}$-convex, a contradiction.)

Definition 3.10.11 We call a closed subset X of Sper A *probasic* if it is of the form $X = \overline{H}_A(F)$ for some subset F of A. Since the intersection of probasic closed sets is again probasic, every subset $X \subseteq$ Sper A has a smallest probasic closed overset that we denote by $\widehat{X}$.

If X is a subset of Sper A we write henceforth

$$P(X) := \big\{ a \in A : a(x) \geq 0 \quad \forall x \in X \big\}.$$

This is a preordering of A and it is precisely the intersection of all elements of X when these are viewed as orderings of A. The preorderings of the form $P(X)$ are thus precisely the saturated preorderings of A. If $X = \{x\}$, then $P(x) := P(\{x\})$ is just x itself, viewed as an ordering of A (cf. Sect. 3.3).

The relationship between preorderings of A and probasic closed subsets of Sper A can be understood formally as follows: We have defined inclusion reversing maps

$$\{\text{Subsets of } A\} \; \underset{P}{\overset{\overline{H}}{\rightleftarrows}} \; \{\text{Subsets of Sper } A\}$$

between the power set of A and Sper A. These maps are such that for every subset $F \subseteq A$ we have

$$P\big(\overline{H}(F)\big) = \widehat{P[F]},$$

the saturation of the preordering of A generated by F, and for every subset X of Sper A we have

$$\overline{H}\big(P(X)\big) = \widehat{X},$$

the smallest probasic closed subset of Sper A that contains X.

If we restrict $\overline{H}$, resp. P, to the saturated preorderings of A, resp. the probasic closed subspaces of Sper A, these operators give rise to inclusion reversing bijections that are inverse to each other. In other words, saturated preorderings of A can be identified with probasic closed subspaces of Sper A via $\overline{H}$ and P. In what follows, we will usually prefer to work with the subspaces of Sper A as these have the advantage of providing greater geometric clarity.

Therefore, given $X \subseteq$ Sper A we will often say X-*convex subgroup* (*ideal, ring*) instead of $P(X)$-convex subgroup, etc. (If $X = \{x\}$ we will also talk of subgroups, convex in x, etc.) Proposition 3.10.10 thus says that for radical ideals $\mathfrak{a}$ and arbitrary preorderings T, $\mathfrak{a}$ is T-convex if and only if $\mathfrak{a}$ is $\overline{H}(T)$-convex. A subgroup of A is X-convex if and only if it is $\widehat{X}$-convex.

Since for subsets $Y \subseteq X \subseteq \operatorname{Sper} A$, every Y-convex ideal is trivially also X-convex, it follows in particular that for every $x \in X$ the ideal $\operatorname{supp}(x)$ is X-convex. We will show next that the converse also holds if X is closed.

Lemma 3.10.12 *If X is a proconstructible subset of* $\operatorname{Sper} A$ *and* $\mathfrak{p}$ *is a prime ideal of A which is not convex in any $x \in X$, then there exists a closed constructible set $Y \supseteq X$ such that $\mathfrak{p}$ is not Y-convex. In particular, $\mathfrak{p}$ is not X-convex.*

Proof For every $x \in X$ there exist elements $a_x, b_x \in A \setminus \mathfrak{p}$ such that $a_x(x) \geq 0$, $b_x(x) \geq 0$ and $a_x + b_x \in \mathfrak{p}$. The sets $Y(x) := \overline{H}(a_x, b_x)$ $(x \in X)$ then constitute a covering of X by closed constructible sets. Since the constructible topology is compact (cf. Corollary 3.4.10(a)) there exist finitely many $x_1, \ldots, x_r \in X$ such that $X \subseteq Y_1 \cup \cdots \cup Y_r =: Y$, where $Y_i := Y(x_i)$. There is no $i \in \{1, \ldots, r\}$ for which $\mathfrak{p}$ is Y_i-convex. It follows from Lemma 3.10.8 that $\mathfrak{p}$ is not Y-convex since $P(Y) = \bigcap_i P(Y_i)$.   □

Theorem 3.10.13 *Let X be a closed subset of* $\operatorname{Sper} A$*. Then the X-convex prime ideals of A are precisely the supports $\operatorname{supp}(x)$ of all points $x \in X$.*

Proof For $x \in X$, the ideal $\operatorname{supp}(x)$ is X-convex, as already remarked. Conversely, consider an X-convex prime ideal $\mathfrak{p}$. By Lemma 3.10.12, $\mathfrak{p}$ is convex in some $x \in X$, i.e., $\mathfrak{p} = \operatorname{supp}(y)$ for some specialization y of x (cf. Theorem 3.7.4). Then $y \in X$ since X is closed.   □

From Theorem 3.10.13 we can deduce a very concrete geometric interpretation of the X-convex radical ideals. If $X \subseteq \operatorname{Sper} A$ is a subset, we call a subset Z of X *algebraic in X* if there exists and ideal $\mathfrak{a} \subseteq A$ such that

$$Z = Z_X(\mathfrak{a}) := X \cap Z_A(\mathfrak{a}) = \{x \in X : a(x) = 0 \text{ for all } a \in \mathfrak{a}\}.$$

Writing

$$I(Y) := \bigcap_{y \in Y} \operatorname{supp}(y) = \{a \in A : a(y) = 0 \text{ for all } y \in Y\},$$

we have:

Theorem 3.10.14 *Let X be closed in* $\operatorname{Sper} A$*. Then the X-convex radical ideals $\mathfrak{a}$ of A correspond bijectively to the subsets Z of X that are algebraic in X via*

$$Z = Z_X(\mathfrak{a}) \quad and \quad \mathfrak{a} = I(Z).$$

Proof Since $Z_X \circ I \circ Z_X = Z_X$, we have $Z_X \circ I(Z) = Z$ for every subset Z that is algebraic in X. Conversely, let $\mathfrak{a}$ be an X-convex radical ideal. It is clear that

$\mathfrak{a} \subseteq I(Z_X(\mathfrak{a}))$. If $f \notin \mathfrak{a}$, then Proposition 3.10.6 yields an X-convex prime ideal $\mathfrak{p} \supseteq \mathfrak{a}$ such that $f \notin \mathfrak{p}$. By Theorem 3.10.13 we have $\mathfrak{p} = \mathrm{supp}(x)$ for some $x \in X$, hence $x \in Z_X(\mathfrak{p}) \subseteq Z_X(\mathfrak{a})$. On the other hand, $f(x) \neq 0$, hence $f \notin I(Z_X(\mathfrak{a}))$. $\square$

What does Theorem 3.10.14 mean in the "geometric setting"? Let V be an affine variety over a real closed field R and let $A := R[V]$ be the associated R-algebra. Consider a closed semialgebraic subset $M \subseteq V(R)$ of the form

$$M = \bigcup_{i=1}^{r} \{x \in V(R) : f_{i1}(x) \geq 0, \ldots, f_{is(i)}(x) \geq 0\} \qquad (3.4)$$

for finitely many $f_{ij} \in A$. (By the—as yet unproven—Finiteness Theorem 3.5.8, every closed semialgebraic M is of the form (3.4). We only require this assumption in order to ensure that $\widetilde{M}$ is closed in $\widetilde{V(R)}$.) The (closed) constructible subset $\widetilde{M}$ of $\mathrm{Sper}\, A$ that corresponds to M is the closure of M in $\mathrm{Sper}\, A$ (with respect to the constructible or Harrison topology). It follows that $P(M) = P(\widetilde{M})$ for the associated (saturated) preorderings of A.

Let $\mathfrak{a}$ be a radical ideal of A and W the associated closed subvariety of V. By Theorem 3.10.14, $\mathfrak{a}$ is M-convex if and only if $\mathfrak{a} = I(Z_{\widetilde{M}}(\mathfrak{a})) = I(\widetilde{M} \cap Z_A(\mathfrak{a}))$. However, $Z_A(\mathfrak{a}) = \widetilde{W(R)}$, and so $\widetilde{M} \cap Z_A(\mathfrak{a}) = \widetilde{M \cap W(R)}$. Since $I(\widetilde{N}) = I(N)$ for semialgebraic $N \subseteq V(R)$ we can formulate Theorem 3.10.14 in the current situation as follows:

Theorem 3.10.15 *Let V be an affine R-variety, M a closed semialgebraic subset of $V(R)$ (of the form (3.4)), $\mathfrak{a}$ a radical ideal of $R[V]$ and W the subvariety of V associated to $\mathfrak{a}$. Then $\mathfrak{a}$ is M-convex if and only if $M \cap W(R)$ is Zariski dense in W.*

We return to our arbitrary commutative ring A and deduce two consequences from Theorems 3.10.13 and 3.10.14. Let T be a preordering and $\mathfrak{a}$ an ideal of A.

Definition 3.10.16 We denote the intersection of all T-convex ideals $\mathfrak{b}$ that contain $\mathfrak{a}$ (i.e., the smallest T-convex ideal that contains $\mathfrak{a}$) by $\mathrm{ci}_T(\mathfrak{a})$. (Here "ci" stands for "convex ideal".)

Proposition 3.10.17 *For every ideal $\mathfrak{a}$ of A we have*

$$\sqrt{\mathrm{ci}_T(\mathfrak{a})} = \{a \in A : \text{there exists } n \geq 0 \text{ and } t \in T \text{ such that } a^{2n} + t \in \mathfrak{a}\}.$$

Proof First of all, $\mathfrak{c} := \sqrt{\mathrm{ci}_T(\mathfrak{a})}$ is the smallest T-convex radical ideal $\supseteq \mathfrak{a}$ by Corollary 3.10.7(a). Let $X := \overline{H}(T)$, i.e., $P(X) = \widehat{T}$. Since $\mathfrak{c}$ is also X-convex by Proposition 3.10.10, $\mathfrak{c}$ is a fortiori the smallest X-convex radical ideal $\supseteq \mathfrak{a}$. Then $\mathfrak{c} = I(Z_X(\mathfrak{a}))$ by Theorem 3.10.14. By the Nullstellensatz Theorem 3.9.11 (applied to $F = \varnothing$, $G = T$, $E = \mathfrak{a}$), $I(Z_X(\mathfrak{a}))$ is exactly the right hand side from the statement of the theorem. $\square$

We note that Proposition 3.10.17 is a generalization of the weak real Nullstellen-satz (Proposition 3.2.10 which deals with the case $T = \Sigma A^2$); the real radical $\sqrt[re]{\mathfrak{a}}$ of $\mathfrak{a}$ is precisely $\sqrt{\mathrm{ci}_{\Sigma A^2}(\mathfrak{a})}$.

The following direct proof of Proposition 3.10.17 can be found in [18, p. 59]: Let $\mathfrak{b} = \{a \in A : a^{2n} + t \in \mathfrak{a} \text{ for some } n \geq 0 \text{ and } t \in T\}$. Obviously, $\mathfrak{b} \subseteq \sqrt{\mathrm{ci}_T(\mathfrak{a})}$ and it suffices to show that $\mathfrak{b} + \mathfrak{b} \subseteq \mathfrak{b}$. Thus, let $b, b' \in \mathfrak{b}$ and $b^{2m} + t = a \in \mathfrak{a}$, $b'^{2n} + t' = a' \in \mathfrak{a}$ with $t, t' \in T$. We may assume that $m = n$. We have

$$\left((b+b')^2 + (b-b')^2\right)^{2n} = (2b^2 + 2b'^2)^{2n} = b^{2n}u + b'^{2n}u'$$

with $u, u' \in T$. Thus,

$$\left((b+b')^2 + (b-b')^2\right)^{2n} + tu + t'u' = au + a'u' \in \mathfrak{a}.$$

Since the left hand side is of the form $(b+b')^{4n} + v$ with $v \in T$, it follows that $b + b' \in \mathfrak{b}$.

The next theorem concerns the relationship between a subset X of Sper A and the probasic closed set $\widehat{X}$ that it generates:

Proposition 3.10.18 *If $X \subseteq$ Sper A is closed, then X and $\widehat{X}$ have the same image under the support map* supp : Sper $A \to$ Spec A.

Proof This follows immediately from Theorem 3.10.13 since $P(X) = P(\widehat{X})$. □

We finish this section with the interesting and arguably difficult problem of determining the relationships between X and $\widehat{X}$ in reasonable situations. For instance, if

$$X = \bigcup_{i=1}^{r} \overline{H}(f_{i1}, \ldots, f_{is(i)})$$

is an explicit closed constructible subset of Sper A, how is $\widehat{X}$ characterized in terms of the f_{ij}? In which cases is $\widehat{X}$ again constructible?

We can give some partial answers to the final question. In the "geometric setting", where A is a finitely generated algebra over a real closed field R, $\widehat{X}$ is never constructible, except in the trivial case where $X = \widehat{X}$, i.e., where X itself is already probasic closed. Indeed, $X \cap V(R) = \widehat{X} \cap V(R)$ by Proposition 3.10.18.

In other cases however, $\widehat{X} \neq X$ can certainly be constructible. For an example, take $A = \mathbb{R}((x))((y))$ and let X consist of three of the four orderings of A.

3.11 Boundedness

In Sect. 2.1 we touched upon the notion of an archimedean field extension with respect to a fixed ordering. We will now generalize this idea in two respects: instead of field extensions we allow arbitrary ring extensions $A \supseteq \Lambda$, and instead of a fixed ordering we consider arbitrary subsets X of Sper A. This leads to the notion of archimedean closure of Λ in A on X, which we will characterize in several ways in this Section. In fact, the archimedean closure can also be viewed as the convex hull of a subring with respect to a preordering, as we will explain below.

In this section $\varphi : \Lambda \to A$ will always denote a ring homomorphism.

Definition 3.11.1 Consider a subset $X \subseteq$ Sper A.

(a) An element $a \in A$ is called *bounded on X over* Λ (or *with respect to* φ) if there exists a $\lambda \in \Lambda$ such that

$$|a(x)| \le (\varphi\lambda)(x)$$

for all $x \in X$. Clearly the elements of A that are bounded on X over Λ form a subring of A that contains $\varphi(\Lambda)$. We denote this subring by $\mathfrak{o}_X(A/\Lambda)$ or $\mathfrak{o}_X(A, \varphi)$. If $X = \{x\}$, we simply write $\mathfrak{o}_x(A/\Lambda) := \mathfrak{o}_{\{x\}}(A/\Lambda)$.

(b) We call A *archimedean on X over* Λ (or *with respect to* φ) if $\mathfrak{o}_X(A/\Lambda) = A$. If Λ is a subring of A and φ the inclusion map, then we call $\mathfrak{o}_X(A/\Lambda)$ the *archimedean closure of Λ in A on X*. In case $\Lambda = \mathfrak{o}_X(A/\Lambda)$, we call Λ *archimedean closed in A on X*.

If we replace φ by the inclusion of the subring $\varphi(\Lambda)$ of A, then $\mathfrak{o}_X(A/\Lambda)$ does not change. Hence, one can always think of Λ as a subring of A (and of φ as the inclusion map).

An element $a \in A$ is (already) bounded on X with respect to φ if and only if there exists a $\mu \in \Lambda$ such that

$$|a(x)| \le |(\varphi\mu)(x)|$$

for all $x \in X$ (take for instance $\lambda = 1 + \mu^2$). The description of $\mathfrak{o}_X(A/\Lambda)$ as archimedean closure (of Λ in A on X) is justified since $\mathfrak{o}_X(A/\Lambda)$ is obviously the largest subring B of A that is archimedean on the image of X in Sper B over Λ.

Note that the condition in Definition 3.11.1 is global with respect to X in the sense that the elements of A are *uniformly* bounded on X by elements of Λ. If X is proconstructible however, we will see in Corollary 3.11.20 that this global condition is equivalent to suitable local ones.

The archimedean closure of a subring of A is actually a convex hull with respect to some preordering. Indeed, consider $X \subseteq$ Sper A and let $T := P(X)$ be the associated preordering as in Sect. 3.10. Again we will say "X-convex" instead of "T-convex". The relation $\le_T$ on A takes the form

$$a \le_T b \Leftrightarrow a(x) \le b(x) \text{ for all } x \in X.$$

Now if $\Lambda \subseteq A$ is a subring, it is immediate that the subring $o_X(A/\Lambda)$ is just the X-convex hull of Λ in A (which is a subring again in the current situation). Therefore this section essentially deals with subrings that are convex with respect to a preordering of the large ring. The subring Λ is archimedean closed in A on X if and only if Λ is X-convex in A.

The construction of the archimedean closure is functorial in the following sense: if

$$
\begin{array}{ccc}
A' & \xrightarrow{\;\alpha\;} & A \\
\uparrow & & \uparrow \\
\Lambda' & \longrightarrow & \Lambda
\end{array}
$$

is a commutative diagram of ring homomorphisms and if $X \subseteq \mathrm{Sper}\,A$ and $X' \subseteq \mathrm{Sper}\,A'$ are subsets such that $(\mathrm{Sper}\,\alpha)(X) \subseteq X'$, then $\alpha\big(o_{X'}(A'/\Lambda')\big) \subseteq o_X(A/\Lambda)$. The proof is clear.

Examples and Remarks 3.11.2

(1) If $A = K$ is a field, Λ a subring of K and $X = \{x\}$ just one ordering of K, then the subring $o_x(K/\Lambda)$ was already introduced in Definition 2.1.8: $o_x(K/\Lambda)$ is the convex hull of Λ in K with respect to x.
(2) In the general situation, the following identities hold for every subset X of $\mathrm{Sper}\,A$:

$$
o_X(A/\Lambda) = o_{\overline{X}}(A/\Lambda) = o_{\widehat{X}}(A/\Lambda),
$$

where $\widehat{X}$ denotes the probasic closed subspace generated by X (cf. Sect. 3.10). Indeed, by definition an element $a \in A$ is in $o_X(A/\Lambda)$ if and only if there exists a $\lambda \in \Lambda$ such that $X \subseteq \overline{H}\big(\varphi(\lambda)^2 - a^2\big)$. (Since the occurring rings are convex hulls with respect to preorderings, the identities follow from $P(X) = P(\overline{X}) = P(\widehat{X})$.)

Definition 3.11.3 If X is a subset of $\mathrm{Sper}\,A$, we define $o_X(A) := o_X(A/\mathbb{Z})$ and call the elements of $o_X(A)$ the *absolutely bounded elements* (*of A*) *on X* (with respect to the unique homomorphism $\mathbb{Z} \to A$ of course). In case $X = \mathrm{Sper}\,A$, the elements of $o(A) := o_{\mathrm{Sper}\,A}(A)$ are called the *absolutely bounded* elements of A. If $A = K$ is a field, $o(K)$ is called the *(real) holomorphy ring of K*. We will study it in more detail in Sect. 3.12.

From the functoriality of the archimedean closure mentioned above it follows that $\alpha\big(o(A)\big) \subseteq o(A')$ for every homomorphism $\alpha \colon A' \to A$. More generally we have $\alpha\big(o_X(A)\big) \subseteq o_{X'}(A')$ for all subsets $X \subseteq \mathrm{Sper}\,A$ and $X' \subseteq \mathrm{Sper}\,A'$ that satisfy $(\mathrm{Sper}\,\alpha)(X') \subseteq X$.

Example 3.11.4 If $a_1, \ldots, a_r \in A$ and $1 + a_1^2 + \cdots + a_r^2$ is a unit in A, then $a_1(1 + a_1^2 + \cdots + a_r^2)^{-1}$ is an element of $o(A)$ and is thus absolutely bounded. In particular, if F is a real field, then all elements of the form $a_1(1 + a_1^2 + \cdots + a_r^2)^{-1}$ with $a_1, \ldots, a_r \in F$ are absolutely bounded. More generally for any preordering T of A, every element $b \in A$ that satisfies an equation of the form $b(1 + a^2 + t) = a$ with $a \in A$ and $t \in T$ is absolutely bounded on $\overline{H}(T)$.

Examples of archimedean ring extensions can be obtained from

Proposition 3.11.5 *Every integral extension of rings $A \supseteq \Lambda$ is archimedean (on the whole of Sper A).*

Proof Every $a \in A$ satisfies an equation of the form

$$a^n + \lambda_1 a^{n-1} + \cdots + \lambda_n = 0$$

for certain $\lambda_i \in \Lambda$. From Proposition 1.7.1 it follows that

$$|a(x)| \le \max\{1, \ |\lambda_1(x)| + \cdots + |\lambda_n(x)|\}$$

for every $x \in$ Sper A. Since the inequality $|\lambda| \le 1 + \lambda^2$ holds in every ordered field we conclude that

$$|a(x)| \le \left(n + 1 + \lambda_1^2 + \cdots + \lambda_n^2\right)(x). \qquad \square$$

We now prove a result that goes substantially beyond Proposition 3.11.5 and which characterizes the archimedean closure of a subring Λ on a subset of the real spectrum. We may assume that char $A = 0$, i.e., that $\mathbb{Z} \subseteq A$, since otherwise Sper $A = \varnothing$ and the concepts become meaningless.

Definition 3.11.6 Let Λ be a ring with $\mathbb{Z} \subseteq \Lambda$. We call a polynomial $f(t) \in \Lambda[t]$ *almost monic* if its leading coefficient is a positive integer.

Theorem 3.11.7 (G.W. Brumfiel [17, Ch. VI]) *Let A be a ring with $\mathbb{Z} \subseteq A$, Λ a subring of A and X a subset of Sper A. For every $a \in A$ the following statements are equivalent:*

(i) *$a \in o_X(A/\Lambda)$, i.e., a is bounded on X over Λ;*
(ii) *there exists an almost monic non-constant polynomial $f(t) \in \Lambda[t]$ such that $f(a^2) \le 0$ on X;*
(iii) *there exists an almost monic non-constant polynomial $f(t) \in \Lambda[t]$ of even degree such that $f(a) \le 0$ on X.*

(For $b \in A$, the expression "$b \le 0$ on X" of course means that $b(x) \le 0$ for all $x \in X$.)

Proof If $a \in o_X(A/\Lambda)$ and $\lambda \in \Lambda$ such that $|a(x)| \le \lambda(x)$ for all $x \in X$, then the polynomial $f(t) = t - \lambda^2$ satisfies (ii). The implication (ii) $\Rightarrow$ (iii) is trivial (just replace $f(t)$ by $f(t^2)$). It remains to show (iii) $\Rightarrow$ (i). For this implication we appeal to the remarkable

Lemma 3.11.8 (G.W. Brumfiel) *Let $n \in \mathbb{N}$ and consider independent variables $t, u_1, \ldots, u_{2n}$ over $\mathbb{Q}$. Let $u := (u_1, \ldots, u_{2n})$. Then there exists $k \in \mathbb{N}$ and polynomials $b(u) \in \mathbb{Q}[u]$ and $h_1(u, t), \ldots, h_k(u, t) \in \mathbb{Q}[u, t]$ such that*

$$t^{2n} + u_1 t^{2n-1} + \cdots + u_{2n} = b(u) + \sum_{i=1}^{k} h_i(u, t)^2.$$

Thus, if Λ is a ring such that $\mathbb{Z} \subseteq \Lambda$ and $f \in \Lambda[t]$ is an almost monic polynomial of even degree $2n$, then there exists a positive integer s, an element $\beta(f) \in \Lambda$ and a polynomial $h(t)$ that is a sum of squares in $\Lambda[t]$ such that $\beta(f) = sf(t) - h(t)$. Returning to the proof of (iii) $\Rightarrow$ (i), if we apply this observation to the polynomials $f(t) + t$ and $f(t) - t$, we obtain a positive integer s and elements $\beta, \widetilde{\beta} \in \Lambda$ such that $\beta = s(f(t) + t) - h(t)$ and $\widetilde{\beta} = s(-f(t) + t) + \widetilde{h}(t)$, where $h, \widetilde{h}$ are sums of squares in $\Lambda[t]$. The substitution $t = a$ then gives

$$\beta(x) \leq sa(x) \leq \widetilde{\beta}(x)$$

for all $x \in X$, from which (i) follows immediately. $\qquad\square$

Proof of Lemma 3.11.8 By induction on n. The case $n = 1$ follows from

$$t^2 + u_1 t + u_2 = \left(t + \tfrac{u_1}{2}\right)^2 + \left(u_2 - \tfrac{u_1^2}{4}\right).$$

If $n > 1$, we apply the "Tschirnhausen transformation" $s := t - \tfrac{u_1}{2n}$ and obtain

$$t^{2n} + u_1 t^{2n-1} + \cdots + u_{2n} = s^{2n} + v_2 s^{2n-2} + v_3 s^{2n-3} + \cdots + v_{2n},$$

where $v_i \in \mathbb{Q}[u]$. Rewriting the right hand side as

$$\left(s^n + \tfrac{v_2-1}{2} s^{n-2}\right)^2 + \left(s^{2n-2} + v_3 s^{2n-3} + \left(v_4 - \left(\tfrac{v_2-1}{2}\right)^2\right) s^{2n-4} + \cdots\right),$$

we can apply the induction hypothesis to the second summand, which proves the claim for any n. $\qquad\square$

Now we come to a different description of the archimedean closure of Λ in A on X, in which a more prominent role is played by the preordering $P(X)$ of elements of A that are non-negative on X. If T is an arbitrary preordering of A, we will also write $\mathfrak{o}_T(A/\Lambda)$ instead of $\mathfrak{o}_{\overline{H}(T)}(A/\Lambda)$ (and $\mathfrak{o}_T(A)$ instead of $\mathfrak{o}_T(A/\mathbb{Z})$). Thus $\mathfrak{o}_T(A/\Lambda)$ is the $\widehat{T}$-convex hull of Λ in A. Because of Example 3.11.2(2), every ring $\mathfrak{o}_X(A/\Lambda)$ is of the form $\mathfrak{o}_T(A/\Lambda)$, for instance with $T = P(X)$.

Proposition 3.11.9 *Let Λ be a subring of A and T a preordering of A.*

(a) *An element $a \in A$ is in $\mathfrak{o}_T(A/\Lambda)$ if and only if there exists $\lambda \in \Lambda$ and $t \in T$ such that*

$$(1+t)(\lambda^2 + a) \in T \quad and \quad (1+t)(\lambda^2 - a) \in T.$$

(b) *An element $a \in A$ is in $\mathfrak{o}_T(A)$ if and only if there exists $n \in \mathbb{N}$ and $t \in T$ such that*

$$(1+t)(n+a) \in T \quad and \quad (1+t)(n-a) \in T.$$

Proof From the existence of λ (resp. n) and t in (a) (resp. (b)), it follows that $-\lambda^2 \leq a \leq \lambda^2$ (resp. $-n \leq a \leq n$) on $X := \overline{H}(T)$. Hence it suffices to verify the converse direction in (a). Given $a \in \mathfrak{o}_T(A/\Lambda)$, we choose $\mu \in \Lambda$ such that $|a(x)| \leq |\mu(x)|$ for all $x \in X$, and set $\lambda := 1+\mu^2$. Since $|a| \leq |\mu| < \lambda \leq \lambda^2$ on X, $\lambda^2 \pm a$ are positive on X and the statement follows from Prestel's Positivstellensatz (Proposition 3.9.5, the (i) $\Rightarrow$ (ii) implication). $\square$

Corollary 3.11.10 *If every element of $1 + T$ is a unit in A (for instance, if A is a field and T is a proper preordering of A), then*

$$\mathfrak{o}_T(A/\Lambda) = \{a \in A : \text{ there exists } \lambda \in \Lambda \text{ such that } \lambda^2 \pm a \in T\}$$

and

$$\mathfrak{o}_T(A) = \{a \in A : \text{ there exists } n \in \mathbb{N} \text{ such that } n \pm a \in T\}.$$ $\square$

Corollary 3.11.11 *If every element of $1 + \Sigma A^2$ is a unit in A, then*

$$\mathfrak{o}\big(\mathfrak{o}(A/\Lambda)/\Lambda\big) = \mathfrak{o}(A/\Lambda).$$

Proof Let $B := \mathfrak{o}(A/\Lambda)$. We need to show that every $b \in B$ is bounded on the whole of Sper B over Λ. Applying Corollary 3.11.10 (with $T = \Sigma A^2$) provides elements $\lambda \in \Lambda$ and $a_1, \ldots, a_r \in A$ and an equation

$$\lambda^4 - b^2 = a_1^2 + \cdots + a_r^2.$$

Every $|a_i|$ is bounded by λ^2 on the whole of Sper A, which shows that the a_i are in B. We conclude that $|b|$ is bounded by λ^2 on the whole of Sper B. $\square$

In the case of fields, Corollary 3.11.10 provides a parametrization of the elements of $\mathfrak{o}_T(A/\Lambda)$:

Proposition 3.11.12 *Consider a field K, a proper preordering T of K and a subring $\Lambda \subseteq K$. Then $\mathfrak{o}_T(K/\Lambda)$ is the set of all elements*

$$\frac{\lambda a}{1 + a^2 + t}$$

such that $\lambda \in \Lambda$, $a \in K$ and $t \in T$.

Proof By Example 3.11.4 every such element is in $\mathfrak{o}_T(K/\Lambda)$. Conversely, given an element $z \neq 0$ in $\mathfrak{o}_T(K/\Lambda)$, Corollary 3.11.10 yields an element $\lambda \in \Lambda$ such that $\lambda^2 - z^2 \in T$, i.e., such that $\left(\frac{\lambda}{z}\right)^2 - 1 =: t \in T$. Setting $a := \frac{\lambda}{z}$, we have $2\frac{\lambda}{z}a = 2a^2 = 1 + a^2 + t$, from which we obtain

$$z = \frac{2\lambda a}{1 + a^2 + t},$$

as required.   □

Corollary 3.11.13 *If K is a field, X a subset of $\operatorname{Sper} K$ and Λ a subring of K, then*

$$\mathfrak{o}_X(K/\Lambda) = \Lambda \cdot \mathfrak{o}_X(K) = \{\lambda z : \lambda \in \Lambda, \, z \in \mathfrak{o}_X(K)\}.$$ □

Corollary 3.11.14 *If K is a field and T a proper preordering of K, then $\mathfrak{o}_T(K)$ is generated as an additive group by the elements $1/(1 + t)$ with $t \in T$.*

Proof This follows from Proposition 3.11.12 and the identity

$$\frac{2a}{1 + a^2 + t} = \frac{(a + 1)^2 - (a - 1)^2}{(a + 1)^2 + (a - 1)^2 + 2t}.$$ □

We return to the case of an arbitrary commutative ring A and want to determine the dependence of the ring $\mathfrak{o}_X(A/\Lambda)$ on the subset $X \subseteq \operatorname{Sper} A$ when the subring Λ is kept fixed. Since $Y \subseteq X$ implies $\mathfrak{o}_Y(A/\Lambda) \supseteq \mathfrak{o}_X(A/\Lambda)$ we always have

$$\mathfrak{o}_X(A/\Lambda) \subseteq \bigcap_{x \in X} \mathfrak{o}_x(A/\Lambda),$$

where the inclusion is strict in general, as the following example shows:

Example 3.11.15 Let R be a real closed field, $A = R[t]$ the polynomial ring in one variable and $\Lambda = R$. Every non-constant polynomial $f \in R[t]$ is in precisely two points of $\operatorname{Sper} R[t]$ unbounded over R, namely in $\pm\infty$ (cf. Example 3.3.14). However, on the open subsets $X := (\operatorname{Sper} R[t]) \setminus \{\pm\infty\}$ of $\operatorname{Sper} R[t]$ only the constant polynomials are archimedean over R. Hence,

$$\mathfrak{o}_X\big(R[t]/R\big) = R \neq R[t] = \bigcap_{x \in X} \mathfrak{o}_x\big(R[t]/R\big).$$

Nevertheless we will shortly see that if X is proconstructible, then the inclusion is an equality. But first:

Proposition 3.11.16 *Let Λ be a subring of A. For every $a \in A$, the "unbounded locus"*

$$\Omega(a/\Lambda) := \{x \in \operatorname{Sper} A : a \notin \mathfrak{o}_x(A/\Lambda)\}$$

of a over Λ is closed in $\operatorname{Sper} A$ and for each point in it, its generalizations are also in it.

Proof We have

$$\Omega(a/\Lambda) = \{x \in \operatorname{Sper} A : \forall \lambda \in \Lambda \ |a(x)| \geq |\lambda(x)|\} = \bigcap_{\lambda \in \Lambda} \overline{H}(a^2 - \lambda^2),$$

and since we may replace "$\geq$" by "$>$", we also have

$$\Omega(a/\Lambda) = \bigcap_{\lambda \in \Lambda} \mathring{H}(a^2 - \lambda^2).$$

The claims then follow from these two representations.   □

Corollary 3.11.17 *If $x, y \in \operatorname{Sper} A$ are such that $y \succ x$, then $\mathfrak{o}_y(A/\Lambda) = \mathfrak{o}_x(A/\Lambda)$.*   □

Example 3.11.18 Consider again a real closed field R and let V be an affine R-variety with associated coordinate algebra $A = R[V]$. Let $u_1, \ldots, u_n$ be generators of $R[V]$ over R and $a := u_1^2 + \cdots + u_n^2 \in A$. Then $\Omega(a/R)$ consists of all elements $x \in \operatorname{Sper} A = \widetilde{V(R)}$ whose "distance to the origin" (with respect to the embedding of V into the affine space $\mathbb{A}^n$ given by $u_1, \ldots, u_n$) is infinitely large compared to R. By Proposition 3.11.16, $\Omega(a/R)$ consists precisely of all the generalizations of the points of

$$\Omega^{\max}(a/R) := \Omega(a/R) \cap (\operatorname{Sper} A)^{\max}.$$

Corollary 3.7.18 then gives

$$\Omega^{\max}(a/R) = \{x \in (\operatorname{Sper} A)^{\max} : k(x) \text{ is not archimedean over } R\}.$$

In particular, $\Omega(a/R)$ is independent of the choice of generators of $R[V]$ and can thus be referred to as the set of *infinitely distant points* in $\widetilde{V(R)}$. This fits well with Remark 3.6.6.

Proposition 3.11.19 *Let Λ be a subring of A and X a proconstructible subset of* Sper A. *If $a \in A$ is bounded in every $x \in X$ over Λ, then there exists an open constructible neighbourhood U of $\overline{X}$ in* Sper A *on which a is bounded over Λ.*

Proof By Proposition 3.11.16, a is also bounded in every $x \in \overline{X}$ over Λ. Hence we may assume that $X = \overline{X}$. For every $x \in X$ there exists $\lambda_x \in \Lambda$ such that $|a(x)| < \lambda_x(x)$. Since X is quasi-compact, there are finitely many $x_1, \ldots, x_N \in X$ such that

$$X \subseteq \bigcup_{i=1}^{N} \{y \in \text{Sper } A : |a(y)| < \lambda_{x_i}(y)\} =: U.$$

The set U is an open constructible neighbourhood of $X = \overline{X}$, and letting $\lambda := 1 + \lambda_{x_1}^2 + \cdots + \lambda_{x_N}^2$ we have $|a(y)| < \lambda(y)$ for all $y \in U$. $\qquad\square$

Corollary 3.11.20 *Let Λ be a subring of A and X a proconstructible subset of* Sper A. *Then*

$$\mathfrak{o}_X(A/\Lambda) = \bigcap_{x \in X} \mathfrak{o}_x(A/\Lambda),$$

where we may restrict the intersection on the right hand side to the points x of $X^{\max}$ or $X^{\min}$.

Proof This follows from Proposition 3.11.19 and Corollary 3.11.17. $\qquad\square$

Remark 3.11.21 Let V be an affine variety over a real closed field R and $A = R[V]$ its coordinate algebra. Given a semialgebraic subset M of $V(R)$ with associated constructible subset $\widetilde{M}$ of Sper A, one might ask if $\mathfrak{o}_{\widetilde{M}}(A/\Lambda)$ is already the intersection of the rings $\mathfrak{o}_x(A/\Lambda)$ with $x \in M$, for instance for $\Lambda = R$. In general the answer to this question is negative. Trivially, $\mathfrak{o}_x(A/R) = A$ for all $x \in V(R)$, whereas only in special cases the R-valued function $x \mapsto a(x)$ is bounded on M for every $a \in A$. In all other cases, $\mathfrak{o}_{\widetilde{M}}(A/R)$ is a proper subring of A. In connection with Corollary 3.11.20, the usefulness of "ideal points" in $\widetilde{M}$ becomes clear at this point once more: For every polynomial function $f \in A$ that is unbounded on M, there exists such a point in $\widetilde{M}$, and in this point f does indeed attain an infinitely large function value compared to R.

3.12 Prüfer Rings and the Real Holomorphy Ring of a Field

We finish Chap. 3 with a detailed study of the holomorphy ring $\mathfrak{o}(F)$ of a (real) field F. The well-developed theory of the holomorphy ring—and more generally the ring $\mathfrak{o}_T(F)$—as well as interesting applications to power sums in fields is mainly due to E. Becker and his student H.-W. Schülting (cf. [7, 8, 106], ...). A key notion in this context is the concept of a Prüfer ring. We will only explain the rudiments of this theory, which for the most part already go back to the work of D.W. Dubois [37]. Our presentation (strongly) emphasizes the use of the real spectrum and differs in this respect from these earlier works. Once again the real spectrum often plays a helpful and clarifying role.

Definition 3.12.1

(a) An integral domain A is called a *Prüfer ring* if every localization $A_\mathfrak{p}$ of A at a prime ideal $\mathfrak{p}$ is a valuation ring. If in addition A is a subring of a field F and if F is the field of fractions of A, then A is called a *Prüfer ring of F*.

(b) The Prüfer ring A is called *residually real* if for all $\mathfrak{p} \in \operatorname{Spec} A$ the residue field $\kappa(\mathfrak{p}) = A_\mathfrak{p}/\mathfrak{p} A_\mathfrak{p}$ is real (i.e., if all valuation rings $A_\mathfrak{p}$ are residually real in the sense of Sect. 2.2). The notion of *residually real Prüfer ring of F* is defined analogously.

Remarks 3.12.2

(1) Every valuation ring B is a Prüfer ring. If in addition $\kappa(B) = B/\mathfrak{m}_B$ is real, then B is convex in $F := \operatorname{Quot}(B)$ with respect to an ordering P of F by Corollary 2.7.2. The same is true for every overring of B in F, and it follows that $B_\mathfrak{p}/\mathfrak{p} B_\mathfrak{p}$ is real for every $\mathfrak{p} \in \operatorname{Spec} B$. This shows that our two definitions of "residually real" (Definitions 3.12.1(b) and 2.2.5) are equivalent.

(2) If A is a Prüfer ring of F, then every overring B of A in F is also a Prüfer ring since $B_\mathfrak{q}$ is an overring of $A_{\mathfrak{q} \cap A}$ in F for every $\mathfrak{q} \in \operatorname{Spec} B$. Similarly, every overring (in F) of a residually real Prüfer ring of F is again a residually real Prüfer ring (cf. Remark 3.12.2(1)).

(3) Let A be an integral domain. If all localizations $A_\mathfrak{m}$ at maximal ideals $\mathfrak{m}$ of A are valuation rings (resp. residually real valuation rings), then A is already a Prüfer ring (resp. a residually real Prüfer ring) since for every $\mathfrak{p} \in \operatorname{Spec} A$ such that $\mathfrak{p} \subseteq \mathfrak{m}$, $A_\mathfrak{p}$ is an overring of $A_\mathfrak{m}$ in the field of fractions of A.

(4) If A is an integral domain and $F = \operatorname{Quot} A$, then $A = \bigcap_\mathfrak{m} A_\mathfrak{m}$ where the intersection is taken over all maximal ideals $\mathfrak{m}$ of A in F. (This is easy to see: Given $\lambda \in \bigcap_\mathfrak{m} A_\mathfrak{m}$, the ideal of all elements a of A such that $a\lambda \in A$ is not contained in any maximal ideal of A, hence must be the unit ideal.) It follows that every Prüfer ring of F is the intersection of valuation rings of F and thus in particular integrally closed.

(5) If A is a Prüfer ring of F and K a further field, then there exists a natural bijection from the set of places $\lambda\colon F \to K \cup \infty$ that are finite on A to $\operatorname{Hom}(A, K)$, namely $\lambda \mapsto \lambda|_A$. (Here and later $\operatorname{Hom}(\cdot, \cdot)$ denotes the set of

ring homomorphisms.) Indeed, every homomorphism $\varphi: A \to K$ has a unique extension to a homomorphism from a valuation ring that contains A (namely $A_{\mathrm{Ker}\,\varphi}$) to K, thus to a place $F \to K \cup \infty$ (that is integral on A).

Exercise 3.12.3 Let A be a Prüfer ring of F. Show that the map $\mathfrak{p} \mapsto A_{\mathfrak{p}}$ is a bijection from Spec A to the set of all valuation rings B of F that contain A and that the inverse map is given by $B \mapsto A \cap \mathfrak{m}_B$.

The importance of Prüfer rings for real algebra is a consequence of the following theorem of A. Dress [36] (see also [70, p. 86]):

Theorem 3.12.4 *Let F be a field of characteristic $\neq 2$. Then the subring A of F that is generated by all elements of the form*

$$\frac{1}{1 + a^2} \quad (a \in F,\ a^2 \neq -1)$$

is a Prüfer ring of F.

Proof Let $\mathfrak{p}$ be a prime ideal of A and $B := A_{\mathfrak{p}}$. Consider $x \in F^*$. We must show that $x \in B$ or $x^{-1} \in B$. First of all, $x^2 \in B$ or $x^{-2} \in B$. This is clear if $x^2 = -1$. Otherwise this follows from the identity

$$\frac{1}{1 + x^2} + \frac{1}{1 + x^{-2}} = 1,$$

since the summands on the left hand side are not both in $\mathfrak{p}$, i.e., since one of those summands is invertible in B. We may assume without loss of generality that $x^2 \in B$. Let $y := 1 + x$. If $y^2 \in B$, then $2x = y^2 - 1 - x^2 \in B$, hence $x \in B$. Otherwise, $y \neq 0$ and $y^{-2} \in B$. Since $(y - 1)^2 = x^2 \in B$ we also have

$$y^{-2}(y - 1)^2 = 1 - 2y^{-1} + y^{-2} \in B,$$

hence $y^{-1} \in B$. From $1 - x = y^{-1}(1 - x^2)$ we then conclude that $x \in B$, as required. $\qquad \square$

We recall that $\mathfrak{o}(F)$ denotes the real holomorphy ring of F (cf. Definition 3.11.3).

Theorem 3.12.5 *Consider a real field F. The following statements are equivalent for every subring A of F:*

(i) $\mathfrak{o}(F) \subseteq A$;
(ii) *A is a residually real Prüfer ring of F;*
(iii) *A is archimedean closed in F, i.e., $\mathfrak{o}(F/A) = A$;*
(iv) *A is a ΣF^2-convex subring of F;*
(v) *there exists a subring Λ of F and a subset X of Sper F such that $A = \mathfrak{o}_X(F/\Lambda)$.*

In particular, the real holomorphy ring $\mathfrak{o}(F)$ *of* F *is the smallest residually real Prüfer ring of* F.

Proof (i) $\Rightarrow$ (ii): Since $\mathfrak{o}(F)$ contains all elements of the form $(1 + a^2)^{-1}$ with $a \in F$, $\mathfrak{o}(F)$ is a Prüfer ring of F by Theorem 3.12.4. Consider a valuation ring B of F such that $\mathfrak{o}(F) \subseteq B$. If B is not residually real, there exist $b_1, \ldots, b_r \in B$ such that $1 + b_1^2 + \cdots + b_r^2 \in \mathfrak{m}_B$, which is impossible since $(1 + b_1^2 + \cdots + b_r^2)^{-1} \in \mathfrak{o}(F) \subseteq B$. Hence A is a residually real Prüfer ring of F.

(ii) $\Rightarrow$ (iii): Since intersections of archimedean closed subrings of F are again archimedean closed in F, it suffices to prove that residually real valuation rings of F are archimedean closed. If B is a residually real valuation ring of F, then B is convex in F with respect to some $x \in \mathrm{Sper}\, F$ (cf. Sect. 2.7). It follows that $\mathfrak{o}_x(F/B) = B$, hence a fortiori $\mathfrak{o}(F/B) = \mathfrak{o}_{\mathrm{Sper}\, F}(F/B) = B$.

The implications (iii) $\Rightarrow$ (iv) $\Rightarrow$ (v) $\Rightarrow$ (i) are trivial. $\square$

Corollary 3.12.6 *The real holomorphy ring* $\mathfrak{o}(F)$ *is the intersection of all residually real valuation rings of* F. $\square$

Remark 3.12.7 Corollary 3.12.6 is already contained in earlier results: By Corollary 3.11.20, $\mathfrak{o}(F)$ is the intersection of the residually real valuation rings $\mathfrak{o}_x(F)$ with $x \in \mathrm{Sper}\, F$, and every residually real valuation ring of F contains such an $\mathfrak{o}_x(F)$ by Sect. 2.7. More generally we know from Sect. 3.11 that every ring $\mathfrak{o}_X(F/\Lambda)$ (with $X \subseteq \mathrm{Sper}\, F$ and Λ a subring of F) is the intersection of the residually real valuation rings $\mathfrak{o}_x(F/\Lambda)$ with $x \in \overline{X}$.

In what follows we fix a field F, assumed to be real without loss of generality, and investigate the real spectrum of its holomorphy ring $\mathfrak{o}(F)$ in some more detail. For simplicity we write $\mathfrak{o} := \mathfrak{o}(F)$.

Proposition 3.12.8 *Let* A *be a residually real Prüfer ring of* F. *Viewing* $\mathrm{Sper}\, F$ *in the usual way as a subspace of* $\mathrm{Sper}\, A$, *we have* $\mathrm{Sper}\, F = (\mathrm{Sper}\, A)^{\min}$.

Proof Let $x \in \mathrm{Sper}\, A$, $\mathfrak{p} := \mathrm{supp}(x)$, $B := A_{\mathfrak{p}}$ and $\overline{x}$ the ordering of $\kappa(\mathfrak{p}) = B/\mathfrak{m}_B$ induced by x. By the Baer–Krull Theorem there exists an ordering y of F that makes B convex and that induces the ordering $\overline{x}$ of $B/\mathfrak{m}_B$. Viewing y and $\overline{x}$ as elements of $\mathrm{Sper}\, B$, it is immediately clear that $y \succ \overline{x}$. Hence the inclusion $\mathrm{Sper}\, B \subseteq \mathrm{Sper}\, A$ shows that y is a generalization of x in $\mathrm{Sper}\, A$ with support (0). $\square$

What is happening with the maximal (= closed) points of $\mathrm{Sper}\, \mathfrak{o}$? Let $x \in \mathrm{Sper}\, \mathfrak{o}$, $\mathfrak{p} := \mathrm{supp}(x)$ and $\overline{x}$ the induced ordering of $\kappa(\mathfrak{p})$. By Corollary 3.7.18, x is a closed point if and only if $\kappa(\mathfrak{p})$ is archimedean over $\mathfrak{o}/\mathfrak{p}$ with respect to $\overline{x}$. Since $\mathfrak{o}/\mathfrak{p}$ is archimedean over $\mathbb{Z}$ with respect to $\overline{x}$ by definition of $\mathfrak{o}$, x being closed is equivalent to $\overline{x}$ being an archimedean ordering of $\kappa(\mathfrak{p})$, i.e., equivalent to the existence of an order preserving embedding $\kappa(\mathfrak{p}) \hookrightarrow \mathbb{R}$. Since this embedding is uniquely determined by $\overline{x}$, we obtain:

Proposition 3.12.9 *The canonical map* $\mathrm{Hom}(\mathfrak{o}, \mathbb{R}) \to \mathrm{Sper}\, \mathfrak{o}$ *is a bijection from* $\mathrm{Hom}(\mathfrak{o}, \mathbb{R})$ *to* $(\mathrm{Sper}\, \mathfrak{o})^{\max}$. $\square$

Corollary 3.12.10 *If F contains* $\mathbb{R}$*, then the support map* supp: $\operatorname{Sper} \mathfrak{o} \to \operatorname{Spec} \mathfrak{o}$
yields a bijection from $(\operatorname{Sper} \mathfrak{o})^{\max}$ *to the space* $(\operatorname{Spec} \mathfrak{o})^{\max}$ *of maximal ideals of* $\mathfrak{o}$*.*

Note that this bijection is not a homeomorphism in general.

Proof Since the field $\mathbb{R}$ does not possess any proper endomorphisms by Corollary 2.1.12, every homomorphism $\varphi\colon \mathfrak{o} \to \mathbb{R}$ is surjective, and φ is determined by its kernel. Thus $\operatorname{supp}(x)$ is a maximal ideal for every $x \in (\operatorname{Sper} \mathfrak{o})^{\max}$ by Proposition 3.12.9. Since all prime ideals of $\mathfrak{o}$ are real reduced, the map supp: $(\operatorname{Sper} \mathfrak{o})^{\max} \to (\operatorname{Spec} \mathfrak{o})^{\max}$ is surjective. In the commutative diagram

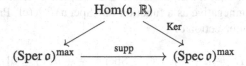

the map Ker is injective. Thus the map supp is also injective by Proposition 3.12.9.
□

By identifying $\operatorname{Hom}(\mathfrak{o}, \mathbb{R})$ with $(\operatorname{Sper} \mathfrak{o})^{\max}$ we thus obtain a compact (Hausdorff) topology on $\operatorname{Hom}(\mathfrak{o}, \mathbb{R})$ (cf. Theorem 3.6.4), which can easily be described without reference to the real spectrum:

Proposition 3.12.11 *The topology induced by* $\operatorname{Hom}(\mathfrak{o}, \mathbb{R}) \subseteq \operatorname{Sper} \mathfrak{o}$ *on* $\operatorname{Hom}(\mathfrak{o}, \mathbb{R})$
coincides with the subspace topology of $\operatorname{Hom}(\mathfrak{o}, \mathbb{R})$ *in* $\mathbb{R}^{\mathfrak{o}} = \prod_{\mathfrak{o}} \mathbb{R}$*.*

Proof Let $\mathcal{T}$, resp. $\mathcal{T}'$, denote the subspace topology of $\operatorname{Hom}(\mathfrak{o}, \mathbb{R})$ in $\operatorname{Sper} \mathfrak{o}$, resp. in $\mathbb{R}^{\mathfrak{o}}$. A subbasis of open sets of $\mathcal{T}$ is given by all the sets

$$S(a) := \{\varphi \in \operatorname{Hom}(\mathfrak{o}, \mathbb{R})\colon \varphi(a) > 0\},$$

where $a \in A$, because $S(a) = \operatorname{Hom}(\mathfrak{o}, \mathbb{R}) \cap \mathring{H}_{\mathfrak{o}}(a)$. Since all $S(a)$ are $\mathcal{T}'$-open, $\mathcal{T}'$ is finer than $\mathcal{T}$. A subbasis of open sets of $\mathcal{T}'$ is given by all the sets

$$S(a, p, q) := \{\varphi \in \operatorname{Hom}(\mathfrak{o}, \mathbb{R})\colon p < \varphi(a) < q\}$$

where $a \in A$ and $p, q \in \mathbb{Q}$ with $p < q$. Since $S(a, p, q) = S(a - p) \cap S(q - a)$ (note that $\mathbb{Q} \subseteq \mathfrak{o}$), $\mathcal{T}$ is also finer than $\mathcal{T}'$. □

In the following we denote the compact space $\operatorname{Hom}(\mathfrak{o}, \mathbb{R})$ simply by Y, and investigate the ring homomorphism

$$\tau\colon \mathfrak{o} \to C(Y, \mathbb{R}), \ a \mapsto \widehat{a},$$

where $C(Y, \mathbb{R})$ denotes the ring of continuous $\mathbb{R}$-valued functions on Y and for all $a \in \mathfrak{o}, \widehat{a}$ is defined by $\widehat{a}(\varphi) := \varphi(a)$ for all $\varphi \in Y$.

It follows immediately from the definition that the functions $\widehat{a}$ separate the points of Y. Furthermore, $\mathbb{Q}$ is contained in $\mathfrak{o}$. The Stone–Weierstrass Theorem [14, Ch. 10,

§4, Th. 3] then implies that $\tau(o)$ is a dense subring of $C(Y, \mathbb{R})$ with respect to the uniform convergence topology.

The subset $C^+(Y, \mathbb{R})$ of nonnegative functions is a preordering of the ring $C(Y, \mathbb{R})$ (identical to the set of squares). Its preimage under τ satisfies

Proposition 3.12.12 *Let $a \in o$. Then $\widehat{a} \in C^+(Y, \mathbb{R})$ if and only if $a + \frac{1}{n}$ is a sum of squares (in F or o) for every $n \in \mathbb{N}$.*

Proof First of all, it is clear that $\Sigma o^2 = o \cap \Sigma F^2$, since if $a_1^2 + \cdots + a_r^2$ is bounded over $\mathbb{Z}$ (on Sper F), then so is each a_i. If $a + \frac{1}{n} \in \Sigma F^2$, then $\widehat{a} + \frac{1}{n} \in C^+(Y, \mathbb{R})$, and if this holds for all $n \in \mathbb{N}$ it follows that $\widehat{a} \in C^+(Y, \mathbb{R})$. Conversely, $\widehat{a} \in C^+(Y, \mathbb{R})$ implies that a is nonnegative as a function on (Sper o)$^{\max}$ (cf. Proposition 3.12.9). For all $n \in \mathbb{N}$, the intersection of

$$\overline{H}_o\left(-a - \tfrac{1}{n}\right) = \{x \in \text{Sper } o : \tfrac{1}{n} + a(x) \le 0\}$$

and (Sper o)$^{\max}$ is thus empty. Hence, $\overline{H}_o\left(-a - \frac{1}{n}\right) = \varnothing$ and so $a + \frac{1}{n} > 0$ on Sper o. In particular, $a + \frac{1}{n}$ is positive on Sper F (cf. Proposition 3.12.8) and hence a sum of squares (cf. Sect. 1.1). $\qquad\square$

Corollary 3.12.13 *The kernel of $\tau : o \to C(Y, \mathbb{R})$ consists precisely of those $a \in F$ for which $\frac{1}{n} \pm a$ is a sum of squares (in o or F) for all $n \in \mathbb{N}$, i.e., those elements of F that are "infinitesimal" with respect to $\mathbb{Q}$ on the whole of Sper F.* $\qquad\square$

Just as for (Sper o)$^{\min}$, there is a purely field theoretic description of (Sper o)$^{\max}$ that makes no reference to o. We use the following notation:

Definition 3.12.14 The set of all $\mathbb{R}$-valued places $F \to \mathbb{R} \cup \infty$ of F is denoted by $M(F)$.

Every $\lambda \in M(F)$ is finite on o by Corollary 3.12.6. There is thus a natural map $M(F) \to \text{Hom}(o, \mathbb{R})$, $\lambda \mapsto \lambda|_o$, which is a bijection by Remark 3.12.2(5). Hence we may identify the set $M(F)$ with $Y = \text{Hom}(o, \mathbb{R})$ in what follows. Note that by Proposition 3.12.9 there is then also a natural bijection between $M(F)$ and (Sper o)$^{\max}$.

In order to transfer the (compact) topology from $\text{Hom}(o, \mathbb{R})$, resp. (Sper o)$^{\max}$ (Proposition 3.12.11) to $M(F)$, we view the elements $\lambda \in M(F)$ as maps from F to the projective line $\mathbb{P}^1(\mathbb{R})$ over $\mathbb{R}$ (which is a compact space), by identifying $c \in \mathbb{R}$ and ∞ with the points with homogeneous coordinates $(c : 1)$ and $(1 : 0)$, respectively. For later use we remark that $(c : d) \mapsto (d : c)$ defines a continuous involution ι on $\mathbb{P}^1(\mathbb{R})$ that exchanges 0 and ∞ and that maps $c \in \mathbb{R}$ to $1/c$.

For $a \in F$ we define the evaluation map $\widehat{a} : M(F) = Y \to \mathbb{R} \cup \infty = \mathbb{P}^1(\mathbb{R})$ by $\widehat{a}(\lambda) = \lambda(a)$. For $a \in o$ this definition coincides with the earlier one ($\widehat{a}$ does not take the value ∞ in this case), and we have:

Proposition 3.12.15 *For every* $a \in F$, *the map* $\widehat{a}: M(F) = Y \to \mathbb{P}^1(\mathbb{R})$ *is continuous (with Y carrying the topology defined earlier). Therefore the topology of $M(F)$ is also the coarsest topology that makes all $\widehat{a}$ ($a \in F$) continuous.*

Proof Without loss of generality we may assume that $a \neq 0$. Let $\lambda \in M(F)$. We prove the continuity of $\widehat{a}$ in λ. Firstly, if $\lambda(a) \neq \infty$, there exist $b, c \in \mathfrak{o}$ such that $a = b/c$ and $\lambda(c) \neq 0$. The continuity of $\widehat{a}$ in λ follows from the fact that $\widehat{a}$ coincides with the continuous function $\widehat{b}/\widehat{c}$ on the open neighbourhood $\{\mu \in M(F): \mu(c) \neq 0\}$ of λ. If on the other hand $\lambda(a) = \infty$, then $\widehat{a} = \iota \circ \widehat{(1/a)}$, where ι denotes the involution on $\mathbb{P}^1(\mathbb{R})$ mentioned earlier. The continuity of $\widehat{a}$ in λ thus follows from the first part of the proof. $\square$

Remark 3.12.16 Let $\rho: \mathrm{Sper}\,\mathfrak{o} \to (\mathrm{Sper}\,\mathfrak{o})^{\max} = M(F)$ denote the (continuous) canonical retraction, which maps every $x \in \mathrm{Sper}\,\mathfrak{o}$ to the tip of the spear $\overline{\{x\}}$ in $\mathrm{Sper}\,\mathfrak{o}$ (i.e., the closed specialization of x in $\mathrm{Sper}\,\mathfrak{o}$), cf. Sect. 3.6. Given $x \in \mathrm{Sper}\,F = (\mathrm{Sper}\,\mathfrak{o})^{\min}$, it is easy to describe $\rho(x)$ without reference to the holomorphy ring $\mathfrak{o}$. Namely, $\rho(x)$ is the "canonical" place $F \to \mathbb{R} \cup \infty$ associated to the ordering x of F, in other words the uniquely determined x-compatible place whose valuation ring is the convex hull $\mathfrak{o}_x(F)$ of $\mathbb{Z}$ in F with respect to x. The restriction $\mathrm{Sper}\,F \to M(F)$ of ρ then makes $M(F)$ into a topological quotient space of $\mathrm{Sper}\,F$ (since this map is closed).

A deeper geometric understanding of the space of places $M(F)$ (for instance in the case of function fields over real closed base fields) is beyond our current abilities, as this would require further knowledge of algebraic geometry beyond what we can cover in this book (e.g., blow-ups). Nevertheless we present two illustrative examples. These are quite special since they are "one-dimensional". Therefore we must emphasize that the holomorphy ring $\mathfrak{o}(F)$ and the space $M(F)$ are also useful for higher-dimensional geometry.

Example 3.12.17 (Schülting, [106, p. 11 ff.]) Consider a one-dimensional real-analytic manifold N and let M be a nonempty compact connected subset of N. Let $\mathcal{O}(M) := \underset{\longrightarrow}{\lim}\, \mathcal{O}(U)$, where U runs through the open neighbourhoods of M and $\mathcal{O}(U)$ denotes the ring of real-analytic functions $U \to \mathbb{R}$. Then $\mathcal{O}(M)$ is an integral domain. Let F denote the field of fractions of $\mathcal{O}(M)$, i.e., the field of "meromorphic functions on M". We assume that the functions of $\mathcal{O}(M)$ separate the points of M (in fact, this is always the case, cf. [106, p. 13]). Then $\mathcal{O}(M) = \mathfrak{o}(F)$ and the canonical evaluation map $M \to \mathrm{Hom}(\mathfrak{o}(F), \mathbb{R})$ is a homeomorphism. Thus we have rediscovered M with its topology as the space of places $M(F)$, and the holomorphy ring $\mathfrak{o}(F)$ as the holomorphic functions on it.

Proof Every point $x \in M$ gives rise to a discrete valuation v_x of F with residue field $\mathbb{R}$, namely the order (of zeroes) at x. We have $\mathcal{O}(M) = \{f \in F: v_x(f) \geq 0 \text{ for all } x \in M\}$, from which $\mathfrak{o}(F) \subseteq \mathcal{O}(M)$ follows. Conversely, every $f \in \mathcal{O}(M)$ is bounded on M, hence bounded on an open neighbourhood U of M since M is compact. If $n \in \mathbb{N}$ is chosen such that $|f(x)| < n$ on U, then $\sqrt{n \pm f}$ are analytic on U. Hence, $n \pm f$ are squares in F, from which $f \in \mathfrak{o}$ follows. Therefore, $\mathfrak{o}(F) = \mathcal{O}(M)$. Now let $\mathfrak{o} := \mathfrak{o}(F) = \mathcal{O}(M)$. The evaluation map $M \to \mathrm{Hom}(\mathfrak{o}, \mathbb{R})$ is trivially continuous and is injective by assumption. Since both spaces are compact, it suffices to verify the surjectivity of this map. Thus, consider a homomorphism $\varphi : \mathfrak{o} \to \mathbb{R}$ with kernel $\mathfrak{p}$. Then $\mathfrak{p}$ is a maximal ideal of $\mathfrak{o}$ since $\mathbb{R} \subseteq \mathfrak{o}$. For $f \in \mathfrak{o}$, let $Z(f) := \{x \in M : f(x) = 0\}$. If the intersection of all $Z(f)$ with $f \in \mathfrak{p}$ is empty, then there exist finitely many $f_1, \ldots, f_r \in \mathfrak{p}$ such that $Z(f_1) \cap \cdots \cap Z(f_r) = \varnothing$ since M is compact. But then $g := f_1^2 + \cdots + f_r^2$ has no zeroes on a neighbourhood of M, meaning that $1/g \in \mathfrak{o}$ which leads to the contradiction $1 = (f_1^2 + \cdots + f_r^2)/g \in \mathfrak{p}$. Thus, consider $x \in M$ such that $f(x) = 0$ for all $f \in \mathfrak{p}$. Since $\mathfrak{p}$ is maximal, it follows immediately that $\varphi(f) = f(x)$ for all $f \in \mathfrak{o}$.   □

This example illustrates the idea that gave rise to the term "holomorphy ring": if F is a field, one views the elements of F as continuous $\mathbb{P}^1(\mathbb{R})$-valued functions on the compact space $M(F)$ ("meromorphic functions"). The elements of $\mathfrak{o}(F)$ are then precisely those functions that never become infinite, in other words the "holomorphic functions".

Example 3.12.18 In this example we use some basic facts from introductory algebraic geometry, see for instance [97], [45, Ch. I]. Consider a smooth projective irreducible curve V, which is defined over $\mathbb{R}$ (i.e., described by homogeneous polynomials with coefficients in $\mathbb{R}$). Let $F = \mathbb{R}(V)$ denote its function field. It is well-known that every point $x \in V(\mathbb{C})$ defines a discrete valuation ring $\mathcal{O}_x$ of F (the ring of regular functions in x), where $\mathcal{O}_x$ is residually real (with residue field $\mathbb{R}$) if and only if $x \in V(\mathbb{R})$. (Furthermore, $\mathcal{O}_x = \mathcal{O}_y$ if and only if x and y are complex conjugate.) Since all the $\mathcal{O}_x$ ($x \in V(\mathbb{C})$) are nontrivial valuation rings of F over $\mathbb{R}$, we have $\mathfrak{o}(F) = \mathcal{O}(V(\mathbb{R}))$, the ring of functions that are regular on the whole of $V(\mathbb{R})$, and there is a canonical bijection $\varepsilon : V(\mathbb{R}) \to M(F)$. If we endow $V(\mathbb{R})$ with its strong (= classical) topology as usual, then $V(\mathbb{R})$ is compact. The functions $f : V(\mathbb{R}) \to \mathbb{R}$ that are regular on $V(\mathbb{R})$ are of course continuous. By Proposition 3.12.11 it then follows that ε is continuous, in other words a homeomorphism. Thus we may identify $V(\mathbb{R})$ with $M(F)$.

The next discussion topic focuses on the relationship between preorderings and (residually real) Prüfer rings with respect to (real) places.

Consider a surjective place $\lambda : F \twoheadrightarrow K \cup \infty$ and a preordering T of F. Then $\lambda(T \cap \mathfrak{o}_\lambda)$ is a preordering of K (cf. Remark 3.9.2(2); $\mathfrak{o}_\lambda$ is the valuation ring of λ), which we—slightly sloppily—denote by $\lambda(T)$.

Definition 3.12.19 The place λ is called *compatible with* T if $-1 \notin \lambda(T)$, i.e., if $\lambda(T)$ is a proper preordering of K.

Remarks 3.12.20

(1) The place λ is compatible with the preordering ΣF^2 of F if and only if λ is real (i.e., K is a real field). In this case we have $\lambda(\Sigma F^2) = \Sigma K^2$.

(2) If $T = P$ is an ordering of F, then λ is compatible with P if and only if $\lambda(P)$ is an ordering of K. This is equivalent to the existence of an ordering Q of K such that λ is compatible with P and Q in the sense of Sect. 2.8 (and then we necessarily have $Q = \lambda(P)$).

We will shortly present a whole list of equivalent characterizations of the compatibility of λ with T, as well as a description of all the places λ of F that satisfy $\mathfrak{o}_T(F) \subseteq \mathfrak{o}_\lambda$ (note that $\mathfrak{o}_T(F)$ is the intersection of all these $\mathfrak{o}_\lambda$). We start with some preliminary results.

Lemma 3.12.21 *Let T be a proper preordering of F and A an overring of $\mathfrak{o}_T(F)$ in F. Then every ideal of A is convex in F with respect to T. In particular, A is a T-convex subring of F.*

Proof Consider $a \in A$ and $b \in F$ such that $0 \leq_T b \leq_T a$. If $a = 0$, then $b = 0$ too. Otherwise we have $0 \leq_T ba^{-1} \leq_T 1$, i.e., $ba^{-1} \in \mathfrak{o}_T(F) \subseteq A$ (since $\mathfrak{o}_T(F)$ is the T-convex hull of $\mathbb{Z}$ in F), and conclude that $b = (ba^{-1})a \in Aa$. □

This shows in particular that the T-convex subrings of F are precisely the overrings of the T-convex hull $\mathfrak{o}_T(F)$ of $\mathbb{Z}$ in F.

Lemma 3.12.22 *Let T be a proper preordering of F, A a Prüfer ring of F and $\mathfrak{p}$ a prime ideal of A. The following statements are equivalent:*

(i) $A_\mathfrak{p}$ *is convex in F with respect to T;*
(ii) $\mathfrak{p}$ *is convex in A with respect to $T \cap A$.*

Proof The T-convexity of $\mathfrak{p}A_\mathfrak{p}$ in F, thus also the $T \cap A$-convexity of $\mathfrak{p} = (\mathfrak{p}A_\mathfrak{p}) \cap A$ in A, follows from (i) by Lemma 3.12.4. Conversely, assume that (ii) holds and consider $a \in A_\mathfrak{p}$ and $b \in F$ such that $0 \leq_T b \leq_T a$. Assume that $b \notin A_\mathfrak{p}$. Then $b^{-1} \in \mathfrak{p}A_\mathfrak{p}$, hence also $ab^{-1} \in \mathfrak{p}A_\mathfrak{p}$, and $0 \leq_T 1 \leq_T ab^{-1}$. There exist $c, s \in A$ such that $c \in \mathfrak{p}$, $s \notin \mathfrak{p}$ and $ab^{-1} = cs^{-1}$. Then $(c - s)s = s^2(ab^{-1} - 1)$ is an element of $T \cap A$ but not of $\mathfrak{p}$, and

$$s^2 + (c - s)s = cs \in \mathfrak{p}$$

contradicts the $T \cap A$-convexity of $\mathfrak{p}$. □

Lemma 3.12.23 (G. Brumfiel) *Consider an integral domain A, a prime ideal $\mathfrak{p}$ of A and a proper preordering T of the field of fractions F of A. The following statements are equivalent:*

(i) $\mathfrak{p}$ *is a $T \cap A$-convex prime ideal of A;*
(ii) *there exists an ordering P of F with $P \supseteq T$ such that $\mathfrak{p}$ is a $P \cap A$-convex prime ideal of A.*

Proof (i) $\Rightarrow$ (ii): By Zorn's Lemma there is a preordering $U \supseteq T$ of F for which $\mathfrak{p}$ is still $U \cap A$-convex and which is maximal with respect to this property. Assume for the sake of contradiction that U is not an ordering of F. Then there exists an

$a \in F$ such that $a \notin U \cup (-U)$. Then $U_1 := U + aU$ and $U_2 := U - aU$ are proper preorderings of F by Lemma 1.1.8. Let us show that $U_1 \cap U_2 = U$. Take $b \in U_1 \cap U_2$ and write $b = u + av = u' - av'$ with $u, u', v, v' \in U$. Then

$$b(v + v') = uv' + u'v \in U,$$

hence $b \in U$ since if $v + v' = 0$, then also $v = v' = 0$. In particular, we have $U \cap A = (U_1 \cap A) \cap (U_2 \cap A)$. By Klapper's Lemma (Lemma 3.10.8) $\mathfrak{p}$ is then convex with respect to one of the $U_i \cap A$, which contradicts the maximality of U. The implication (ii) $\Rightarrow$ (i) is trivial. $\square$

Now we can present the list of equivalent characterizations of the compatibility of a place with a preordering, announced earlier. As before, $\mathfrak{o}_\lambda$ denotes the valuation ring of a place λ of F and $\mathfrak{m}_\lambda$ denotes its maximal ideal.

Theorem 3.12.24 *Let $\lambda\colon F \twoheadrightarrow K \cup \infty$ be a surjective place of F and T a proper preordering of F. Denote the holomorphy ring of F by $\mathfrak{o} := \mathfrak{o}(F)$. The following statements are equivalent:*

  (i) *λ is compatible with T;*
 (ii) *$\mathfrak{m}_\lambda \cap (1 + T) = \varnothing$;*
(iii) *λ is finite on $\mathfrak{o}_T(F)$ (i.e., $\mathfrak{o}_T(F) \subseteq \mathfrak{o}_\lambda$);*
 (iv) *$\mathfrak{o}_\lambda$ is T-convex;*
  (v) *there exists an ordering $P \supseteq T$ of F such that $\mathfrak{o}_\lambda$ is P-convex;*
 (vi) *there exists an element $x \in \overline{H}_F(T)$ such that λ is compatible with x;*
(vii) *there exists a coarsening μ of λ in $M(F)$ (i.e., $\mathfrak{o}_\mu \subseteq \mathfrak{o}_\lambda$) which is compatible with T.*

For the remaining statements we assume that λ is real. Then $\mathfrak{o} \subseteq \mathfrak{o}_\lambda$, and the prime ideal $\mathfrak{p} := \mathfrak{o} \cap \mathfrak{m}_\lambda$ of $\mathfrak{o}$ satisfies $\mathfrak{o}_\lambda = \mathfrak{o}_\mathfrak{p}$. The following statements are then equivalent to those above:

(viii) *The prime ideal $\mathfrak{p}$ of $\mathfrak{o}$ is $T \cap \mathfrak{o}$-convex;*
 (ix) *there exists an element $y \in \overline{H}_\mathfrak{o}(T \cap \mathfrak{o})$ such that $\operatorname{supp}(y) = \mathfrak{p}$;*
  (x) *$\mathfrak{p} \cap (1 + T \cap \mathfrak{o}) = \varnothing$.*

Proof The hard work has already been done. We just need to put everything together.

The equivalence of (i) and (ii) follows directly from Definition 3.12.19, while the equivalence of (iii) and (iv) follows from Lemma 3.12.4. By Corollary 3.11.14, $\mathfrak{o}_T(F)$ is generated by all elements of the form $(1 + t)^{-1}$ with $t \in T$, which gives the equivalence of (ii) and (iii). The equivalence of (v) and (vi) is clear by Remark 3.12.20(2) (cf. Proposition 2.8.10), and (v) $\Rightarrow$ (iv) is trivial. Conversely, assuming (iv), it follows from Lemma 3.12.5 that $\mathfrak{m}_\lambda$ is convex with respect to $T \cap \mathfrak{o}_\lambda$. By Lemma 3.12.23 there is then an ordering $P \supseteq T$ of F such that $\mathfrak{m}_\lambda$ is convex with respect to $P \cap \mathfrak{o}_\lambda$. A further application of Lemma 3.12.5 gives (v). (v) $\Rightarrow$ (vii): If we choose P as in (v) and μ as a place with valuation ring $\mathfrak{o}_P(F)$, then μ is compatible with T, coarsens λ, and satisfies $\mu \in M(F)$. Conversely, (vii) $\Rightarrow$

(iii) follows by applying the proven implication (i) $\Rightarrow$ (iii) to μ. All this establishes the equivalence of statements (i) to (vii).

Assume now that λ is real, i.e., $\mathfrak{o} \subseteq \mathfrak{o}_\lambda$ (cf. Theorem 3.12.5). The equality $\mathfrak{o}_\lambda = \mathfrak{o}_\mathfrak{p}$ was already established in Exercise 3.12.3, and the equivalence of (iv) and (viii) follows from Lemma 3.12.5. The equivalence of (viii) and (ix) follows from Proposition 3.10.10 and Theorem 3.10.13, and (ii) $\Rightarrow$ (x) is trivial. Conversely, assume (x). If $\mathfrak{q}$ is an ideal of $\mathfrak{o}$ that is maximal with respect to the property $\mathfrak{q} \cap (1 + T \cap \mathfrak{o}) = \varnothing$, then $\mathfrak{q}$ is prime and convex in $\mathfrak{o}$ with respect to $T \cap \mathfrak{o}$ by Proposition 3.10.5. Hence, $\mathfrak{o}_\mathfrak{q}$ is T-convex in F by the proven implication (viii) $\Rightarrow$ (iv). From $\mathfrak{o}_\mathfrak{q} \subseteq \mathfrak{o}_\mathfrak{p}$ and Lemma 3.12.4 then follows the T-convexity of $\mathfrak{o}_\mathfrak{p} = \mathfrak{o}_\lambda$, in other words, (iv). This proves the theorem. $\qquad\square$

Proposition 3.12.25 *Consider a surjective place* $\lambda \colon F \twoheadrightarrow K \cup \infty$ *and a preordering* T *of* F. *If* λ *is compatible with* T, *then*

$$\lambda(\mathfrak{o}_T(F)) = \mathfrak{o}_{\lambda(T)}(K).$$

In particular, $\lambda(\mathfrak{o}(F)) = \mathfrak{o}(K)$ *if* λ *is real.*

Proof By Corollary 3.11.14, $\mathfrak{o}_T(F)$ is generated by all elements of the form $(1 + t)^{-1}$ with $t \in T$. The assumption signifies that λ is finite on $\mathfrak{o}_T(F)$ (cf. Theorem 3.12.24). However, for $t \in T$ we have

$$\lambda\left(\tfrac{1}{1+t}\right) = \begin{cases} \frac{1}{1+\lambda(t)} & \text{if } t \in T \cap \mathfrak{o}_\lambda, \\ 0 & \text{if } t \notin \mathfrak{o}_\lambda. \end{cases}$$

The first claim then follows since $\lambda(T) = \lambda(T \cap \mathfrak{o}_\lambda)$, while the second claim is the special case $T = \Sigma F^2$ (i.e., $\lambda(T) = \Sigma K^2$), cf. Remark 3.12.20(1). $\qquad\square$

Since the prime ideals $\mathfrak{p}$ of $\mathfrak{o}(F)$ correspond to the residually real valuation rings of F (via $\mathfrak{p} \mapsto \mathfrak{o}(F)_\mathfrak{p}$, cf. Exercise 3.12.3 and Theorem 3.12.5), the second claim of Proposition 3.12.25 can also be formulated as follows:

If $\mathfrak{p}$ *is a prime ideal of* $\mathfrak{o}(F)$, *then* $\mathfrak{o}(F)/\mathfrak{p}$ *is the holomorphy ring of the field of fractions* $\kappa(\mathfrak{p})$ *of* $\mathfrak{o}(F)/\mathfrak{p}$.

Corollary 3.12.26 *If* A *is a residually real Prüfer ring of* F *and* $\lambda \colon F \twoheadrightarrow K \cup \infty$ *a surjective place which is integral on* A, *then* $\lambda(A)$ *is a residually real Prüfer ring of* K.

Proof By Theorem 3.12.5 we have $\mathfrak{o}(F) \subseteq A$, hence $\mathfrak{o}(K) \subseteq \lambda(A)$ by Proposition 3.12.25. Then apply Theorem 3.12.5 again. $\qquad\square$

Proposition 3.12.27 *Consider a preordering* T *of* F, *a surjective place* $\lambda \colon F \twoheadrightarrow K \cup \infty$ *that is compatible with* T *and an ordering* Q *of* K. *Then* $Q \supseteq \lambda(T)$ *if and*

only if there exists an ordering $P \supseteq T$ of F that is compatible with λ and such that
$Q = \lambda(P)$.

Proof For the nontrivial direction, let $Q \supseteq \lambda(T)$ and write $Q^* := Q \setminus \{0\}$. If P is an
ordering of F such that $T \cup \lambda^{-1}(Q^*) \subseteq P$, then P clearly satisfies the claim. Hence
we must show that $T \cup \lambda^{-1}(Q^*)$ generates a proper preordering of F. Assume for
the sake of contradiction that this is not the case. Then there exist $s_i \in \lambda^{-1}(Q^*)$,
$t_i \in T$ $(i = 1, \ldots, N)$ such that

$$1 + s_1 t_1 + \cdots + s_N t_N = 0.$$

Hence there must be at least one i for which $\lambda(t_i) = \infty$. Let v denote the valuation
associated to λ and assume without loss of generality that $v(t_1) \leq \cdots \leq v(t_N)$.
Then we obtain

$$t_1^{-1} + s_1 + s_2 t_2' + \cdots + s_N t_N' = 0,$$

where t_1^{-1} and all $t_i' = t_i/t_1$ are in $T \cap \mathfrak{o}_\lambda$. Applying λ to this identity gives a
contradiction.   □

Corollary 3.12.28 *Let A be a residually real Prüfer ring of F and T a preordering
of F. Then $\overline{H}_A(T \cap A)$ is the closure of $\overline{H}_F(T)$ in Sper A (where we view Sper F =
(Sper A)^{min} as a subspace of Sper A).*

Proof Since $\overline{H}_F(T) \subseteq \overline{H}_A(T \cap A)$, we must show that every $y \in \overline{H}_A(T \cap A)$ has
a generalization in $\overline{H}_F(T)$. Let $\mathfrak{p} := \mathrm{supp}(y)$ and $B := A_\mathfrak{p}$. We view y also as a
point in Sper B (with support $\mathfrak{m}_B$). Then $y \in \overline{H}_B(T \cap B)$. The claim follows by
applying Proposition 3.12.27 to the place $F \to \kappa(B) \cup \infty$ associated to B.   □

We conclude this section with a discussion of the relationships between the real
spectra of $\mathfrak{o} := \mathfrak{o}(F)$ and other residually real Prüfer rings of F.

Proposition 3.12.29 *Let A be a residually real Prüfer ring of F, thus an overring
of $\mathfrak{o}$ in F, and let Y(A) be the image of Sper A under the restriction map
r_A: Sper A → Sper $\mathfrak{o}$.*

(a) *Y(A) is a proconstructible subspace of Sper $\mathfrak{o}$ that is stable under generalization in Sper $\mathfrak{o}$, and r_A is a homeomorphism from Sper A to Y(A);*

(b) *Y(A) consists of all $x \in$ Sper $\mathfrak{o}$ that satisfy $\mathfrak{o}_{\mathrm{supp}(x)} \supseteq A$, or equivalently, such
that $A \cdot \mathrm{supp}(x) \neq A$ (where $A \cdot \mathrm{supp}(x)$ is the usual notation for the ideal of A
generated by $\mathrm{supp}(x)$);*

(c) *Y(A) is the union of all Y(B) where B runs through the valuation rings of F
for which $A \subseteq B$.*

Proof Let $\mathfrak{p}$ be a prime ideal of A. From Exercise 3.12.3 it follows that $\mathfrak{o}_{\mathfrak{p}\cap\mathfrak{o}} = A_{\mathfrak{p}}$ (both valuation rings have the same centre $\mathfrak{o}$). This implies the commutativity of the diagram

$$
\begin{array}{ccc}
\text{Spec } A & \longrightarrow & \{\text{valuation rings of } F/A\} \\
\downarrow & & \cap \\
\text{Spec } \mathfrak{o} & \longrightarrow & \{\text{valuation rings of } F/\mathfrak{o}\}
\end{array}
$$

where the horizontal arrows are the bijections of Exercise 3.12.3 (localization). It follows that the restriction map Spec $A \to$ Spec $\mathfrak{o}$ is injective and that it preserves the residue fields, from which already follows that r_A is injective and that $Y(A) = \{x \in \text{Sper } \mathfrak{o} : A \subseteq \mathfrak{o}_{\text{supp}(x)}\}$. For this reason (c) is clear, as is the stability of $Y(A)$ under generalization. A priori $Y(A)$ is proconstructible. If $a = b/c \in A$ with $0 \neq b$ and $c \in \mathfrak{o}$, then $r_A(\mathring{H}_A(a)) = \mathring{H}_\mathfrak{o}(bc) \cap Y(A)$. It follows that r_A is a homeomorphism to $Y(A)$. It remains to show the second claim of (b). If $\mathfrak{q} \in$ Spec $\mathfrak{o}$ and $A \subseteq \mathfrak{o}_{\mathfrak{q}}$, then $A\mathfrak{q} \subseteq A \cap \mathfrak{q}\,\mathfrak{o}_{\mathfrak{q}} \neq A$. Conversely, if $A\mathfrak{q} \subseteq \mathfrak{p}$ with $\mathfrak{p} \in$ Spec A, then $A \subseteq A_{\mathfrak{p}} = \mathfrak{o}_{\mathfrak{p}\cap\mathfrak{o}} \subseteq \mathfrak{o}_{\mathfrak{q}}$. $\square$

Corollary 3.12.30 *If A and A' are residually real Prüfer rings of F with $Y(A) = Y(A')$, then $A = A'$.*

Proof From Proposition 3.12.29 it follows that given a prime ideal $\mathfrak{q}$ of $\mathfrak{o}$, $\mathfrak{o}_{\mathfrak{q}}$ contains A if and only if $\mathfrak{o}_{\mathfrak{q}}$ contains A', and that A resp. A' is the intersection of all such $\mathfrak{o}_{\mathfrak{q}}$. $\square$

By Proposition 3.12.29 we may identify Sper A with the subspace $Y(A)$ of Sper $\mathfrak{o}$. This space can be described quite explicitly as follows. Let $y \in$ Sper $F = (\text{Sper } \mathfrak{o})^{\min}$. The specializations x of y in Sper $\mathfrak{o}$ correspond *uniquely* to the subrings B of F that are convex with respect to y, via the map $x \mapsto \mathfrak{o}_{\text{supp}(x)}$. (This follows from Proposition 3.12.29 and Exercise 3.12.3.) This correspondence is order reversing. Hence y corresponds to the largest such subring (namely $B = F$) and the closed specialization $\overline{y}$ of y in Sper $\mathfrak{o}$ corresponds to the smallest such subring (namely $B = \mathfrak{o}_y(F)$). Given any specialization x of y, it follows from Proposition 3.12.29(b) that x is in $Y(A) =$ Sper A if and only if A is contained in $\mathfrak{o}_{\text{supp}(x)}$. In particular, the support ideal of the closed specialization of y in Sper A is precisely the centre of $A \cdot \mathfrak{o}_y(F)$ (the y-convex hull of A) in $\mathfrak{o}$.

Results that are similar to Propositions 3.12.9–3.12.15 for the holomorphy ring $\mathfrak{o}(F)$ can also be established for the ring $A = \mathfrak{o}_T(F)$, where T is any proper preordering of F. We will just indicate these results and leave the proofs to the reader.

The canonical map $\text{Hom}(A, \mathbb{R}) \to$ Sper A is a bijection onto the subset

$$
\text{Sper } A \cap (\text{Sper } \mathfrak{o})^{\max}
$$

of $(\text{Sper } A)^{\max}$. This subset contains the tips of all the spears $\overline{\{x\}}$ in $\text{Sper } \mathfrak{o}$ for $x \in$ $\overline{H}_F(T)$, but is larger in general. There is a natural bijective correspondence between $\text{Hom}(A, \mathbb{R})$ and the closed (hence compact) subspace

$$M(F/T) := \{\lambda \in M(F) : \lambda \text{ and } T \text{ are compatible}\}$$

of $M(F)$. A result similar to Proposition 3.12.11 then also holds for A, and Proposition 3.12.12 can be generalized as follows:

If $a \in A$ and $\widehat{a} \colon \text{Hom}\,(A, \mathbb{R}) \to \mathbb{R}$ denotes the evaluation map, then $\widehat{a} \geq 0$ if and only if

$$a + \tfrac{1}{n} \in T \quad \text{for all } n \in \mathbb{N}.$$

Chapter 4
Recent Developments

In this short chapter we briefly discuss a number of important developments and advances that mostly occurred after the 1989 publication of *Einführung in die reelle Algebra*, and that are directly related to topics covered in Chapters 1, 2 and 3.

4.1 Counting Real Solutions

The results from Sects. 1.7–1.10 on counting real zeroes of univariate polynomials are just the classical tip of a modern iceberg, that has emerged in the past decades. We can merely sketch the general idea, for more details we have to refer to the literature. The support $\mathrm{supp}(f)$ of a polynomial f in n variables is the finite set of n-tuples $(\alpha_1, \ldots, \alpha_n) \in \mathbb{Z}^n$ for which the monomial $x_1^{\alpha_1} \cdots x_n^{\alpha_n}$ appears in f with nonzero coefficient. Let a system of n real polynomials $f_i(x)$ $(i = 1, \ldots, n)$ in n variables $x = (x_1, \ldots, x_n)$ be given. The number of nondegenerate complex solutions of the system $f_1(x) = \cdots = f_n(x) = 0$ is at most $\prod_{i=1}^{n} \deg(f_i)$ (Bézout), and for the number of solutions in $(\mathbb{C}^*)^n$ there is the Bernstein-Khovanskiĭ-Kushnirenko (BKK) upper bound in terms of the mixed volume of the convex hulls of the supports $\mathrm{supp}(f_i)$. When it comes to *real* zeroes of *real* polynomials, Khovanskiĭ discovered around 1980 that what matters for their count is just the number of different monomials in the f_i, not their degrees or the volume of their convex hull. More precisely, he gave an explicit number $\varphi(n, N)$ with the following property: Given any real polynomials $f_1, \ldots, f_n$ in n variables that together have at most N monomials, the system $f_1(x) = \cdots = f_n(x)$ has at most $\varphi(n, N)$ nondegenerate real solutions in the open positive orthant. Note how in the univariate case such a statement follows from Descartes' rule of sign (cf. Corollary 1.10.3(a)). In more than one variable, proofs for results of this kind become a lot more difficult.

This chapter is co-authored by Thomas Unger.

© Springer Nature Switzerland AG 2022

M. Knebusch, C. Scheiderer, *Real Algebra*, Universitext,
https://doi.org/10.1007/978-3-031-09800-0_4

This discovery gave rise to the term *fewnomials*, and it has meanwhile grown into a large theory with many ramifications. The bound $\varphi(n, N)$ itself is huge and is generally believed to be far from best possible. For small values of $N - n$, much better or even sharp bounds have been proved. For example, it has been shown for $n = 2$ that two real trinomials $f_1(x, y)$, $f_2(x, y)$ can have at most 5 nondegenerate positive common zeroes, and that this bound is sharp.

For more details we refer to the books by Khovanskiĭ [58] and Sottile [110].

4.2   Quadratic Forms

Related to Sect. 3.8, some remarkable developments and applications are to be reported. The Witt ring of quadratic forms, as defined for fields of characteristic $\neq 2$ in Sect. 1.2, can be defined (and is an important object) for arbitrary rings (or even schemes) on which 2 is invertible [61]. Let A be a ring with $\frac{1}{2} \in A$. A regular quadratic space over A is a pair $\varphi = (M, b)$ consisting of a finitely generated projective A-module M and a regular symmetric bilinear form $b: M \times M \to A$. The sum and product of quadratic spaces are defined as in the field case, using orthogonal sum and tensor product, respectively. Let $\widehat{W}(A)$ be the Witt-Grothendieck ring of isomorphism classes of regular quadratic spaces over A. For any finitely generated projective module M, the hyperbolic space $\mathbb{H}(M)$ of M is $h_M = (M \oplus M^\vee, b_M)$ where b_M is the canonical symmetric bilinear form on $M \oplus M^\vee$. By definition, the Witt ring $W(A)$ is the quotient ring $\widehat{W}(A)/\mathcal{H}$ where $\mathcal{H}$ is the ideal in $\widehat{W}(A)$ consisting of (the classes of) all hyperbolic spaces $\mathbb{H}(M)$. Every element of $W(A)$ is represented by a regular quadratic space $\varphi = (M, b)$.

Over general rings A, Witt's Cancellation Theorem (see Theorem 1.2.1 in the field case) usually fails, and Witt decomposition (see loc. cit.) has no analogue. For semilocal rings A however, Witt cancellation holds, and much of the theory passes from the field case to this more general situation. See [63] for general background.

We define a signature of A to be any ring homomorphism $\sigma: W(A) \to \mathbb{Z}$, cf. Definition 1.2.14. By functoriality of the Witt ring, every element $\alpha \in \mathrm{Sper}(A)$ gives a signature sign_α of A. It maps the Witt class of (M, b) to the signature of the regular quadratic form $(M, b) \otimes_A k(\alpha)$ over the real closed field $k(\alpha)$.

Fixing $\varphi \in W(A)$, the map $\mathrm{sign}(\varphi): \mathrm{Sper}(A) \to \mathbb{Z}$ is clearly continuous, and hence constant on connected components of $\mathrm{Sper}(A)$. As in Proposition 3.8.2 over fields, we get a ring homomorphism $\mathrm{sign}: W(A) \to C(\mathrm{Sper}(A), \mathbb{Z})$, the *global signature* of A. It turns out that Proposition 3.8.6 remains true in general as well, but this becomes much harder to prove. In fact, Mahé [78] showed for arbitrary A that the cokernel of the global signature is a 2-primary torsion group. As a consequence it follows that every signature $\sigma: W(A) \to \mathbb{Z}$ comes from an ordering, i.e., $\sigma = \mathrm{sign}_\alpha$ for some $\alpha \in \mathrm{Sper}(A)$, and that $\mathrm{sign}: \mathrm{Sper}(A) \to \mathrm{Hom}(W(A), \mathbb{Z})$ induces a bijection from the set of connected components of $\mathrm{Sper}(A)$ to the set of all signatures.

Calculating or estimating the exponent of the cokernel of the global signature $\mathrm{sign}: W(A) \to C(\mathrm{Sper}(A), \mathbb{Z})$ is an important problem. For general rings it is very hard. For the coordinate rings $A = \mathbb{R}[V]$ of affine algebraic varieties V over $\mathbb{R}$, Mahé [79] proved an explicit (although very large) upper bound: There exists a function $w\colon \mathbb{N} \to \mathbb{N}$ such that the cokernel of the global signature of $\mathbb{R}[V]$ is annihilated by $2^{w(d)}$ whenever $\dim(V) \le d$.

In the case when A is a field, or more generally a semilocal ring, the integer $s \ge 0$ (or $s = \infty$) for which 2^s is the exponent of the cokernel of sign is called the stability index $\mathrm{st}(A)$ of A (see Definition 3.8.7). Via the theory of fans, the stability index is closely related to real valuations of A, and can be understood via powerful local–global principles. Marshall's abstract framework of *spaces of orderings* provides strong tools for analyzing the stability index and related invariants.

The stability index has found attractive applications in geometry, and Corollary 3.8.13 (the equivalence of (i) and (ii)) already points in this direction. Let V be an affine $\mathbb{R}$-variety. If $M \subset V(\mathbb{R})$ is a basic open semialgebraic set, let $s(M)$ denote the minimal number r such that $M = \{x \in V(\mathbb{R})\colon f_1(x) > 0, \ldots, f_r(x) > 0\}$ for suitable $f_i \in \mathbb{R}[V]$, and put $s(V) = \sup\{s(M)\colon M \subset V(\mathbb{R})$ basic open$\}$. For basic closed sets M, define $\bar{s}(M)$ similarly using nonstrict inequalities $f_i(x) \ge 0$, and let $\bar{s}(V) = \sup\{\bar{s}(M)\colon M \subset V(\mathbb{R})$ basic closed$\}$.

Theorem 4.2.1 *Let V be an affine $\mathbb{R}$-variety of dimension $d \ge 1$.*

(a) (Bröcker–Scheiderer) $s(V) \le d$,
(b) $\bar{s}(V) \le \frac{1}{2}d(d+1)$.

In fact, equality holds in (a) and (b) if $V(\mathbb{R})$ is Zariski-dense in V.

See [16, 99] where much more general versions are proved. Basic closed sets that actually require $\frac{d}{2}(d+1)$ inequalities are somewhat pathological; Averkov–Bröcker [6] proved later that $\bar{s}(P) \le d$ holds for any d-dimensional polyhedron P in $\mathbb{R}^n$. Note however that it remains very difficult to produce an explicit description of minimal (or just small) length, for a general basic (open or closed) set M. Not surprisingly, the price one has to pay for a short description is that the inequalities tend to have very large degrees. See e.g., Burési–Mahé [21] for some explicit degree bounds.

To allow for more flexible applications to geometrical questions, Marshall's spaces of orderings were later generalized to abstract real spectra [80] and to spaces of signs [2]. The book [2] presents an extensive account of such applications, at varying levels of abstractness (description of semialgebraic, semianalytic and constructible sets by few inequalities).

4.3 Stellensätze

In Sect. 3.9 we discussed various *Stellensätze*. This is another topic where remarkable progress was made since the German edition appeared. The results in Propositions 3.9.5 and 3.9.8 and Theorem 3.9.11 are of a general nature and hold for arbitrary preorderings in arbitrary rings. Their common feature is that if a ring element f takes a particular sign on $\mathrm{Sper}(A)$ (e.g., is positive, or nonnegative, or zero throughout), they guarantee the existence of a "certificate" for this sign.

The *Positivstellensatz* is a certificate "with denominator", in the sense that it has the form $sf = 1 + t$, where s and t are sums of squares, and thus expresses f as a quotient of two elements that are strictly positive. Similarly, the *Nichtnegativstellensatz* is a certificate with denominator $sf = t'$, where s and t' are sums of squares, and expresses f as a quotient of two elements that are nonnegative.

Making more specific assumptions, one gets much more powerful statements. A door-opener and breakthrough in this direction was Schmüdgen's Positivstellensatz:

Theorem 4.3.1 (Schmüdgen) *Let polynomials $g_1, \ldots, g_r \in \mathbb{R}[x] = \mathbb{R}[x_1, \ldots, x_n]$ be given such that the basic closed set $K = \{x \in \mathbb{R}^n : g_i(x) \geq 0 \ (i = 1, \ldots, r)\}$ is compact. Then every polynomial f with $f|_K > 0$ is contained in the preordering $P = P[g_1, \ldots, g_r]$.*

(See Examples and Remarks 3.9.2(5) for the notation; nowadays people usually write $PO(g_1, \ldots, g_r)$ instead of $P[g_1, \ldots, g_r]$.)

Schmüdgen's primary interest came from analysis rather than real algebra, more precisely from the (multivariate) K-moment problem. Given a closed set $K \subset \mathbb{R}^n$, the question is to characterize all linear forms $L \colon \mathbb{R}[x_1, \ldots, x_n] \to \mathbb{R}$ that are integration with respect to a suitable (Borel) measure supported on K. Clearly, such L has to satisfy $L(f) \geq 0$ whenever $f|_K \geq 0$. Conversely, Theorem 4.3.1 implies that this condition is already sufficient for K as in the theorem.

In his original paper [103], Schmüdgen deduced Theorem 4.3.1 from his solution to the K-moment problem. For this he combined operator-theoretic arguments with Stengle's Positivstellensatz. Later it was realized that there exists a purely algebraic approach to Theorem 4.3.1. The key for this is the archimedean property. A convex subcone M of an $\mathbb{R}$-algebra A is called *archimedean* if $\{0, 1\} \subset M$ and $\mathbb{Z} + M = A$. Thinking of the elements of M as being nonnegative in a suitably general sense, this property requires that every element of A is exceeded by some positive integer, whence the terminology archimedean. The hypotheses of the theorem imply that the preordering $P[g_1, \ldots, g_r]$ in $A = \mathbb{R}[x]$ is archimedean, which makes Theorem 4.3.1 a consequence of the general *Archimedean Positivstellensatz*. For the precise statement of this latter result we refer to the literature. It allows a variety of substantially different variations on the question of finding *denominator-free* certificates for strict positivity. The archimedean Positivstellensatz is essentially equivalent to the so-called Representation Theorem, usually attributed to Stone, Kadison, Dubois and Krivine. See [92, Theorem 5.2.6] or [101, 1.5.9], and compare also [22] for a different approach.

Seeing Theorem 4.3.1 in the archimedean context allows for useful generalizations. In the situation of the theorem, the quadratic module $M := QM(g_1, \ldots, g_r)$ generated by the g_i consists of all polynomials $f = s_0 + \sum_{i=1}^r s_i g_i$ where the s_i are sums of squares of polynomials. Note that M is a subset (usually proper) of the preordering P. In contrast to P, the cone M need not be archimedean when K is compact. But when it is, M contains every polynomial f strictly positive on K. The smaller complexity of a representation in M, compared with one in P (only $r + 1$

generators instead of 2^r) is an important point for applications. From Theorem 4.3.1 it follows that M is archimedean if and only if M contains a polynomial g for which $\{x \in \mathbb{R}^n : g(x) \geq 0\}$ is compact. This fact is often referred to as Putinar's Positivstellensatz.

Jacobi and Prestel used the local–global principle for weakly isotropic quadratic forms (due to Bröcker and Prestel, around 1973) to prove various refinements and variations of these results. For example, if $K = \{x \in \mathbb{R}^n : g_1(x) \geq 0, \ g_2(x) \geq 0\}$ is compact (the case $r = 2$), it is always true that $M = QM(g_1, g_2)$ is archimedean. For a detailed account of these results we refer to [92], in particular to [92, Chapters 6–8].

Archimedean Nichtnegativstellensätze (as opposed to Positivstellensätze) were studied by Scheiderer, see e.g., [101, Sect. 3] for an overview. An important tool in this context is the archimedean local–global principle. It is used, for example, to show for nonsingular affine $\mathbb{R}$-varieties V with $\dim(V) \leq 2$ and $V(\mathbb{R})$ compact, that every nonnegative polynomial on V is a sum of squares in $\mathbb{R}[V]$, cf. [100]. In contrast, as soon as the dimension is ≥ 3, there always exist nonnegative polynomials that are not sums of squares. Similar results hold more generally for basic closed sets under suitable regularity assumptions.

Such theoretical results have important applications of various kinds. For consequences on multivariate moment problems, see e.g. the overview in [101, Sect. 5]. An area where they are omnipresent is polynomial optimization [74]. Roughly since the year 2000, semidefinite programming techniques started to be used to optimize polynomial functions under polynomial side conditions. The moment relaxation method, suggested by Lasserre and Parrilo, allows optimization of a polynomial over a compact (basic) semialgebraic set $K \subset \mathbb{R}^n$ up to arbitrary precision, at least in theory. The basic idea is to relax the positivity condition $f|_K \geq c$ by the condition $f - c \in QM(g_1, \ldots, g_r)$, if $K = \{x : g_1(x) \geq 0, \ldots, g_r(x) \geq 0\}$. In other words, being nonnegative is replaced by being a weighted sum of squares. If we bound the degrees $\deg(s_i)$ of the sums of squares in representations

$$f - c = s_0 + \sum_i s_i g_i \qquad (*)$$

as $\deg(s_i) \leq 2d$, then finding the supremum $f_{(d)}$ of all numbers c for which $(*)$ exists becomes a standard semidefinite program. Semidefinite programs (SDP) are solved very efficiently by interior path methods, due to ground-breaking ideas of Karmarkar, Nesterov and Nemirenko in the 1980s and 1990s. Observe how the Schmüdgen and Putinar Positivstellensätze imply that the restricted optima converge to the true optimum when the degree d is taken to infinity. In other words, $f_* = \min f(K) = \lim_{d \to \infty} f_{(d)}$. In practice the semidefinite programs become more and more complex when d is increased. Still, these ideas have become very important and are now used in a large variety of situations.

4.4 Noncommutative Stellensätze

We should also mention the appearance of various kinds of *noncommutative*
Stellensätze. To explain the setting, let A be a not necessarily commutative algebra
with unit over a real field F (in many instances $F = \mathbb{R}$) and with an involution
$x \mapsto x^*$. In other words, F is contained in the centre of A and $x \mapsto x^*$ is an anti-
automorphism of order two, i.e., satisfies $(x + y)^* = x^* + y^*$, $(x \cdot y)^* = y^* \cdot x^*$
and $(x^*)^* = x$ for all $x, y \in A$. Such an algebra is often called a $*$-*algebra*. An
element $a \in A$ is called *symmetric* (or *hermitian*) if $a^* = a$. The analogue of sums
of squares in this context are sums of hermitian squares, i.e., finite sums $\sum_i x_i^* x_i$
with $x_i \in A$. We denote the set of sums of hermitian squares by ΣA^2. There exist
several approaches to defining what being positive means for elements of A. As this
subject is witnessing an explosive growth (we refer to Schmüdgen's survey [105]
for an excellent overview of results in this area up to 2009), we can only highlight a
number of representative examples.

The prototypical example of a $*$-algebra is a full matrix algebra $A = M_n(F)$
with transpose involution $* = T$. More generally one can consider central simple
algebras with involution. For example, in their 1976 paper [94] Procesi and Schacher
investigated central simple algebras A with involution $*$, centre a field K and fixed
field $F := K \cap \{a \in A : a^* = a\}$ a real field. Given an ordering P of F, the
involution $*$ is defined to be *positive at* P if the quadratic form $x \mapsto \mathrm{Trd}_A(x^*x)$
is positive semidefinite at P, where Trd_A denotes the reduced trace of A, cf. [64,
§1.A]. In this situation P is called a $*$-*ordering* of F. Consider such a $*$-ordering
P. A symmetric element $a \in A$ is defined to be *positive at* P if the quadratic
form $x \mapsto \mathrm{Trd}_A(x^*ax)$ is positive semidefinite at P. A symmetric element $a \in A$ is
defined to be *positive* if it is positive at all $*$-orderings P of F. Procesi and Schacher
proved a noncommutative version of Artin's Theorem [94, Thm. 5.4]: Let $a \in A$ be
any symmetric element. Then a is positive if and only if it is a weighted sum of
hermitian squares,

$$a = \sum_{\varepsilon \in \{0,1\}^n} \alpha_1^{\varepsilon_1} \cdots \alpha_n^{\varepsilon_n} \Big(\sum_i x_i^2 \Big),$$

where the (finitely many) $x_i \in A$ are symmetric and $\alpha_1, \ldots, \alpha_n$ are the elements of
F appearing in a diagonalization of the form $x \mapsto \mathrm{Trd}_A(x^*x)$.

They showed that if $\deg A = 2$, i.e., $\dim_K A = 4$, i.e., A is a quaternion algebra,
the weights are superfluous ([94, Cor. 5.5]). In general they are not, cf. [59], [60].
See [5] for an approach via signatures of hermitian forms. The weights are also
superfluous if every ordering of F is a $*$-ordering, cf. [94, Prop. 5.3]. This happens
for example in the "split" case where $A = M_n(F)$ and $* = T$, as already observed
in 1974 by Gondard and Ribenboim, cf. [41].

In the context of real multivariate *matrix polynomials* (or *polynomial matrices*)
$A = M_n(\mathbb{R}[x_1, \ldots, x_m])$, analogues have been obtained of the Krivine strict
Positivstellensatz by Cimprič [27] in 2011 and the Schmüdgen and Putinar strict

Positivstellensätze by Scherer and Hol [102] in 2006. These are the equivalences (i) $\Leftrightarrow$ (ii), (i) $\Leftrightarrow$ (ii') and (i) $\Leftrightarrow$ (ii''), respectively, in [27, Thm. 2]: Consider a finite set of polynomials $S = \{g_1, \ldots, g_d\}$ in $\mathbb{R}[x] = \mathbb{R}[x_1, \ldots, x_m]$ and let $\widehat{S} = \{g_1^{\alpha_1} \cdots g_d^{\alpha_d} : \alpha \in \{0, 1\}^d\}$. Consider the basic closed set $K_S = \{x \in \mathbb{R}^m : g_1(x) \geq 0, \ldots, g_d(x) \geq 0\}$ and the quadratic modules

$$M_S^n = \left\{ C_0 + \sum_{i=1}^d C_i g_i : C_0, \ldots, C_d \in \Sigma M_n(\mathbb{R}[x])^2 \right\}, \quad T_S^n = M_{\widehat{S}}^n, \quad \text{and } T_S = M_{\widehat{S}}^1.$$

Let $M \in M_n(\mathbb{R}[x])$ be any symmetric matrix. Then: (i) $M(x)$ is strictly positive definite for every $x \in K_S$ if and only if (ii) there exist $t \in T_S$ and $V \in T_S^n$ such that

$$(1 + t)M = I_n + V.$$

Furthermore, if K_S is compact, then (i) and (ii) are equivalent to: (ii') there exists $\varepsilon > 0$ such that $M - \varepsilon I_n \in T_S^n$. Finally, if one of the sets $K_{\{g_i\}} = \{x \in \mathbb{R}^m : g_i(x) \geq 0\}, i = 1, \ldots, d$, is compact, then (i), (ii) and (ii') are equivalent to: (ii'') there exists $\varepsilon > 0$ such that $M - \varepsilon I_n \in M_S^n$.

A generalization is obtained by dropping the requirements that the variables $x_1, \ldots, x_m$ commute and that the matrix size is fixed. For example, in his seminal 2002 paper [48], Helton considered the free $*$-algebra $A = \mathbb{R}\langle x, x^* \rangle$ generated by m noncommuting variables $x = (x_1, \ldots, x_m)$ and their "transposes" $x^* = (x_1^*, \ldots, x_m^*)$. It is common practice to refer to the elements of A as *noncommutative* (or *NC*) *polynomials* (or simply *polynomials*). When substituting matrices for the variables of a noncommutative polynomial, the involution $*$ is replaced by matrix transposition T. A symmetric noncommutative polynomial $p \in A$ is *matrix-positive* provided the matrix $p(X, X^T)$ is positive semidefinite for all matrix tuples $X \in M_r(\mathbb{R})^m$ and all $r \in \mathbb{N}$. With this definition of positivity, Helton proved the following noncommutative analogue of Artin's Theorem [48, Thm. 1.1]: Let p be any symmetric noncommutative polynomial. Then p is matrix-positive if and only if it is a sum of hermitian squares.

In [82] McCullough and Putinar pointed out that Helton's Theorem remains valid for $\mathbb{C}\langle x, x^* \rangle$ and that Helton's proof can easily be adapted to show this. They also presented a new proof, based upon Carathéodory's Theorem from convex geometry and a Hahn–Banach separation argument.

In 2004, Helton and McCullough [50] established a strict Positivstellensatz for three classes of matrix-valued noncommutative polynomials. To simplify the exposition we only describe the case of symmetric noncommutative polynomials. Given a collection $\mathcal{P}$ of symmetric polynomials in noncommuting variables $x = (x_1, \ldots, x_m)$, its *positivity domain* $\mathcal{D}_\mathcal{P}$ is defined as follows: let H be a separable real Hilbert space and let $\mathcal{D}_\mathcal{P}(H)$ denote the collection of tuples $X = (X_1, \ldots, X_m)$ such that each X_j is a symmetric operator on H and $p(X)$ is positive semidefinite for each $p \in \mathcal{P}$. Then $\mathcal{D}_\mathcal{P}$ is defined to be the collection of tuples X such that $X \in \mathcal{D}_\mathcal{P}(H)$ for some H. The positivity domain $\mathcal{D}_\mathcal{P}$ is *bounded* if there exists a constant $C > 0$ such that $C^2 - X_j^T X_j$ is positive semidefinite whenever $X \in \mathcal{D}_{\mathcal{P}}$. Helton and McCullough proved [50, Thm. 1.2]: Assume that $\mathcal{D}_{\mathcal{P}}$ is bounded with constant $C > 0$ and let q be any symmetric polynomial that is strictly positive definite on $\mathcal{D}_{\mathcal{P}}$. Then there exist polynomials $p_i \in \mathcal{P}$ and polynomials $s_i, r_j, t_{k,\ell}$ such that q can be represented as a finite weighted sum of hermitian squares,

$$q = \sum_i s_i^T p_i s_i + \sum_j r_j^T r_j + \sum_{k,\ell} t_{k,\ell}^T (C^2 - x_k^2) t_{k,\ell}.$$

By changing some assumptions, this result was turned into a Nichtnegativstellensatz by Helton, Klep and McCullough in 2012. They also improved the weighted sum of squares representation by giving optimal degree bounds, cf. [49]. Pascoe derived some Positivstellensätze for noncommutative rational expressions from the Helton–McCullough theorem in 2018, cf. [87].

An exciting aspect of Positivstellensätze for free $*$-algebras is their applications to engineering problems, cf. [30].

Our final example is Schmüdgen's 2005 strict Positivstellensatz for the Weyl algebra $\mathcal{W}(d)$ using methods from operator theory and functional analysis, cf. [104]. The set-up is as follows: for $d \in \mathbb{N}$, the *Weyl algebra* $\mathcal{W}(d)$ is the complex $*$-algebra with unit element 1, generators $a_1, \ldots, a_d, a_{-1}, \ldots, a_{-d}$, defining relations

$$a_k a_{-k} - a_{-k} a_k = 1 \text{ for } k = 1, \ldots, d,$$

$$a_k a_\ell - a_\ell a_k = 0 \text{ for } k, \ell = -d, \ldots, -1, 1, \ldots, d, \ k \neq -\ell,$$

and involution given by $a_k^* = a_{-k}$ for $k = 1, \ldots, d$. The Weyl algebra $\mathcal{W}(d)$ has a natural filtration with corresponding graded algebra the polynomial algebra $\mathbb{C}[z, \overline{z}]$ in $2d$ complex variables $z = (z_1, \ldots, z_d)$, $\overline{z} = (\overline{z_1}, \ldots, \overline{z_d})$, where z_j and $\overline{z_j}$ correspond to a_j and a_j^*, respectively. If $c \in \mathcal{W}(d)$ is an element of degree n, the polynomial associated to the n-th component of c is denoted $c_n(z, \overline{z})$. Let α be a fixed positive number which is not an integer, and let $\mathcal{N}$ denote the set of all finite products of the elements $a_1^* a_1 + \cdots + a_d^* a_d + (\alpha + n)1$, where $n \in \mathbb{Z}$. Finally, consider the positive cone

$$\mathcal{W}(d)_+ := \{x \in \mathcal{W}(d) : \langle \pi_0(x)\varphi, \varphi \rangle \geq 0 \text{ for all } \varphi \in \mathcal{S}(\mathbb{R}^d)\},$$

where π_0 denotes the Schrödinger representation, acting on the Schwartz space $\mathcal{S}(\mathbb{R}^d)$, considered as a dense domain of the Hilbert space $L^2(\mathbb{R}^d)$ with inner product $\langle \cdot, \cdot \rangle$. Schmüdgen proved [104, Thm. 1.1]: Let $c \in \mathcal{W}(d)$ be any symmetric element of even degree $2m$. Assume that there exists $\varepsilon > 0$ such that $c - \varepsilon \cdot 1 \in \mathcal{W}(d)_+$, and that $c_{2m}(z, \overline{z}) > 0$ for all $z \in \mathbb{C}^d$, $z \neq 0$. If m is even, then there exists $b \in \mathcal{N}$ such that $bcb \in \Sigma \mathcal{W}(d)^2$. If m is odd, then there exists $b \in \mathcal{N}$ such that $\sum_{j=1}^d ba_j ca_j^* b \in \Sigma \mathcal{W}(d)^2$.

References

1. Andradas, C., Bröcker, L., Ruiz, J.M.: Minimal generation of basic open semianalytic sets. Invent. Math. **92**(2), 409–430 (1988). https://doi.org/10.1007/BF01404461
2. Andradas, C., Bröcker, L., Ruiz, J.M.: Constructible sets in real geometry, Ergebnisse der Mathematik und ihrer Grenzgebiete (3) [Results in Mathematics and Related Areas (3)], vol. 33. Springer, Berlin (1996). https://doi.org/10.1007/978-3-642-80024-5
3. Artin, E.: Über die Zerlegung definiter Funktionen in Quadrate. Abh. Math. Sem. Univ. Hamburg **5**(1), 100–115 (1927). https://doi.org/10.1007/BF02952513
4. Artin, E., Schreier, O.: Algebraische Konstruktion reeller Körper. Abh. Math. Sem. Univ. Hamburg **5**(1), 85–99 (1927). https://doi.org/10.1007/BF02952512
5. Astier, V., Unger, T.: Signatures of Hermitian forms, positivity, and an answer to a question of Procesi and Schacher. J. Algebra **508**, 339–363 (2018). https://doi.org/10.1016/j.jalgebra.2018.05.004
6. Averkov, G., Bröcker, L.: Minimal polynomial descriptions of polyhedra and special semi-algebraic sets. Adv. Geom. **12**(3), 447–459 (2012). https://doi.org/10.1515/advgeom-2011-059
7. Becker, E.: The real holomorphy ring and sums of $2n$th powers. In: Real Algebraic Geometry and Quadratic Forms (Rennes, 1981). Lecture Notes in Math., vol. 959, pp. 139–181. Springer, Berlin-New York (1982)
8. Becker, E.: Valuations and real places in the theory of formally real fields. In: Real Algebraic Geometry and Quadratic Forms (Rennes, 1981). Lecture Notes in Math., vol. 959, pp. 1–40. Springer, Berlin-New York (1982)
9. Becker, E., Köpping, E.: Reduzierte quadratische Formen und Semiordnungen reeller Körper. Abh. Math. Sem. Univ. Hamburg **46**, 143–177 (1977). https://doi.org/10.1007/BF02993018
10. Becker, E., Spitzlay, K.J.: Zum Satz von Artin-Schreier über die Eindeutigkeit des reellen Abschlusses eines angeordneten Körpers. Comment. Math. Helv. **50**, 81–87 (1975). https://doi.org/10.1007/BF02565735
11. Bochnak, J., Coste, M., Roy, M.F.: Géométrie algébrique réelle. Ergebnisse der Mathematik und ihrer Grenzgebiete (3) [Results in Mathematics and Related Areas (3)], vol. 12. Springer, Berlin (1987)
12. Bourbaki, N.: Algèbre. Masson, Paris
13. Bourbaki, N.: Algèbre Commutative. Masson, Paris
14. Bourbaki, N.: Topologie Générale. Masson, Paris
15. Bröcker, L.: Zur Theorie der quadratischen Formen über formal reellen Körpern. Math. Ann. **210**, 233–256 (1974). https://doi.org/10.1007/BF01350587

© Springer Nature Switzerland AG 2022
M. Knebusch, C. Scheiderer, *Real Algebra*, Universitext,
https://doi.org/10.1007/978-3-031-09800-0

16. Bröcker, L.: On the stability index of Noetherian rings. In: Real Analytic and Algebraic Geometry (Trento, 1988). Lecture Notes in Math., vol. 1420, pp. 72–80. Springer, Berlin (1990). https://doi.org/10.1007/BFb0083912

17. Brumfiel, G.W.: Partially ordered rings and semi-algebraic geometry. London Mathematical Society Lecture Note Series, vol. 37. Cambridge University Press, Cambridge-New York (1979)

18. Brumfiel, G.W.: Real valuation rings and ideals. In: Real Algebraic Geometry and Quadratic Forms (Rennes, 1981). Lecture Notes in Math., vol. 959, pp. 55–97. Springer, Berlin-New York (1982)

19. Brumfiel, G.W.: Witt Rings and K-Theory. pp. 733–765 (1984). https://doi.org/10.1216/RMJ-1984-14-4-733. Ordered fields and real algebraic geometry (Boulder, Colo., 1983)

20. Brumfiel, G.W.: The real spectrum compactification of Teichmüller space. In: Geometry of Group Representations (Boulder, CO, 1987). Contemp. Math., vol. 74, pp. 51–75. Amer. Math. Soc., Providence (1988). https://doi.org/10.1090/conm/074/957511

21. Burési, J., Mahé, L.: Reducing inequalities with bounds. Math. Z. **227**(2), 231–243 (1998). https://doi.org/10.1007/PL00004371

22. Burgdorf, S., Scheiderer, C., Schweighofer, M.: Pure states, nonnegative polynomials and sums of squares. Comment. Math. Helv. **87**(1), 113–140 (2012). https://doi.org/10.4171/CMH/250

23. Carral, M., Coste, M.: Normal spectral spaces and their dimensions. J. Pure Appl. Algebra **30**(3), 227–235 (1983). https://doi.org/10.1016/0022-4049(83)90058-0

24. Cassels, J.W.S., Ellison, W.J., Pfister, A.: On sums of squares and on elliptic curves over function fields. J. Number Theory **3**, 125–149 (1971).https://doi.org/10.1016/0022-314X(71)90030-8

25. Chaillou, J.: Hyperbolic differential polynomials and their singular perturbations. Mathematics and its Applications, vol. 3. D. Reidel Publishing Co., Dordrecht-Boston (1979). Translated from the French by J. W. Nienhuys

26. Choi, M.D., Lam, T.Y.: An old question of Hilbert. In: Conference on Quadratic Forms—1976 (Proc. Conf., Queen's Univ., Kingston, Ont., 1976), pp. 385–405. Queen's Papers in Pure and Appl. Math., No. 46 (1977)

27. Cimprič, J.: Strict Positivstellensätze for matrix polynomials with scalar constraints. Linear Algebra Appl. **434**(8), 1879–1883 (2011). https://doi.org/10.1016/j.laa.2010.11.046

28. Coste, M., Roy, M.F.: La topologie du spectre réel. In: Ordered Fields and Real Algebraic Geometry (San Francisco, Calif., 1981). Contemp. Math., vol. 8, pp. 27–59. Amer. Math. Soc., Providence (1982)

29. Coste-Roy, M.F.: Faisceau structural sur le spectre réel et fonctions de Nash. In: Real Algebraic Geometry and Quadratic Forms (Rennes, 1981). Lecture Notes in Math., vol. 959, pp. 406–432. Springer, Berlin-New York (1982)

30. de Oliveira, M.C., Helton, J.W., McCullough, S.A., Putinar, M.: Engineering systems and free semi-algebraic geometry. In: Emerging Applications of Algebraic Geometry. IMA Vol. Math. Appl., vol. 149, pp. 17–61. Springer, New York (2009). https://doi.org/10.1007/978-0-387-09686-5_2

31. Delfs, H., Knebusch, M.: Semialgebraic topology over a real closed field. I. Paths and components in the set of rational points of an algebraic variety. Math. Z. **177**(1), 107–129 (1981). https://doi.org/10.1007/BF01214342

32. Delfs, H., Knebusch, M.: Semialgebraic topology over a real closed field. II. Basic theory of semialgebraic spaces. Math. Z. **178**(2), 175–213 (1981). https://doi.org/10.1007/BF01262039

33. Delfs, H., Knebusch, M.: Locally semialgebraic spaces. Lecture Notes in Mathematics, vol. 1173. Springer-Verlag, Berlin (1985). https://doi.org/10.1007/BFb0074551

34. Delzell, C.N., Madden, J.J.: A completely normal spectral space that is not a real spectrum. J. Algebra **169**(1), 71–77 (1994). https://doi.org/10.1006/jabr.1994.1272

35. Dickmann, M., Schwartz, N., Tressl, M.: Spectral spaces. New Mathematical Monographs, vol. 35. Cambridge University Press, Cambridge (2019). https://doi.org/10.1017/9781316543870

36. Dress, A.: Lotschnittebenen mit halbierbarem rechtem Winkel. Arch. Math. (Basel) **16**, 388–392 (1965). https://doi.org/10.1007/BF01220047
37. Dubois, D.W.: Infinite primes and ordered fields. Dissertationes Math. (Rozprawy Mat.) **69**, 43 (1970)
38. Elman, R., Lam, T.Y., Wadsworth, A.R.: Orderings under field extensions. J. Reine Angew. Math. **306**, 7–27 (1979)
39. Fuller, A.T.: Aperiodicity determinants expressed in terms of roots. Int. J. Control **47**(6), 1571–1593 (1988). https://doi.org/10.1080/00207178808906122
40. Gantmacher, F.R.: Matrizentheorie. Springer-Verlag, Berlin (1986). https://doi.org/10.1007/978-3-642-71243-2. With an appendix by V. B. Lidskij, With a preface by D. P. Želobenko, Translated from the second Russian edition by Helmut Boseck, Dietmar Soyka and Klaus Stengert
41. Gondard, D., Ribenboim, P.: Le 17e problème de Hilbert pour les matrices. Bull. Sci. Math. (2) **98**(1), 49–56 (1974)
42. Gorin, E.A.: Asymptotic properties of polynomials and algebraic functions of several variables. Uspehi Mat. Nauk **16**(1 (97)), 91–118 (1961)
43. Grätzer, G.: General Lattice Theory. Birkhäuser Verlag, Basel-Stuttgart (1978). Lehrbücher und Monographien aus dem Gebiete der Exakten Wissenschaften, Mathematische Reihe, Band 52
44. Harrison, D.K.: Witt Rings. Lecture Notes. Univ. of Kentucky, Lexington (1970)
45. Hartshorne, R.: Algebraic Geometry. Springer-Verlag, New York-Heidelberg (1977). Graduate Texts in Mathematics, No. 52
46. Helmke, U.: Rational functions and Bezout forms: a functorial correspondence. Linear Algebra Appl. **122/123/124**, 623–640 (1989). https://doi.org/10.1016/0024-3795(89)90669-1
47. Helmke, U., Fuhrmann, P.A.: Bezoutians. Linear Algebra Appl. **122/123/124**, 1039–1097 (1989). https://doi.org/10.1016/0024-3795(89)90684-8
48. Helton, J.W.: "Positive" noncommutative polynomials are sums of squares. Ann. Math. (2) **156**(2), 675–694 (2002). https://doi.org/10.2307/3597203
49. Helton, J.W., Klep, I., McCullough, S.: The convex Positivstellensatz in a free algebra. Adv. Math. **231**(1), 516–534 (2012). https://doi.org/10.1016/j.aim.2012.04.028
50. Helton, J.W., McCullough, S.A.: A Positivstellensatz for non-commutative polynomials. Trans. Am. Math. Soc. **356**(9), 3721–3737 (2004). https://doi.org/10.1090/S0002-9947-04-03433-6
51. Hochster, M.: Prime ideal structure in commutative rings. Trans. Am. Math. Soc. **142**, 43–60 (1969). https://doi.org/10.2307/1995344
52. Hölder, O.: Die Axiome der Quantität und die Lehre vom Maß. Leipz. Ber. **53**, 1–64 (1901)
53. Hörmander, L.: On the division of distributions by polynomials. Ark. Mat. **3**, 555–568 (1958). https://doi.org/10.1007/BF02589517
54. Jacobson, N.: Lectures in Abstract Algebra (3 volumes). D. Van Nostrand Co., Inc., Toronto, New York, London (1951–1954)
55. Jacobson, N.: Basic Algebra (2 volumes). W. H. Freeman and Co., San Francisco, Calif. (1974 and 1980)
56. Johnstone, P.T.: Stone Spaces. Cambridge Studies in Advanced Mathematics, vol. 3. Cambridge University Press, Cambridge (1982)
57. Kato, K.: A Hasse principle for two-dimensional global fields. J. Reine Angew. Math. **366**, 142–183 (1986). https://doi.org/10.1515/crll.1986.366.142. With an appendix by Jean-Louis Colliot-Thélène
58. Khovanskiĭ, A.G.: Fewnomials. Translations of Mathematical Monographs, vol. 88. American Mathematical Society, Providence (1991). https://doi.org/10.1090/mmono/088. Translated from the Russian by Smilka Zdravkovska
59. Klep, I., Unger, T.: The Procesi-Schacher conjecture and Hilbert's 17th problem for algebras with involution. J. Algebra **324**(2), 256–268 (2010). https://doi.org/10.1016/j.jalgebra.2010.03.022

60. Klep, I., Špenko, v., Volčič, J.: Positive trace polynomials and the universal Procesi-Schacher conjecture. Proc. Lond. Math. Soc. (3) **117**(6), 1101–1134 (2018). https://doi.org/10.1112/plms.12156

61. Knebusch, M.: Symmetric bilinear forms over algebraic varieties. In: Conference on Quadratic Forms—1976 (Proc. Conf., Queen's Univ., Kingston, Ont., 1976), pp. 103–283. Queen's Papers in Pure and Appl. Math., No. 46 (1977)

62. Knebusch, M., Kolster, M.: Wittrings. Aspects of Mathematics, vol. 2. Friedr. Vieweg & Sohn, Braunschweig (1982)

63. Knus, M.A.: Quadratic and Hermitian Forms Over Rings. Grundlehren der mathematischen Wissenschaften [Fundamental Principles of Mathematical Sciences], vol. 294. Springer-Verlag, Berlin (1991). https://doi.org/10.1007/978-3-642-75401-2. With a foreword by I. Bertuccioni

64. Knus, M.A., Merkurjev, A., Rost, M., Tignol, J.P.: The Book of Involutions. American Mathematical Society Colloquium Publications, vol. 44. American Mathematical Society, Providence (1998). https://doi.org/10.1090/coll/044. With a preface in French by J. Tits

65. Kreĭn, M.G., Naĭmark, M.A.: The method of symmetric and Hermitian forms in the theory of the separation of the roots of algebraic equations. Linear and Multilinear Algebra **10**(4), 265–308 (1981). https://doi.org/10.1080/03081088108817420. Translated from the Russian by O. Boshko and J. L. Howland

66. Krivine, J.L.: Anneaux préordonnés. J. Anal. Math. **12**, 307–326 (1964). https://doi.org/10.1007/BF02807438

67. Krull, W.: Allgemeine Bewertungstheorie. J. Reine Angew. Math. **167**, 160–196 (1932). https://doi.org/10.1515/crll.1932.167.160

68. Kunz, E.: Einführung in die kommutative Algebra und algebraische Geometrie. Vieweg Studium: Aufbaukurs Mathematik [Vieweg Studies: Mathematics Course], vol. 46. Friedr. Vieweg & Sohn, Braunschweig (1980). With an English preface by David Mumford

69. Lam, T.Y.: The algebraic theory of quadratic forms. W. A. Benjamin, Inc., Reading (1973). Mathematics Lecture Note Series

70. Lam, T.Y.: Orderings, Valuations and Quadratic Forms. CBMS Regional Conference Series in Mathematics, vol. 52. Published for the Conference Board of the Mathematical Sciences, Washington, DC; by the American Mathematical Society, Providence (1983). https://doi.org/10.1090/cbms/052

71. Lam, T.Y.: An introduction to real algebra. Rocky Mount. J. Math. **14**(4), 767–814 (1984). https://doi.org/10.1216/RMJ-1984-14-4-767. Ordered fields and real algebraic geometry (Boulder, Colo., 1983)

72. Lam, T.Y.: Introduction to Quadratic Forms Over Fields. Graduate Studies in Mathematics, vol. 67. American Mathematical Society, Providence (2005)

73. Lang, S.: Algebra. Graduate Texts in Mathematics, vol. 211, 3rd edn. Springer, New York (2002). https://doi.org/10.1007/978-1-4613-0041-0

74. Lasserre, J.B.: An introduction to polynomial and semi-algebraic optimization. Cambridge Texts in Applied Mathematics. Cambridge University Press, Cambridge (2015). https://doi.org/10.1017/CBO9781107447226

75. Leicht, J.B.: Zur Charakterisierung reell abgeschlossener Körper. Monatsh. Math. **70**, 452–453 (1966). https://doi.org/10.1007/BF01300449

76. Łojasiewicz, S.: Sur le problème de la division. Stud. Math. **18**, 87–136 (1959). https://doi.org/10.4064/sm-18-1-87-136

77. Lorenz, F., Leicht, J.: Die Primideale des Wittschen Ringes. Invent. Math. **10**, 82–88 (1970). https://doi.org/10.1007/BF01402972

78. Mahé, L.: Signatures et composantes connexes. Math. Ann. **260**(2), 191–210 (1982). https://doi.org/10.1007/BF01457236

79. Mahé, L.: Théorème de Pfister pour les variétés et anneaux de Witt réduits. Invent. Math. **85**(1), 53–72 (1986). https://doi.org/10.1007/BF01388792

80. Marshall, M.A.: Spaces of Orderings and Abstract Real Spectra. Lecture Notes in Mathematics, vol. 1636. Springer, Berlin (1996). https://doi.org/10.1007/BFb0092696

81. Matsumura, H.: Commutative Algebra. Mathematics Lecture Note Series, vol. 56, 2nd edn. Benjamin/Cummings Publishing Co., Inc., Reading (1980)
82. McCullough, S., Putinar, M.: Noncommutative sums of squares. Pac. J. Math. **218**(1), 167–171 (2005). https://doi.org/10.2140/pjm.2005.218.167
83. Milnor, J.: Algebraic K-theory and quadratic forms. Invent. Math. **9**, 318–344 (1969/1970). https://doi.org/10.1007/BF01425486
84. Motzkin, T.S.: The arithmetic-geometric inequality. In: Inequalities (Proc. Sympos. Wright-Patterson Air Force Base, Ohio, 1965), pp. 205–224. Academic Press, New York (1967)
85. Obreschkoff, N.: Verteilung und Berechnung der Nullstellen reeller Polynome. VEB Deutscher Verlag der Wissenschaften, Berlin (1963)
86. Ostrowski, A.: Über einige Lösungen der Funktionalgleichung $\varphi(x) \cdot \varphi(x) = \varphi(xy)$. Acta Math. **41**(1), 271–284 (1916). https://doi.org/10.1007/BF02422947
87. Pascoe, J.E.: Positivstellensätze for noncommutative rational expressions. Proc. Am. Math. Soc. **146**(3), 933–937 (2018). https://doi.org/10.1090/proc/13773
88. Pfister, A.: Zur Darstellung definiter Funktionen als Summe von Quadraten. Invent. Math. **4**, 229–237 (1967). https://doi.org/10.1007/BF01425382
89. Pourchet, Y.: Sur la représentation en somme de carrés des polynômes à une indéterminée sur un corps de nombres algébriques. Acta Arith. **19**, 89–104 (1971). https://doi.org/10.4064/aa-19-1-89-104
90. Prestel, A.: Lectures on Formally Real Fields. Lecture Notes in Mathematics, vol. 1093. Springer, Berlin (1984). https://doi.org/10.1007/BFb0101548
91. Prestel, A.: Einführung in die Mathematische Logik und Modelltheorie. Vieweg Studium: Aufbaukurs Mathematik [Vieweg Studies: Mathematics Course], vol. 60. Friedr. Vieweg & Sohn, Braunschweig (1986). https://doi.org/10.1007/978-3-663-07641-4
92. Prestel, A., Delzell, C.N.: Positive Polynomials. Springer Monographs in Mathematics. Springer, Berlin (2001). https://doi.org/10.1007/978-3-662-04648-7. From Hilbert's 17th problem to real algebra
93. Prieß-Crampe, S.: Angeordnete Strukturen. Ergebnisse der Mathematik und ihrer Grenzgebiete [Results in Mathematics and Related Areas], vol. 98. Springer, Berlin (1983). https://doi.org/10.1007/978-3-642-68628-3. Gruppen, Körper, projektive Ebenen. [Groups, fields, projective planes]
94. Procesi, C., Schacher, M.: A non-commutative real Nullstellensatz and Hilbert's 17th problem. Ann. Math. (2) **104**(3), 395–406 (1976). https://doi.org/10.2307/1970962
95. Robson, R.O.: The Ideal Theory of Real Algebraic Curves and Affine Embeddings of Semi-Algebraic Spaces and Manifolds. Ph.D. Thesis, Stanford University (1981)
96. Roman, S.: Field Theory. Graduate Texts in Mathematics, vol. 158, 2nd edn. Springer, New York (2006)
97. Schafarewitsch, I.R.: Grundzüge der algebraischen Geometrie. Friedr. Vieweg+Sohn, Braunschweig (1972). Übersetzung aus dem Russischen von Rudolf Fragel, Logik und Grundlagen der Mathematik, Band 12
98. Scharlau, W.: Quadratic and Hermitian Forms. Grundlehren der Mathematischen Wissenschaften [Fundamental Principles of Mathematical Sciences], vol. 270. Springer, Berlin (1985). https://doi.org/10.1007/978-3-642-69971-9
99. Scheiderer, C.: Stability index of real varieties. Invent. Math. **97**(3), 467–483 (1989). https://doi.org/10.1007/BF01388887
100. Scheiderer, C.: Sums of squares on real algebraic surfaces. Manuscr. Math. **119**(4), 395–410 (2006). https://doi.org/10.1007/s00229-006-0630-5
101. Scheiderer, C.: Positivity and sums of squares: a guide to recent results. In: Emerging Applications of Algebraic Geometry. IMA Vol. Math. Appl., vol. 149, pp. 271–324. Springer, New York (2009). https://doi.org/10.1007/978-0-387-09686-5_8
102. Scherer, C.W., Hol, C.W.J.: Matrix sum-of-squares relaxations for robust semi-definite programs. Math. Program. **107**(1-2, Ser. B), 189–211 (2006). https://doi.org/10.1007/s10107-005-0684-2

103. Schmüdgen, K.: The K-moment problem for compact semi-algebraic sets. Math. Ann. **289**(2), 203–206 (1991). https://doi.org/10.1007/BF01446568

104. Schmüdgen, K.: A strict Positivstellensatz for the Weyl algebra. Math. Ann. **331**(4), 779–794 (2005). https://doi.org/10.1007/s00208-004-0604-4

105. Schmüdgen, K.: Noncommutative real algebraic geometry—some basic concepts and first ideas. In: Emerging Applications of Algebraic Geometry. IMA Vol. Math. Appl., vol. 149, pp. 325–350. Springer, New York (2009). https://doi.org/10.1007/978-0-387-09686-5_9

106. Schülting, H.W.: Über reelle Stellen eines Körpers und ihren Holomorphiering. Ph.D. Thesis, Univ. Dortmund (1979)

107. Schwartz, N.: The Basic Theory of Real Closed Spaces. Regensburger Mathematische Schriften [Regensburg Mathematical Publications], vol. 15. Universität Regensburg, Fachbereich Mathematik, Regensburg (1987)

108. Schwartz, N.: The basic theory of real closed spaces. Mem. Am. Math. Soc. **77**(397), viii+122 (1989). https://doi.org/10.1090/memo/0397

109. Serre, J.P.: Groupes algébriques et corps de classes. Hermann, Paris (1975). Deuxième édition, Publication de l'Institut de Mathématique de l'Université de Nancago, No. VII, Actualités Scientifiques et Industrielles, No. 1264

110. Sottile, F.: Real Solutions to Equations from Geometry. University Lecture Series, vol. 57. American Mathematical Society, Providence (2011). https://doi.org/10.1090/ulect/057

111. Stengle, G.: A nullstellensatz and a positivstellensatz in semialgebraic geometry. Math. Ann. **207**, 87–97 (1974). https://doi.org/10.1007/BF01362149

112. Swan, R.G.: Topological examples of projective modules. Trans. Am. Math. Soc. **230**, 201–234 (1977). https://doi.org/10.2307/1997717

113. van der Waerden, B.L.: Algebra (2 volumes). Heidelberger Taschenbücher. Springer, Berlin-Heidelberg-New York (1971 and 1967)

114. Weber, H.: Lehrbuch der Algebra (3 Volumes). Reprint: Chelsea Publ. Comp., New York (1963)

Symbol Index

The set of natural numbers $\{1, 2, 3, \ldots\}$ is denoted $\mathbb{N}$, while $\mathbb{N}_0 := \mathbb{N} \cup \{0\}$. As usual, $\mathbb{Z}$, $\mathbb{Q}$, $\mathbb{R}$ and $\mathbb{C}$ denote the sets of integers, rational numbers, real numbers and complex numbers, respectively. For a prime power q, the field with q elements is denoted $\mathbb{F}_q$.

Given a set X, the notation $\#X$ is used for the number of elements of X in a naive sense: $\#X \in \mathbb{N}_0 \cup \{\infty\}$. The power set of X is denoted 2^X. If Y and Z are subsets of X, we let $Y \setminus Z := \{y \in Y : y \notin Z\}$.

Rings are always assumed commutative and unitary. Subrings contain the identity and all ring homomorphisms preserve the identity. Given a ring A, the group of units is denoted A^* and the subset $\{a^2 : a \in A\}$ of A is denoted A^2. If A is an integral domain, we write $\mathrm{Quot}\, A$ for the field of fractions of A. The transcendence degree of a field extension L/K is denoted $\mathrm{tr.deg.}(L/K) \in \mathbb{N}_0 \cup \{\infty\}$, see also Sect. 2.11.

© Springer Nature Switzerland AG 2022
M. Knebusch, C. Scheiderer, *Real Algebra*, Universitext,
https://doi.org/10.1007/978-3-031-09800-0

Index

© Springer Nature Switzerland AG 2022
M. Knebusch, C. Scheiderer, *Real Algebra*, Universitext,
https://doi.org/10.1007/978-3-031-09800-0

203

Printed in the United States
by Baker & Taylor Publisher Services